Chemical Substitutes from Agricultural and Industrial By-Products

Chemical Substitutes from Agricultural and Industrial By-Products

Bioconversion, Bioprocessing, and Biorefining

Edited by Suraini Abd-Aziz, Misri Gozan, Mohamad Faizal Ibrahim, and Lai-Yee Phang

WILEY-VCH

The Editors

Prof Ts Dr Suraini Abd-Aziz
Universiti Putra Malaysia
Faculty of Biotechnology &
Biomolecular Sciences
43400 Serdang, Selangor
Malaysia

Prof Dr-Ing Misri Gozan
Universitas Indonesia
Faculty of Engineering
16424 Depok
Indonesia

Associate Prof Ts Dr Mohamad Faizal Ibrahim
Universiti Putra Malaysia
Faculty of Biotechnology &
Biomolecular Sciences
43400 Serdang, Selangor
Malaysia

Associate Prof Ts Dr Lai-Yee Phang
Universiti Putra Malaysia
Faculty of Biotechnology &
Biomolecular Sciences
43400 Serdang, Selangor
Malaysia

Cover Image: © ARTFULLY PHOTOGRAPHER/Shutterstock

All books published by **WILEY-VCH** are carefully produced. Nevertheless, authors, editors, and publisher do not warrant the information contained in these books, including this book, to be free of errors. Readers are advised to keep in mind that statements, data, illustrations, procedural details or other items may inadvertently be inaccurate.

Library of Congress Card No.: applied for

British Library Cataloguing-in-Publication Data
A catalogue record for this book is available from the British Library.

Bibliographic information published by the Deutsche Nationalbibliothek
The Deutsche Nationalbibliothek lists this publication in the Deutsche Nationalbibliografie; detailed bibliographic data are available on the Internet at <http://dnb.d-nb.de>.

© 2024 WILEY-VCH GmbH, Boschstr. 12, 69469 Weinheim, Germany

All rights reserved (including those of translation into other languages). No part of this book may be reproduced in any form – by photoprinting, microfilm, or any other means – nor transmitted or translated into a machine language without written permission from the publishers. Registered names, trademarks, etc. used in this book, even when not specifically marked as such, are not to be considered unprotected by law.

Print ISBN: 978-3-527-35186-2
ePDF ISBN: 978-3-527-84112-7
ePub ISBN: 978-3-527-84113-4
oBook ISBN: 978-3-527-84114-1

Typesetting   Straive, Chennai, India

Contents

Preface *xv*
About the Editors *xvii*

1 **A Glance on Biorefinery of Chemical Substitutes from Agriculture and Industrial By-products** *1*
  Suraini Abd-Aziz, Misri Gozan, Mohamad F. Ibrahim, Lai-Yee Phang, and Mohd A. Jenol
1.1 Introduction *1*
1.2 Analysis of Feedstocks for Composition and Potential for Chemical Substitutes *3*
1.2.1 Different Types of Agricultural Wastes and Associated Risks *4*
1.2.2 Waste Utilization Routes *5*
1.2.2.1 Fertilizer Application *5*
1.2.2.2 Fibers for Textile Industry *5*
1.2.2.3 Mushroom Cultivation *6*
1.2.2.4 Organic Acids *7*
1.2.2.5 Industrial Enzymes *7*
1.2.3 Industrial By-products *8*
1.2.3.1 Agriculture, Horticulture, and Landscaping *8*
1.2.3.2 Use as Raw Material or Additive of New Products *8*
1.3 Potential Application of Chemical Substitute Extracted from Selected Agricultural Wastes and Industrial By-products *9*
1.4 Conclusions *13*
  References *13*

2 **Antioxidants from Agricultural Wastes and their Potential Applications** *19*
  Mohd A. Jenol, Yazmin Hussin, Pei H. Chu, Suraini Abd-Aziz, and Noorjahan B. Alitheen
2.1 Introduction to Antioxidants and their Usages *19*
2.2 Sources of Antioxidants *21*
2.3 Alternative Antioxidants Sources from Agricultural Wastes *22*
2.4 Extraction of Antioxidants from Selected Agricultural Waste *22*

2.4.1 Maceration 23
2.4.2 Pressurized Liquid Extraction 26
2.4.3 Microwave-assisted Extraction 27
2.4.4 Ultrasounds-assisted Extraction 28
2.4.5 Supercritical Fluid Extraction 29
2.5 Potential Applications of Antioxidants Extracted from Selected Agricultural Wastes 30
2.5.1 Food 30
2.5.2 Cosmetics 32
2.5.3 Therapeutics 33
2.6 Future Direction of Antioxidants from Agriculture Wastes 34
2.7 Conclusions 35
References 35

3 Lemongrass Oleoresin in Food Flavoring 39
Madihah Md Salleh, Shankar Ramanathan, and Rohaya Mohd Noor
3.1 Introduction 39
3.2 Types of Lemongrass and Their Components 40
3.3 Potential Chemical Substitutes from Lemongrass 42
3.3.1 Essential Oil 42
3.3.2 Phytoconstituents 43
3.3.3 Oleoresins 43
3.4 Characteristics and Properties of Oleoresin 44
3.5 Lemongrass Oleoresin Composition and Function 44
3.6 Extraction Technique of Lemongrass Oleoresin 46
3.6.1 Chemical Extraction 46
3.6.2 Steam Distillation 49
3.6.3 Pressurized Liquid Extraction (PLE) 50
3.7 Application of Lemongrass Oleoresin as Food Flavoring 51
3.8 Oleoresin Prospect 53
3.9 Conclusions 53
References 54

4 Nanocarbon Material and Chemicals from Seaweed for Energy Storage Components 59
Tirto Prakoso, Hary Devianto, Heri Rustamaji, Praswasti PDK Wulan, and Misri Gozan
4.1 Introduction 59
4.2 Source of Seaweed 62
4.2.1 Red Seaweed 62
4.2.2 Brown Seaweed 62
4.2.3 Green Seaweed 63
4.3 Potential Material Substitute from Seaweed 64

4.3.1	Activated Carbon from Seaweed	*64*
4.3.2	Graphene from Seaweed	*66*
4.4	Utilization of Seaweed-based Material for Energy Storage Component	*76*
4.4.1	Seaweed-derived Carbon Material for Supercapacitor Component	*76*
4.4.2	Seaweed-derived Chemical Materials for Battery Component	*79*
4.5	Future Prospects and Challenges	*82*
4.6	Conclusions	*83*
	References	*83*

5   Spent Mushroom Substrate as Alternative Source for the Production of Chemical Substitutes   *87*
Vikineswary Sabaratnam, Chia Wei Phan, Hariprasath Lakshmanan, and Jegadeesh Raman

5.1	Introduction	*87*
5.2	Spent Mushroom Substrate (SMS) as Source of Bulk Enzymes	*90*
5.2.1	Enzymes Extracted from SMS for Bioremediation	*91*
5.2.2	Enzymes Extracted from SMS for Green Fuel Feedstock Production: Case Study	*92*
5.3	Various Challenges and Future Prospects in the Use of SMS	*94*
5.3.1	Challenges in the Use of Enzymes in SMS	*94*
5.3.2	Future Prospects for Use of SMS for Production of Green Chemicals	*95*
5.4	Conclusions	*97*
	References	*98*

6   Essential Oil from Pineapple Wastes   *103*
Mohamad F. Ibrahim, Nurshazana Mohamad, Mariam J. M. Fairus, Mohd A. Jenol, and Suraini Abd-Aziz

6.1	Introduction	*103*
6.2	Pineapple Wastes	*104*
6.3	Pineapple Essential Oil	*105*
6.4	Extraction of Essential Oils	*106*
6.4.1	Distillation	*106*
6.4.1.1	Hydro-distillation	*107*
6.4.1.2	Soxhlet Extraction	*107*
6.4.2	Enzyme-assisted Extraction	*109*
6.4.3	Supercritical Fluid Extraction	*109*
6.5	Extracted Essential Oil Compounds	*112*
6.5.1	Essential Oils and Hydrosols	*114*
6.5.2	Applications of Essential Oils and Hydrosols	*114*
6.5.3	Market Analysis of Essential Oils and Hydrosols	*116*
6.6	Conclusions	*117*
	References	*118*

7	**Chicken Feather as a Bioresource to Produce Value-added Bioproducts** *123*
	Kai L. Sim, Radin S. R. Yahaya, Suriana Sabri, and Lai-Yee Phang
7.1	Introduction *123*
7.2	Valorization of Chicken Feathers *124*
7.2.1	Feather Composition *125*
7.2.2	Types of Feathers Treatment *125*
7.3	Bioprocessing of Chicken Feathers into Chemical Substitutes *128*
7.3.1	Feather Meal *129*
7.3.2	Bioplastic *130*
7.3.3	Biofertilizer *130*
7.3.4	Keratinase *131*
7.4	Molecular Approaches to Improve Keratinolytic Propensity of Native Host *132*
7.4.1	Overexpression of Keratinase from Native Host *132*
7.4.2	Strain Engineering *134*
7.4.3	Enzymatic Consortium *134*
7.5	Molecular Approaches to Improve Recombinant Keratinase Production and Characteristics *135*
7.5.1	Propeptide Engineering *136*
7.5.2	Promoter Engineering *136*
7.5.3	Signal Peptide Engineering *137*
7.5.4	Directed Evolution *137*
7.5.5	Alteration of Protein Domains *138*
7.6	Challenges and Future Perspectives *138*
7.7	Conclusions *140*
	References *140*
8	**Bio-bleaching Agents Used for Paper and Pulp Produced from the Valorization of Corncob, Wheat Straw, and Bagasse** *145*
	Kanya C. H. Alifia, Tjandra Setiadi, Ramaraj Boopathy, Hendro Risdianto, Muhammad Irfan, and Ibnu M. Hidayatullah
8.1	Introduction *145*
8.2	Characteristics of Biomass Substrate for Bio-bleaching Enzyme Production *146*
8.3	Microbial Sources of Bio-bleaching Enzymes *148*
8.3.1	Fungi *148*
8.3.2	Yeast *149*
8.3.3	Bacteria *149*
8.4	Bio-bleaching Enzymes and Their Usage in Pulp and Paper Industry *150*
8.4.1	Xylanase *150*
8.4.2	Cellulase *151*
8.4.3	Laccase *152*

8.5	Bioprocessing of Agricultural Wastes for Bio-bleaching Enzyme Production *154*
8.5.1	General Block Flow Diagram *154*
8.5.2	Upstream Processing *157*
8.5.3	Downstream Processing *158*
8.6	Techno-economic Evaluation *159*
8.6.1	Technical Analysis *159*
8.6.2	Economic Analysis *162*
8.7	Challenges and Future Outlooks *165*
8.8	Conclusions *167*
	References *168*

9 Recovery of Industrially Useful Enzymes from Rubber Latex Processing By-products *173*
Tan W. Kit, Yong Y. Seng, Siti N. Azlan, Nurulhuda Abdullah, and Fadzlie W. F. Wong

9.1	Introduction *173*
9.2	Processing of Natural Rubber Latex for the Production of Rubber Products *175*
9.2.1	NRL Overview *175*
9.2.2	NRL Structure *176*
9.2.3	NRL Preservation *176*
9.2.4	Deproteinization of NRL *176*
9.3	General Characteristics of Plant-derived Lysozymes and Chitinases *177*
9.4	Conventional and Alternative Activity Assays for Lysozymes and Chitinases *178*
9.5	Potential Application of Plant-derived Lysozymes and Chitinases *182*
9.6	Potential Strategy for Recovering Lysozymes and Chitinases from NRL *182*
9.7	Conclusions *187*
	References *188*

10 Sago Wastes as a Feedstock for Biosugar, Precursor for Chemical Substitutes *193*
Mohd A. Jenol, Muhd N. Ahmad, Dayang S. A. Adeni, Micky Vincent, and Nurashikin Suhaili

10.1	Introduction *193*
10.2	Current Status of Sago Starch Industry *194*
10.2.1	Sago Palm Cultivation *194*
10.2.2	Sago Starch Industry in Malaysia *195*
10.2.3	Sago Starch Processing in Sarawak *195*
10.3	Sago Wastes Biomass *196*
10.3.1	Sago Wastewater *196*
10.3.2	Sago Bark *197*
10.3.3	Sago Hampas *199*

10.3.4	Sago Frond 200
10.4	Bioconversion of Sago Wastes into Biosugars and its Derivative Precursors 200
10.5	Bioprocessing Sago Wastes Fermentable Sugar for Chemicals Substitute 202
10.5.1	L-Lactic Acid 202
10.5.2	Antimicrobial and Prebiotic Sugar (Cellobiose) from Sago Frond 203
10.5.3	Sago Frond Silage 204
10.5.4	Enzymes 205
10.5.5	Kojic Acid and its Derivatives 206
10.6	Challenges and Prospect of Sago Wastes Biorefinery 207
10.6.1	Challenges 207
10.6.2	Future Direction of Sago Waste Utilization 207
10.7	Conclusions 209
	References 209

11 Biofertilizer and Other Chemical Substitutes from Sugarcane By-products 213

Is Fatimah, Ganjar Fadillah, Tatang S. Julianto, Rudy Syahputra, and Habibi Hidayat

11.1	Introduction 213
11.2	Sugarcane By-products Conversion into Biofertilizer 216
11.3	Sugarcane Bagasse as Raw Material for Soil Improver: Phenol Degradation 218
11.4	Sugarcane By-products Conversion into Chemical 223
11.5	Sugarcane By-product as Material for Biocomposites 227
11.6	Future Perspective of Sugarcane By-products Conversion in the Sugarcane Industrial Cycle 228
11.7	Future Usage and Applications of Sugarcane By-products 229
11.8	Conclusions 230
	References 230

12 Cocoa Butter Substitute from Tengkawang (*Shorea stenoptera*) 235

Muhammad A. Darmawan, Suraini Abd-Aziz, and Misri Gozan

12.1	Introduction 235
12.2	Composition and Characteristics of Tengkawang Butter 237
12.2.1	Fatty Acids Profile of Tengkawang Butter 237
12.2.2	Quality Parameters of Tengkawang Butter 238
12.2.3	Solid Fat Content of Tengkawang Butter 239
12.2.4	Thermal Properties of Tengkawang Butter 240
12.3	Traditional Treatment Process 241
12.4	Extraction and Purification Process of Tengkawang 242
12.4.1	Chemical Purification of Tengkawang 243

12.4.2	Physical Purification of Tengkawang	*244*
12.5	Economic Feasibility Based on Process Simulation	*244*
12.6	Benefits and Future Outlook of Tengkawang Butter	*249*
12.7	Conclusions	*250*
	References	*251*

13 Bio-succinic Acid Production from Biomass and their Applications *255*

Abdullah A. I. Luthfi, Jian P. Tan, Wen X. Woo, Nurul A. Bukhari, and Hikmah B. Hariz

13.1	Introduction	*255*
13.1.1	Background of Succinic Acid Production	*255*
13.1.2	Global Market and Demands for Bio-succinic Acid	*256*
13.2	Valorization of Biomass to Bio-succinic Acid	*257*
13.2.1	Compositional Attributes of the Biomass	*258*
13.2.2	Size Reduction	*258*
13.2.3	Pretreatment	*259*
13.2.4	Hydrolysis	*259*
13.3	Bio-succinic Acid as Fermentative Metabolite	*260*
13.3.1	Microbial Workhorse as Potential Biocatalyst	*260*
13.3.1.1	Bacteria	*260*
13.3.1.2	Fungi	*261*
13.3.2	Fermentation Conditions	*261*
13.3.3	Media Composition	*263*
13.3.3.1	Carbon Sources	*263*
13.3.3.2	Nitrogen Sources	*264*
13.3.3.3	Carbon Dioxide	*264*
13.3.3.4	Additives	*264*
13.3.4	Fermentation Configuration and Strategies	*264*
13.3.4.1	Separate Hydrolysis and Fermentation	*265*
13.3.4.2	Simultaneous Saccharification and Fermentation/Co-fermentation	*265*
13.3.4.3	Consolidated Bioprocessing	*265*
13.3.4.4	Operational Modes	*266*
13.3.4.5	Immobilization	*266*
13.4	Purification and Recovery of Succinic Acid	*267*
13.4.1	Fermentation Broth Constituents	*267*
13.4.2	Downstream Processing Method	*269*
13.4.2.1	Reactive Extraction	*270*
13.4.2.2	Forward Osmosis	*271*
13.4.2.3	Ion Exchange	*271*
13.4.2.4	Crystallization	*271*
13.5	Application of Bio-succinic Acid	*272*
13.6	Conclusions	*273*
	References	*273*

14	**Furfural and Derivatives from Bagasse and Corncob** 279
	Muryanto Muryanto, Yanni Sudiyani, Andre F. P. Harahap, and Misri Gozan
14.1	Introduction 279
14.2	Furfural as a Building Block Material 280
14.2.1	5-Chloromethylfurfural 280
14.2.2	HMFCA and Esther 281
14.3	Furfural Derivatives 281
14.3.1	Furfuryl Alcohol 281
14.3.2	Furan 283
14.3.3	Tetrahydrofuran 284
14.3.4	Furoic Acid 284
14.4	Lignocellulosic Biomass as Raw Material for Furfural Production 285
14.4.1	Sugarcane Bagasse 286
14.4.2	Corncob and Corn Stover 286
14.4.3	Other Agriculture and Industrial Biomass 287
14.4.3.1	Rice Straw 287
14.4.3.2	Oil Palm Biomass 287
14.4.3.3	Forest/Wood Biomass 287
14.5	Furfural Production 288
14.5.1	Furfural Production Process 288
14.5.1.1	Pretreatment 288
14.5.1.2	Hydrolysis Process 289
14.5.1.3	Dehydration Processes 291
14.5.1.4	Purification 292
14.5.2	Factors Affecting the Furfural Production 293
14.5.2.1	Catalyst 293
14.5.2.2	Solvent 294
14.6	Techno-economical Aspect 295
14.7	Future Trends 296
14.8	Conclusions 297
	References 297

15	**Levulinic and Formic Acids from Rice Straw and Sugarcane Bagasse** 301
	Jabosar R. H. Panjaitan and Misri Gozan
15.1	Introduction 301
15.2	Potential of Biomass Source for the Production of Levulinic and Formic Acids 304
15.2.1	Rice Straw 304
15.2.2	Sugarcane Bagasse 304
15.3	Levulinic Dan Formic Acids Formation 306
15.4	Pretreatment and Production Technologies 306
15.4.1	Size Reduction 306
15.4.2	Biological Pretreatment 307
15.4.3	Delignification 308

15.5	Purification Technologies *309*
15.6	Economic Feasibilities *312*
15.7	Case Studies *313*
15.8	Conclusions *314*
	References *314*

16 Cellulase as Biocatalyst Produced from Agricultural Wastes *319*
Wichanee Bankeeree, Suraini Abd-Aziz, Sehanat Prasongsuk, Pongtharin Lotrakul, Syahriar NMM Ibrahim, and Hunsa Punnapayak

16.1	Introduction *319*
16.2	Cellulases Diversity *320*
16.2.1	Functional Types of Cellulases *320*
16.2.2	Cellulase Structures *321*
16.2.3	Catalytic Mechanisms *322*
16.3	Cellulase-producing Microorganisms *323*
16.3.1	Aerobic Microorganisms *323*
16.3.2	Anaerobic Microorganisms *323*
16.4	Cellulase Properties *325*
16.5	Strategies to Improve Cellulase Production *326*
16.5.1	Utilization of Agricultural Waste *326*
16.5.2	Production Processes *328*
16.5.3	Consolidated Bioprocessing *330*
16.6	Techno-economic Analysis to Produce Biofuels *331*
16.7	Conclusions *332*
	References *332*

17 Conversion of Glycerol Derived from Biodiesel Production to Butanol and 1,3-Propanediol *337*
Prawit Kongjan, Alissara Reungsang, and Sureewan Sittijunda

17.1	Introduction *337*
17.2	Crude Glycerol Characteristics and Impurities *338*
17.3	Bioconversion of Crude Glycerol into Butanol and 1,3-Propanediol *340*
17.4	Purification and Recovery of 1,3-Propanediol and Butanol *343*
17.4.1	1,3-Propanediol *343*
17.4.1.1	Ion-exchange Resin-based Separation *343*
17.4.1.2	Hydrodistillation-based Separation *343*
17.4.2	Butanol Separation *345*
17.4.2.1	In Situ Gas Stripping *345*
17.4.2.2	In Situ Pervaporation *346*
17.5	Applications of 1,3-Propanediol and Butanol *346*
17.6	Challenges and Future Perspective *348*
17.7	Conclusions *349*
	Acknowledgment *350*
	References *350*

18 Sustainability of Chemical Substitutes from Agricultural and Industrial By-products *355*
Lai-Yee Phang, Suraini Abd-Aziz, Misri Gozan, and Mohamad F. Ibrahim
18.1 Introduction *355*
18.2 Sustainable Development Strategies, Policies and Regulations in Indonesia and Malaysia *358*
18.2.1 Indonesia's National Sustainable Development Strategy *358*
18.2.2 Malaysia's Policies and Regulations for Sustainable Development *359*
18.3 Case Study 1: Techno-economic Analysis for the Production of Cellulase *360*
18.3.1 Cost Analysis Using SuperPro Designer *360*
18.3.2 Environmental Impact Analysis *363*
18.3.3 Social Aspect *364*
18.4 Case Study 2: Techno-economic Analysis for the Production of Biofertilizer *364*
18.4.1 Cost Analysis Using SuperPro Designer *364*
18.4.2 Environmental Impact Analysis *367*
18.4.3 Social Aspect *367*
18.5 Challenges and Market Opportunities *368*
18.5.1 Availability and Reliability of By-products as Feedstocks *368*
18.5.2 Production Cost *368*
18.5.3 Technical Challenges *368*
18.5.4 Market Opportunities *369*
18.6 Conclusions *369*
References *370*

Index *375*

Preface

The emerging field of exploitation for agricultural wastes and industrial by-products is the alternative way to find potential feedstocks for chemical substitutes. This is our second book with Wiley-VCH-GmbH that discusses the bioconversion, bioprocessing, and biorefining of chemical substitutes from agricultural and industrial by-products. Some biomass can be directly extracted or processed and purified as a chemical substitute for other platforms. However, biomass may require various pretreatment before bioconversion and purification. Chemical substitution focuses on finding new and less hazardous solutions for a particular process or product through a biorefinery approach. The advanced biorefinery is capable of transforming agricultural wastes and industrial by-products into various chemical substitutes or value-added products. Therefore, the development of sustainable biorefining approaches will contribute to cost-effective technology.

Our recently published book in January 2022 with Wiley-VCH-GmbH entitled *Biorefinery of Oil Producing Plants for Value-Added Products* is a two-volume set that delivers a comprehensive exploration of oil-producing plants, from their availability to their pretreatment, bioenergy generation, chemical generation, and bioproduct generation, concluded with an insightful analysis of the economic effects of oil-producing plants.

This book presents several substitutes for chemicals obtained from biomass of agricultural wastes and industrial by-products covering upstream to downstream perspectives including antioxidant, oleoresin, nanocarbon materials, enzymes, essential oil, biobleaching agents, biosugars, biofertilizers, cocoa butter substitute, bio-succinic acids, furfural derivatives, levulinic acids, cellulases, and 1,3-propanediol. In each chapter, the individual aspect of bioconversion, bioprocessing, and downstream process of chemical substitutes produced from selected agricultural and industrial by-products to selected chemical substitutes is discussed.

We thank and express our appreciation of the multidisciplinary team of authors for the discussion and communication, especially for their scientific contribution to this book. Thanks to the reviewers for the improvement of the book's contents. We also thank reviewers whose suggestions immensely helped to improve the quality of the chapters. We sincerely thank the Wiley-VCH team composed of Dr. Lifen

Yang, Project Manager – Ms. Lesley Fenske, and their production and typesetting teams for constantly supporting us during the editorial process. We firmly believe that the information in this book will enhance the interdisciplinary scientific skills of readers, while also deepening their fundamental knowledge of the chemical substitutes that can be derived from agricultural and industrial by-products.

April 2023
Malaysia

Suraini Abd-Aziz

About the Editors

Editor
SURAINI ABD-AZIZ
Universiti Putra Malaysia

Suraini Abd-Aziz is a professor at the Department of Bioprocess Technology, Faculty of Biotechnology and Biomolecular Sciences, Universiti Putra Malaysia, Malaysia. She is currently the Asian Federation of Biotechnology (AFOB) Vice President of Malaysia region, President of AFOB Malaysia Chapter, and AFOB Co-chair of Bioenergy and Biorefinery Division. She graduated with a BSc (Honors) in Clinical Biochemistry from Universiti Kebangsaan Malaysia (the National University of Malaysia) in 1992, MSc (Biochemical Engineering) from the University of Wales Swansea, United Kingdom in 1994, and PhD (Biochemical Engineering) from the University of Wales Swansea, United Kingdom in 1997. In research, she developed her research interest in the field of Industrial Biotechnology specializing in bioenergy and bio-based chemicals, which focused on the utilization of agro-wastes for bioenergy and value-added products toward zero waste emission. She is one of the Malaysia's Top Research Scientists (TRSM 2018). She has been recognized as World's Top 2% Scientist in 2021 and 2022 (Career Achievement) published by Stanford University and Elsevier BV. In 2021, she received the Research Exchange Award 2021 by The Korean Society for Biotechnology and Bioengineering. Apart from being the project leader of her own research grants, she also actively contributed to other projects in her capacity as an expert in enzyme technology, fermentation technology, bioprocess engineering, and environmental biotechnology. With an h-Index of 37 (total citations more than 3800), she has published more than 200 refereed journals internationally.

About the Editors

Editor
MISRI GOZAN
Universitas Indonesia

Misri Gozan is a professor at the Bioprocess Engineering Program, Department of Chemical Engineering, Faculty of Engineering, Universitas Indonesia, Indonesia. He is the former Asian Federation of Biotechnology (AFOB) Vice President of Indonesia region. He completed his Bachelor of Engineering Education in the Department of Gas and Petrochemical Engineering (later Chemical Engineering) at Universitas Indonesia in 1993. He earned a Master in Technology (M.Tech.) in Process and Environmental Engineering from Massey University, Palmerston North, New Zealand, in 1998. In 2004, he finished his Doctor of Ingineuer (Dr.-Ing., Cumlaude) at the Faculty of Geology, Forestry, and Hydrology, TU Dresden, Germany. With an *h*-Index of 15 (total citations of 723), his research works were published in international articles related to bioprocess engineering, especially in the efficacy of biochemicals extracted from *Nicotiana tabacum*, biochemical synthesis from agricultural solid waste, and oil extraction of indigenous plants. He has been an Honorary Member of the ASEAN Federation of Engineering Organization (AFEO) since 2017. He received a research award from The Korean Society of Biotechnology and Bioengineering in 2016. In 2014 he received an award as one of the Indonesian Outstanding National Lecturers.

Editor
MOHAMAD FAIZAL IBRAHIM
Universiti Putra Malaysia

Mohamad Faizal Ibrahim is an associate professor at the Department of Bioprocess Technology, Faculty of Biotechnology and Biomolecular Sciences, Universiti Putra Malaysia, Malaysia. He obtained his Bachelor of Science (Biotechnology) in 2009 and PhD (Environmental Biotechnology) in 2013, both from UPM before being appointed as a senior lecturer at UPM in 2013. His works are mostly on the utilization of biomass for value-added products, such as enzymes, biofuels, biosugars, animal feed, activated carbon, essential oils, and other biobased chemicals, which cover various bioprocessing approaches, such as pretreatment, extraction, fermentation, recovery and purification, adsorption, carbonization, and activation. With an *h*-Index of 16 (total citations of more than 900), he has published more than 48 refereed journals internationally, 9 book chapters, 2 edited books, and 2 patents. He is a board member of the Asian Federation of Biotechnology, AFOB Bioenergy, and Biorefinery Academic Division, currently the Vice President of the AFOB Malaysia Chapter and Cochair of Science Policy and Governance of Young Scientist Network, Academy of Science Malaysia.

Editor
LAI-YEE PHANG
Universiti Putra Malaysia

Lai-Yee Phang is an associate professor at the Department of Bioprocess Technology, Faculty of Biotechnology and Biomolecular Sciences, Universiti Putra Malaysia, Malaysia. She graduated with a Bachelor of Science (Biotechnology) in 1998 from the Department of Biotechnology, Faculty of Food Science and Biotechnology, UPM, and obtained a Master's degree in Science in the field of Environmental Biotechnology at the same department in 2001. She pursued her PhD study at Kyushu Institute of Technology (KIT), Japan. Her research interest is in Environmental Biotechnology, focusing on the conversion of waste materials into valuable products, such as fine chemicals and biofuels, which includes screening of microorganisms, fermentation, optimization, and pretreatment. She is also working on the phytoremediation of contaminated soil. With an *h*-Index of 25 (total citations more than 1800), she has authored and coauthored over 70 papers in refereed journals and 7 book chapters. She is a member of the AFOB and is currently the Treasurer of the AFOB Malaysia Chapter.

1

A Glance on Biorefinery of Chemical Substitutes from Agriculture and Industrial By-products

Suraini Abd-Aziz[1], Misri Gozan[2], Mohamad F. Ibrahim[1], Lai-Yee Phang[1], and Mohd A. Jenol[1]

[1] Universiti Putra Malaysia, Faculty of Biotechnology and Biomolecular Sciences, Department of Bioprocess Technology, Serdang, 43400, Selangor, Malaysia
[2] Universitas Indonesia, Kampus UI, Faculty of Engineering, Bioprocess Engineering Program, Department of Chemical Engineering, Depok, 16424, Indonesia

1.1 Introduction

The agricultural sector is among the important biological sectors that contributed to the production of biomass. This is vital to the bioeconomy development as the biomass produced is used as feedstock for various processes of value-added products. Koul et al. [1] characterized and listed the various sources of agricultural waste as shown in Figure 1.1:

The emerging field on the utilization of agricultural wastes and industrial by-products is the alternative way to find potential feedstocks for chemical substitutes. Chemical substitution focuses on finding new and less hazardous solutions for a particular process or product through advanced biorefinery approach. Due to the exponential rise in the demand for worldwide energy, the depletion of existing fossil fuel reserves is deemed as the vital issue, thus increasing the global interest in creating a new energy pathway via the bioprocessing of agricultural wastes as well as agro-industrial biomass [2]. Agricultural waste is defined as unwanted waste as a result of agricultural activities. Industrial biomass is known to be a solid inorganic residue, which may come from various sources, including mining and metal industry, chemical industry or building industry, energy production, and forest industry [3]. These biomass are potential feedstocks with abundant organic and inorganic nutrient composition.

Every year agricultural-based industries contribute a huge amount of residues, which may result in various environmental problems due to improper waste management. This situation will further have a negative impact on animal and human health. Unfortunately, the disposal of untreated agro-industrial wastes has been reported [4] to create other concerning problems associated with climate change. The common waste management procedures applied by the related agro-industry

---

Chemical Substitutes from Agricultural and Industrial By-Products: Bioconversion, Bioprocessing, and Biorefining,
First Edition. Edited by Suraini Abd-Aziz, Misri Gozan, Mohamad Faizal Ibrahim, and Lai-Yee Phang.
© 2024 WILEY-VCH GmbH. Published 2024 by WILEY-VCH GmbH.

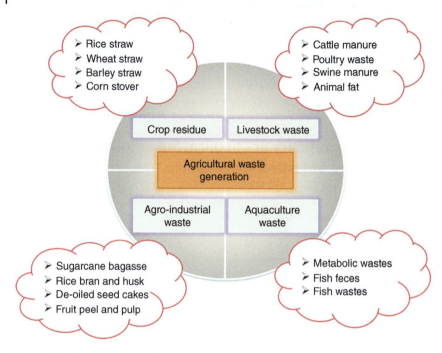

Figure 1.1 Various sources of agricultural waste. Source: Adapted from Koul et al. [1].

sectors, including burning, open dumping, and improper landfilling, should be improved with the availability and advancement of current technology.

The by-product from an industrial, commercial, mining, or agricultural operation is known as an "industrial by-product" if it is not a primary product and was not created independently throughout the process. The determination of the inorganic industrial by-product's composition and generic properties is specified by factors such as type of industry and the process. In depth, the environmental and technical properties of by-products are influenced by the equipment, process condition, and the feedstock. This further implies the suitability of the by-product for recycling material. As an example, reclaimed asphalt, waste concrete, and slag are recyclable materials, which have been used all over the world. However, the recycling rates of certain by-products vary among nations, and it can be challenging to locate precise data on the rates. The latter issue is likely caused by inadequate documentation of the amounts and end locations of by-products as well as differences in how trash is classified and how it is managed (e.g. whether filling mines is considered recycling) [5].

The evolution of the biorefinery concept was started during the late 1990s. Biorefinery covers spectrum of processes, as it defines as an overall concept of biomass processing plant conversion and/or extraction into various valuable products [6]. National Renewable Energy Laboratory (NREL) describes a biorefinery concept as the integration of biomass conversion processes as well as equipment to refine value-added products, including chemicals, fuels, and power. The International Energy Agency (IEA) Bioenergy Task 42 has defined a biorefinery as "the sustainable processing of biomass into a spectrum of marketable products (food, feed, materials, chemicals) and energy (fuels, power, heat)." Thus, a biorefinery can be a process, a facility, a plant, or a cluster of facilities for the biomass conversion [7].

An inventive and effective method of using biomass resources for the synergistic co-production of power, heat, and biofuels alongside ingredients for food and feed, pharmaceuticals, chemicals, materials, minerals, and short-cycle CO_2 is biorefining. This process involves processing biomass sustainably to create a variety of marketable biobased products and bioenergy/biofuels. Biorefining is one of the vital technologies of the circular economy, closing loops of raw biomass materials, minerals, water, and carbon. Biorefining is the optimal strategy for large-scale sustainable utilization of biomass in the bioeconomy [8].

1.2 Analysis of Feedstocks for Composition and Potential for Chemical Substitutes

Agricultural wastes are vast, which cover the by-products generated from various agricultural products, including crops, dairy products, fruits, meat, poultry, and vegetables. They are the non-product outputs of the production and processing of agricultural products that may contain material that can benefit humans but whose economic value is less than the cost of collection, transportation, and processing for beneficial use. The physicochemical composition of the wastes varies based on the type of agricultural activities as well as the system, which further results in a different form of wastes, such as solid, liquid, and slurry [1].

Agricultural-based residues can be classified into two distinct types based on the point of waste generated, such as field waste and process waste. The field wastes are mainly generated from harvesting activity, which includes leaves, stems, seedpods, and stalks. Meanwhile, the process wastes are generally the by-products produced during processing activity of the harvested products into alternate valuable products [8].

Nowadays, the raising public concern toward major categories of agricultural waste has been deemed to be the main issue. Concurrently, these wastes also have been recognized to threaten the sustainability of agricultural regimes. In fact, several agricultural activities that are affected include livestock wastes (dung, urine, residual milk, wash water, and waste feed), slaughterhouse wastes (blood, bones, flesh, hair, hides, etc.), crop residues (leaf litter, seed pods, stalks, stems, straws, husks, and weeds), poultry wastes (bedding material, droppings, spilled feed, and feathers), agro-industrial wastes (bagasse, molasses, peels [cassava, orange, and potato], pulps [apple, orange, guava, mango, papaya, pineapple, pomegranate, tomato, etc.], oil-seed cakes (coconut, groundnut, palm kernel cake, soybean, mustard, etc.), and aquaculture wastes (fecal waste, uneaten feed) [9, 10].

Through advancement in bioprocessing technology, agricultural wastes, which is also known as biomass, have been determined to be the promising feedstock and alternative renewable resource that helps to reduce existing high dependence on fossil fuel-based sector. This is deemed as impactful approach as it can help in conserving the natural resources as well as alleviate the environmental impact. The major component of agricultural biomass is lignocellulosic biomass that has shown a promising platform in various bioprocesses, including biofuels, platform chemicals, and bioproducts [11]. Lignocellulosic biomass is made up of cellulose, hemicellulose, and lignin, as cellulose is identified as the most abundant biopolymer. Cellulose is accounted to make up 30–50% of the total biomass, which is determined to be a vital substrate in microbial transformation. Various efforts have been explored to provide the strategic management and valorization of lignocellulosic biomass for the production of value-added products, which further resulted in elevating the value chain of the biomass.

1.2.1 Different Types of Agricultural Wastes and Associated Risks

The agricultural sector is urge to discover well sustainable waste management due to its vital impact on the biodiversity, human societies and world economy [12]. Due to the rapidly increasing demand, sustainable resource management is mandatory. Agricultural waste production is not only limited to farming activities, but it also covers other activities associated with food chain and farming. In each and every stage, the waste generated poses a significant risk to the environment and humans, generally, due to poor management. Oluseun Adejumo and Adebiyi [13] explained that there are seven associated agricultural wastes, which are animal wastes, food and meat processing wastes, crop production wastes, on-farm medical wastes, horticultural production wastes, industrial agricultural wastes, and chemical wastes.

The demand for using pesticides for growing crops to protect the yield of the land from weeds and insects is increasing. This is due to the impact of 42% reduction in global food production in the case of completely stopping pesticide application [14]. The main concerns in regard to this situation are the overexploitation of the pesticides, the leftover chemicals in the containers, and packaging in disposal manner. Extra measure should be taken for the disposal of these sources of waste in order to avoid a negative impact on the environment. This is even worse if they are thrown into ponds or fields because these types of agricultural wastes can contaminate the farmland with their lasting chemicals, and even lead to food poisoning. Alternatively, the usage of fertilizer to maintain the productivity of the land and the health of plants is one of the approaches used, in which organic fertilizer can be produced from the remains of farm animals using appropriate equipment. Animal waste in agriculture mainly comprises the carcasses of manure, livestock, and wastewater that come from the sanitation process of slaughterhouses. Proper management with the suitable equipment for carcass storage and disposal is necessary to avoid air pollution and bad odors. Additionally, if animal waste is not properly disposed of, it can pollute water sources, harm soil fertility, and produce greenhouse gases. One of the best ways to cut costs and waste is to reprocess animal waste into feed.

Improper management of agricultural wastes can contribute to several events, which pose a threat to living being. Based on the study reported by Adejumo and Adebiyi [13], there are three types of risks associated with improper management of agricultural wastes, which are food security, health and environmental implication, and flood. One of the main problems in the shortage of food supply is agricultural waste, associated with food wastage. This can be overcome by the recycling approach, subsequently improving food security by increasing animal protein production [15]. Apart from that, improper agricultural waste management subjected to indiscriminate burning and dumping results in various pollution, which subsequently endangers human health. Also, the indiscriminate dumping of agricultural wastes may result in the blockage of waterways, thus eventually causing floods.

1.2.2 Waste Utilization Routes

1.2.2.1 Fertilizer Application

The utilization of agricultural wastes has benefited various sectors and applications. One of them is the agricultural activity itself. The use of fertilizer from agricultural wastes, such as animal and organic (crop residue) manure, could enhance the yield of the crop. According to Wu et al. [16], the soil richness is affected by the organic fertilizer, due to the encouragement of rhizospheric microorganisms and the conditioning of the soil, such as texture and stability. In addition, the slow decomposition of organic residue is beneficial during vegetative growth process because of the slow nutrient release [17]. This could avoid the utilization of chemical fertilizer, which is associated with exposure to high-risk problems, including soil infertility, water pollution, and plant toxicity. In the case of post-chemical fertilizer-based water and land pollution, the treatment using integrated nutrient management (INM) is deemed to be vital in sustaining agricultural activity by reducing environmental pollution and elevating soil fertility [18].

In several case studies, the feasibility of agricultural wastes has been observed to give a positive impact on crop cultivation. Badar and Qureshi [19] have demonstrated that utilization of wheat bran composted with various microorganisms that has boosted soil fertility as well as enhanced the growth of sunflower cultivation. In another study conducted by Nisa et al. [20], the combination of agricultural waste biochar with straw compost as fertilizer has significantly improved the yield of paddy cultivation. These situations provide evidence of a promising approach to agricultural wastes in fertilizer application, which further enhance the productivity of crops and improve the soil condition. This further helps in conserving the environment by reducing the impact of chemically synthesized fertilizers, which contributed to various environmental pollutions.

1.2.2.2 Fibers for Textile Industry

Agricultural wastes are rich in valuable components that can be utilized in various industries, such as textiles. In the textile industry, the fiber component is one of the most important elements. The potential of agricultural wastes to be applied in the textile sector is vast due to main component of cellulosic fiber in most agricultural wastes. According to Das et al. [21], the most predominant features of textile's fiber are high cellulose and low lignin content. Therefore, the pretreatment of lignocellulosic biomass is necessary to remove the impurities of polymers, which are lignin and hemicellulose [22]. There are various agricultural wastes that have been identified to be ideal alternative fiber for the textile industry, including bagasse, cornhusk, oil palm biomass, pineapple, and leaves [1, 23, 24]. Among all the lignocellulosic biomass, bagasse has been found to have the highest cellulose content, which is determined as 65–70% in crude bagasse. The treatment of sugarcane bagasse fiber using alkali–H_2O_2 is found to have enhanced softness of fiber and improved mechanical properties [25]. In addition, corn husk fiber has been found to have a rough surface with a hollow cross-sectional view, which can be potentially applied as thermal insulation materials [26].

Apart from that, over the years, there has been an increase in interest in utilizing pineapple leaf fiber (PALF) due to its properties, which are beneficial in various applications including textiles. Asim et al. [24] summarized that the PALF has tensile strength ranging from 126.6 to 1627.0 MPa and Young's modulus of 4.41–82.5 GPa. The development of PALF-reinforced composite enhanced the properties of the composite to be versatile to be applied in the textile industry. The reinforcement of the composite is done by cooperating PALF with other composites, including epoxy [27], polypropylene [28], polyester [29], and polycarbonate [30].

1.2.2.3 Mushroom Cultivation

Agricultural wastes contain rich nutrients, and one of them is lignocellulosic content, which is a promising component for mushroom cultivation. Various types of edible mushrooms have been known to be effectively grown using agricultural wastes, including Button – *Agaricus*, Chinese mushroom – *Ganoderma*, Oyster – *Pleurotous*, Shiitake – *Lentinula edodes*, and straw – *Volvariella volvacea* [31]. Apart from helping in waste management, these mushrooms contain high nutritional values, which benefit the consumers. Several studies have been conducted to assess the feasibility of various types of agricultural wastes in mushroom cultivation. Goswami et al. [32] revealed that paddy straw has been commonly utilized as a substrate in paddy mushroom cultivation. In other cases, Kwon and Kim [33] explained that various sources of agricultural residues, including rice straw, sugarcane straw and bagasse, maize straw, wheat bran and straw, and soybean husk, have been used in *Calocybe indica*, oyster, and *Volvariella*.

Zakil et al. [34] have demonstrated the combination of various agricultural biomass, including oil palm biomass, sugarcane bagasse, corncob, and rubber tree sawdust, in cultivation of oyster mushroom, *Pleurotus ostreatus*. Based on the results obtained, this biomass contains a rich amount of potassium, which is vital in mushroom growth. In addition, the optimum condition determined to enhance the growth of *P. ostreatus* was the combination of sugarcane bagasse and rubber tree sawdust with a 1:1 ratio. In another study, Baktemur et al. [35] determined the feasibility of several other agricultural wastes, including corncob, peanut shell, poplar sawdust, wheat stalk, and vine pruning waste, in the cultivation of shiitake mushroom (*L. edodes* [Berk.] Pegler). It was concluded that the type of agricultural waste used has a significant impact on the composition of volatile aroma of the mushroom. This situation is due to the sulfur content of the biomass, which is subsequently important in the odor formation for shiitake mushrooms. All in all, the utilization of agricultural wastes in mushroom cultivation has been widely studied for their feasibility and corelation in the growth of mushrooms.

1.2.2.4 Organic Acids

The production of organic acids from agricultural wastes has been documented for the past decades, including acetic acid, citric acid, formic acid (FA), succinic acid, and a few other organic acids [36]. Kumar et al. [37] explained that the utilization of lignocellulosic biomass is one of the promising approaches for the production of various types of organic acids. In acid catalysis of lignocellulosic biomass, the formation of acetic acid, levulinic acid (LA), and FA can be observed. Citric acid, which is known as the second-largest fermentation product, is mainly produced by *Aspergillus niger* [38]. Various agricultural wastes have been utilized in citric acid production, including apple pomace [39], cassava bagasse [36], oil palm empty fruit bunch [40], and sugarcane bagasse [38].

Succinic acid is one of the most valuable chemicals in various industries, including agricultural, food, and pharmaceutical. The utilization of agricultural wastes is deemed to be a potential substitute for the current petroleum-based chemical processes. Various studies have been conducted to explore the potential of agricultural wastes as an alternative substrate for the production of succinic acid. Liu et al. [41] have demonstrated that the production of succinic acid achieved is 39.3 g/l using sugarcane bagasse hydrolysate by *Escherichia coli* BA305. The feasibility of cornstalk hydrolysate in succinic acid production has been successfully demonstrated with a yield of 0.85 and 23.1 g/l of succinic acid produced [42]. The utilization of agricultural wastes into organic acids is highly promising for sustainable production.

1.2.2.5 Industrial Enzymes

Bioconversion of agricultural wastes also benefited enzyme-based related industry. In the production of enzymes from agricultural wastes, there are two modes of fermentation involved, solid-state and submerged fermentation. Despite having gains and losses, solid-state fermentation is the most popular technique of choice in the larger scale for enzyme production. This situation is mainly due to the higher production of enzymes obtained, resulting from enhanced interaction between microorganisms and substrate as well as a supportive medium for the growth of microorganisms [43]. According to Koul et al. [1], several enzymes have been recognized in high market demand, including cellulases, lignases, proteases, pectinases, and xylanases. In the literature reviews, several studies have successfully demonstrated the production of these enzymes using bioconversion of various agricultural wastes. Ferreira da Silva et al. [44] have utilized banana pseudostem, jatropha, and coconut fiber in an attempt to produce lignin peroxidase and cellulases. The results showed that the highest production of lignin peroxidase and cellulases was 49 916 U/g and 19.5 FPU/g using jatropha and banana pseudostem, respectively. In another study, other agricultural wastes have been documented to be potentially used as substrates for the production of cellulases and xylanase. According to Dhillon et al. [45], the highest enzyme activity was recorded in the fermentation using the combination of rice straw and wheat bran. These situations have elevated the value chain of agricultural wastes, as the processes using negative-cost agricultural wastes benefited various industries as well as the environment.

1.2.3 Industrial By-products

In term of chemical substitutes from industrial by-products, various waste materials can be utilized based on the source.

1.2.3.1 Agriculture, Horticulture, and Landscaping

The use of a by-product in agriculture or horticulture presumes that it has properties favorable to plant growth. The positive effect can be related to the supply of nutrients or conditioning soil by altering its chemical, physical, or biological composition. In the former case, the nutrients need to be soluble or transformable to a form that is available to plants. By-products that have liming properties, which can correct soil acidity and increase crop yield, have been used for ground improvement. Blast furnace slag, steelmaking slag, and ash from wood combustion are the most common by-products used for this purpose. Kiln dust is also suitable for agricultural purposes.

According to US Environmental Protection Agency (USEPA) [46], industrial by-products containing potentially toxic metals, including nickel, lead, and cadmium, have been used in the production of zinc fertilizer, which is closely monitored by the USEPA of their maximum contamination limit. The calcium- and magnesium-based by-products contained in steelmaking slag and sludge are observed to provide a positive impact on herbage cultivation. The motivation for using wood ash is to avoid depleting essential soil nutrients and reduce the harmful effects of acidification of forest soils and surface waters. Wood ash also releases potassium, sodium, boron, and sulfur. In other cases, the utilization of red mud from mining industry has been extensively studied for its potential. According to Hua et al. [47], the positive responses of soil contaminated with metal/metalloid with the addition of red mud as an immobilizer have revealed its potential use. Seawater-neutralized red mud has also shown a good capacity to immobilize soluble acid and metals, particularly aluminum, zinc, and copper, from acid-sulfate soil solutions. This feature makes it a desirable alternative for lime if leaching is an issue.

1.2.3.2 Use as Raw Material or Additive of New Products

Some by-products have proven to be practical raw materials for new products. For example, blast furnace slag (BFS) has been widely used in the manufacture of concrete owing to properties similar to Portland cement, whereas fiber sludge with high inorganic content has been proven to be a suitable raw material in the manufacture of cement. Fiber sludge has also been used as the base raw material in some industrial sorbent and animal bedding products [48]. Blast furnace slag can be combined with silica or alumina and can be converted to fibers for rock wool. Spent foundry sand can be used as a source of silica in this process.

Fly ash can be mixed with the main raw material to produce ceramics, floor and wall tiles, sound insulation panels, fillers in polymers and rubbers, and zeolites and inorganic fibers [38], among others. Besides ash, foundry sand, in particular, is used as an additive to provide fines in cement and concrete. Some applications, such as the manufacture of cement, may require mechanical activation that enables the use of higher proportions of ash and the attainment of improved quality for the new product. At the same time, the potential effect on the durability of concrete structures has raised some discussion about the suitability of ash for this purpose.

1.3 Potential Application of Chemical Substitute Extracted from Selected Agricultural Wastes and Industrial By-products

The book consists of 17 chapters that discuss the selected agricultural wastes and industrial by-products as precursors or substrates for chemical substitutes or themselves as the chemical substitute.

It begins with a glance at biorefinery of chemical substitutes from agriculture and industrial by-products (Abd-Aziz et al.: Chapter 1). Agricultural waste is defined as unwanted waste as a result of agricultural activities, while industrial by-products are leftover from industrial activities with abundant organic and inorganic nutrient composition. The advanced biorefinery is capable of utilizing agricultural wastes and industrial by-products to various chemical substitutes or value-added products. This chapter, in general, will cover the consolidated bioconversion, bioprocessing, and downstream process of the significant chemical substitutes produced from agriculture and industrial by-products about the specific chemical substitutes that will be discussed in each chapter.

Antioxidant bioactive molecules can be extracted from agro-industrial wastes and can potentially be alternative nutrient sources to produce dietary supplements for health benefits (Jenol et al.: Chapter 2). In view of this, the chapter provides an overview of the various antioxidant bioactive molecules found in numerous agro-industrial wastes and biological approaches for their recovery from agricultural wastes. Also, their potential applications in pharmaceutical industries and food production as food and animal feed additives are incorporated in this chapter.

One of the components in lemongrass (*Cymbopogon citratus*) that shows high potential in bioactive activities is oleoresin (Md Salleh et al.: Chapter 3). Oleoresins are a natural mixture of essential oils and resin extracted from lemongrass leaves. Oleoresins in lemongrass provide a number of advantages over traditional spices as flavoring agents. Oleoresins can also be standardized for flavor and have a long shelf life for application in the food industry. Nowadays, the increase in demand for natural food flavors in the European market has opened more opportunities for suppliers of oleoresins in developing countries.

Seaweed is a promising biomass resource for producing high-added-value materials such as food or usable flavoring to improve the nutritious quality of food preparation and active components with antioxidant and anti-excitant properties, an alternative renewable resource for biofuel production, chemicals, and nanocarbon materials (such as activated carbon, graphene, and carbon nanotube). This chapter focuses on the options of developing those materials from seaweeds

(Prakaso et al.: Chapter 4), which these structures have encountered applications in various technological fields, such as adsorption, catalysis, hydrogen storage, and electronics.

Spent Mushroom Substrate (SMS) is the by-product of mushroom cultivation (Sabaratnam et al.: Chapter 5). It is the unutilized lignocellulosic substrate, which contains lignin, cellulose, hemicellulose, mycelia, sugars, enzymes, and chemical entities left after the harvesting of commercial mushrooms. As the mushroom industry is rapidly growing, it is facing challenges in disposing SMS. The potentials of using SMS for the production of value-added products include the production of bulk hydrolytic enzymes such as laccase, xylanase, lignin peroxidase, cellulase, and hemicellulose from SMS, and their applications can be explored. This chapter reviews the scientific research and practical applications of SMS as a readily available and cheap source of enzymes for the bioremediation of harmful soil and water contaminants, as well as for the hydrolysis of complex components into simple components for fermentable feedstock production.

Pineapple is a popular tropical fruit consumed worldwide. After processing pineapple for human consumption, large amounts of waste materials are generated, including the crown, peel, and core. These wastes are usually disposed of in landfills, causing environmental problems. However, pineapple wastes contain valuable compounds, including essential oils, that can be extracted and utilized for various applications. There are various techniques that can be used to extract pineapple essential oil, including steam distillation or hydrodistillation, solvent extraction, and supercritical fluid extraction. However, further research is required to optimize the extraction methods and explore the full potential of these essential oils. The use of essential oils derived from pineapple waste can also contribute to sustainable waste management practices and environmental preservation (Ibrahim et al.: Chapter 6).

Poultry processing plants produce huge amounts of chicken feathers. The common feathers disposal methods cause environmental pollution. As feathers are composed of 91% keratin, they form feather hydrolysate after being subjected to chemical, biological, or combined treatments. This chapter provides the treatments available for the conversion of feathers into protein hydrolysate that can be further used to produce feather meals, biofertilizers, and bioplastics. Keratinase enzyme plays an important role in the hydrolysis of feathers and has been applied in leather, feed, detergent, and cosmetics industries. The upstream process that applies molecular approaches to improve the keratinase characteristics will be discussed in this chapter. In addition, the challenges of the bioprocessing of feathers and the upstream process are highlighted (Lai-Yee et al.: Chapter 7).

Agricultural wastes, such as corncob, wheat straw, and bagasse, are available as cheap and natural carbon sources for enzyme production. The pulp and paper industry has started to phase out chemical bleaching agents with enzymes, such as cellulases, xylanases, and laccases, specifically for the pre-bleaching step and the delignification step. Using enzymes will be beneficial not only economically but also environmentally as the process contributes to less toxic waste. Multiple studies have researched producing cellulase, xylanase, or laccase in varying production methods. This chapter aims to highlight the potential of agricultural wastes with high lignocellulosic content for enzyme production, explain the usage of bio-bleaching enzymes, and analyze the techno-economical aspect through a case study (Alifia et al.: Chapter 8).

Natural rubber latex (NRL), the white sap that comes from the *Hevea brasiliensis* tree, is an important raw material for the production of various rubber products, such as gloves, tires, and tubes. However, during the processing of NRL, a protein-rich fraction (the serum) is generated and is treated as waste. The fraction generally consists of significant quantities of enzymes, namely lysozymes and chitinases, which can be industrially useful. This chapter draws attention to the general characteristics and potential applications of the enzymes and assay developments for the enzymes' activity. This chapter subsequently explores the recoveries of plant-derived lysozymes and chitinases and proposes a strategy for recovering the enzymes from the NRL. The strategy ultimately promotes the transition toward circular and sustainable bioeconomy through waste valorization (Tan et al.: Chapter 9).

In Southeast Asia, the sago palm serves as one of the most important starch providers, which has been utilized as food for centuries. In Malaysia, specifically in Sarawak, sago starch-based agro-industry is one of the major revenue sources for the state. In the near future, the sago industry will eventually be facing major challenges related to the waste management due to the rising demand of the products. The processing of sago palm into starch generates a huge amount of several types of waste, including bark, hampas, wastewater, and sago frond. The bioconversion of these wastes into value-added products (biosugar and bioproducts derivation, including L-Lactic acid, cellobiose, silage, enzymes, and kojic acid) will be reviewed. The prospects and challenges in the sago wastes biorefinery will be discussed (Jenol et al.: Chapter 10).

Sugarcane is one of the most important food crops in Asian countries for more than a millennial. As part of the industrial crops, the strategy for improving the efficiency of the agricultural sector for a vast and sustainable industry is required. Sugarcane bagasse is a main residue produced by sugar industries as the milling by-product. Chemically, it is a fibrous residue of sugarcane bagasse that contains mainly cellulose, hemicellulose, and lignin. The utilization of sugarcane bagasse and other residues in supporting industrial processes has been developed. Various chemicals are generated by conversion and bioconversion of sugarcane by-products, including biofertilizers, soil improvers, and biocomposite. In addition, the future prospects of sugarcane by-product development to support sugarcane industrial activity are presented (Fatimah et al.: Chapter 11).

Tengkawang (*Shorea stenoptera*) is a potential endemic plant of Kalimantan. However, its production process is currently still carried out traditionally. Tengkawang butter has the potential to be a cocoa butter equivalent (CBE) because it has a fatty acid composition and melting point close to cocoa butter (CB). The fatty acid profile of the tengkawang butter is dominated by palmitic acid (16–24%), stearic acid (40–47%), and oleic acid (31–33%). The melting point of tengkawang butter is at the human mouth (body) temperature of 35–37 °C. Physical and chemical processing of tengkawang is carried out to obtain the quality of tengkawang butter according to the standard butter material. This chapter also discusses the economic feasibility based on process simulation. The utilization of tengkawang butter as a substitute for CB has drawn attention as a strategy to achieve cost-effective food material production and conservation of endemic plants in Kalimantan (Darmawan et al.: Chapter 12).

Bio-succinic acid production is gaining attraction globally due to fossil fuel depletion and environmental deterioration. The production process also offers excellent performance, is less environmentally damaging, allows the use of abundant biomass resources, and is closely related to the sustainability agenda, i.e. responsible consumption and production. This chapter provides an overview of the recent insights on bio-succinic acid and the importance of its production through biochemical routes (Luthfi et al.: Chapter 13). Technologies related to the lignocellulosic valorization in bio-succinic acid production, including pretreatment, hydrolysis, and fermentation of various microbial workhorses, configurations, and strategies, as well as the purification and recovery of succinic acid are some of the recent interesting features highlighted in this chapter. Finally, the application of bio-succinic acid as a precursor to various products is also presented.

Furfural is one of the building block materials that can react to produce various fine chemicals. Furfural is widely used as a solvent in the petroleum processing industry, agrochemicals, and pharmaceuticals. It can be further converted into derivatives, such as furfuryl alcohol, furan, and tetrahydrofuran. Furfural is produced from the dehydration of xylose (Muryanto et al.: Chapter 14). Furfural production depends on the condition of hemicellulose/xylose content, catalyst used, solvent for process, and the condition of hydrolysis and dehydration process. Xylose is derived from lignocellulosic biomass. This chapter discusses furfural and its derivatives, its hydrolysis and purification processes, and the bagasse and corncob feedstock for furfural production. Techno-economic analysis is discussed based on process simulation and selected case studies.

Lignocellulose biomass can be converted to FA and LA. LA has been identified as a promising platform chemical for the next generation, such as biodiesel additives. FA represents one of the most promising modern fuel sources for membrane fuel cells. LA production commences with depolymerizing the biomass cellulose fraction into oligosaccharides and glucose. Next, pentose is hydrolyzed to 5-hydroxymethylfurfural (HMF) and dehydrated into LA and FA. The temperature and pressure of the reaction are susceptible to the products. Sugarcane bagasse and rice straw are among industrial and agricultural waste biomass. This chapter discusses the availability of these biomass and alternative production processes of LA and FA. This chapter also discusses the economic feasibility of process simulation and available studies (Panjaitan and Gozan: Chapter 15).

Cellulases are a group of diverse enzymes that are significant for a number of industrial applications, especially the conversion of lignocellulosic feedstock into monomeric sugars in biorefinery process. In recent years, the market demand for these enzymes has been continually raised due to the growth of the biorefinery industry, which also caused an increase in the commercial enzyme price. Several strategies have been proposed to reduce the cost of cellulase, while the utilization of agricultural wastes as an inexpensive source for cellulase production is mainly described in this chapter. Furthermore, cellulase classification, potential strains with high-level cellulase production, co-culture (mixed) system, and consolidated bioconversion are also discussed (Bankeeree et al.: Chapter 16).

Biodiesel is a renewable energy alternative to diesel, and during its production, abundant crude glycerol is generated as a by-product. As a waste product, crude glycerol needs to be treated before disposal. Alternatively, it can be converted into valuable products, such as 1,3-propanediol (1,3-PDO; $CH_2(CH_2OH)_2$) and butanol (C_4H_9OH). The present chapter focuses on the utilization of microbes for the production of 1,3-PDO and butanol from crude glycerol. Pathways associated with the production of 1,3-PDO and butanol, as well as characteristics and impurities in crude glycerol, are highlighted. In addition, applications of 1,3-PDO and butanol and the downstream purification and recovery processes are examined. Finally, challenges and future research perspectives on the production and application of 1,3-PDO and butanol are also noted (Kongjan et al.: Chapter 17).

The overall highlight of the book discussed the sustainable production of chemical substitutes from agricultural and industrial by-products as a way to preserve resources and protect the environment. Hence, significant attention is directed toward developing green technologies to valorize these by-products. This increase in value can make the economics of chemical substitute production from by-products more feasible and workable. However, the cost of synthesizing chemical substitutes and productivity barriers is still very challenging. This chapter provides the policies and initiatives for the sustainable development of selected countries, as well as case studies on assessing the sustainability of selected chemical substitute production strategies, considering three perspectives, i.e. environmental, economic, and social. This chapter presents two case studies on cellulase and biofertilizer production. The case studies discuss the cost and pinch analysis using the process simulation software Aspen Plus. The challenges and market opportunities will be covered (Lai-Yee et al.: Chapter 18).

1.4 Conclusions

The development of sustainable solutions for the management of by-products and food waste is one of the main challenges in our society. These solutions must be able to take full advantage of the biological potential of biomaterials and achieve economic, social, and environmental benefits. With the nutritional problems faced by society today (hunger indicators and the growing world population), the use of food waste for human food should be a priority. Wastes and by-products produced in developing countries have a powerful nutritional and functional use in their formulation and a powerful tool in minimizing hunger. In addition, the added value generated by the diversification of the productive chains can create job opportunities for the residents, generating an additional social benefit.

References

1 Koul, B., Yakoob, M., and Shah, M.P. (2022). Agricultural waste management strategies for environmental sustainability. *Environ. Res.* 206: 112285.

2 Torres-León, C., Ramírez-Guzman, N., Londoño-Hernandez, L. et al. (2018). Food waste and byproducts: an opportunity to minimize malnutrition and hunger in developing countries. *Front. Sustain. Food Syst.* 2: 52.
3 Obi, F.O., Ugwuishiwu, B.O., and Nwakaire, J.N. (2016). Agricultural waste concept, generation, utilization and management. *Niger. J. Technol.* 35: 957–964.
4 Sadh, P.K., Duhan, S., and Duhan, J.S. (2018). Agro-industrial wastes and their utilization using solid state fermentation: a review. *Bioresour. Bioprocess.* 5: 1–15.
5 Sorvari, J. and Wahlström, M. (2014). Industrial by-products. In: *Handbook of Recycling* (ed. E. Worrell and M.A. Reuter), 231–253. Elsevier https://doi.org/10.1016/B978-0-12-396459-5.00017-9.
6 Kamm, B., Schneider, B.U., Hüttl, R.F. et al. (2006). Lignocellulosic feedstock biorefinery–combination of technologies of agroforestry and a biobased substance and energy economy. *Forum der Forsch.* 19: 53–62.
7 de Jong, E. and Jungmeier, G. (2015). Biorefinery concepts in comparison to petrochemical refineries. In: *Industrial Biorefineries & White Biotechnology* (ed. A. Pandey, R. Höfer, M. Taherzadeh, et al.), 3–33. Elsevier.
8 IEA Bioenergy. (2022). Newsletter IEA Bioenergy Task42. Task42 Biorefining a Futur BioEconomy 2017:1–25. http://task42.ieabioenergy.com/publications/newsletter-number-2-august-2017 (accessed 1 October 2022).
9 Fadzil, N.F. and Othman, S.A. (2021). The growing biorefinery of agricultural wastes: a short review. *J. Sustainable Nat. Res.* 2: 46–51.
10 Tripathi, N., Hills, C.D., Singh, R.S., and Atkinson, C.J. (2019). Biomass waste utilisation in low-carbon products: harnessing a major potential resource. *Npj Clim Atmos Sci* 2: 35. https://doi.org/10.1038/s41612-019-0093-5.
11 Duque-Acevedo, M., Belmonte-Ureña, L.J., Cortés-García, F.J., and Camacho-Ferre, F. (2020). Agricultural waste: review of the evolution, approaches and perspectives on alternative uses. *Global Ecol. Conserv.* 22: e00902. https://doi.org/10.1016/j.gecco.2020.e00902.
12 WWF. (2023). Impact of sustainable agriculture and farming practices. Sustainainable Agriculture. https://www.worldwildlife.org/industries/sustainable-agriculture (accessed 3 January 2023).
13 Oluseun Adejumo, I. and Adebukola, A.O. (2021). Agricultural solid wastes: causes, effects, and effective management. In: *Strategies of Sustainable Solid Waste Management* (ed. H.M. Saleh). IntechOpen https://doi.org/10.5772/intechopen.93601.
14 Northerly. (2019). Top causes of agricultural waste, and how Northerly works to combat them. https://northerly.ag/causes-of-agricultural-waste (accessed 3 January 2023).
15 Adejumo, I., Adetunji, C., and Adeyemi, O. (2017). Influence of UV light exposure on mineral composition and biomass production of mycomeat produced from different agricultural substrates. *J. Agric. Sci. Belgrade* 62: 51–59. https://doi.org/10.2298/JAS1701051A.
16 Wu, L., Jiang, Y., Zhao, F. et al. (2020). Increased organic fertilizer application and reduced chemical fertilizer application affect the soil properties and bacterial

communities of grape rhizosphere soil. *Sci. Rep.* 10: 9568. https://doi.org/10.1038/s41598-020-66648-9.

17 Shaji, H., Chandran, V., and Mathew, L. (2021). Organic fertilizers as a route to controlled release of nutrients. In: *Controlled Release Fertilizers for Sustainable Agriculture* (ed. F.B. Lewu, T. Volova, S. Thomas, and K.R. Rakhimol), 231–245. Elsevier https://doi.org/10.1016/B978-0-12-819555-0.00013-3.

18 Wu, W. and Ma, B. (2015). Integrated nutrient management (INM) for sustaining crop productivity and reducing environmental impact: a review. *Sci. Total Environ.* 512–513: 415–427. https://doi.org/10.1016/j.scitotenv.2014.12.101.

19 Badar, R. and Qureshi, S.A. (2015). Utilization of composted agricultural waste as organic fertilizer for the growth promotion of sunflower plants. *J Pharmacogn. Phytochem.* 3: 184–187.

20 Nisa, K., Siringo-Ringo, L., and Zaitun, M. (2019). The utilization of agricultural waste biochar and straw compost fertilizer on paddy plant growth. *IOP Conf. Ser. Mater. Sci. Eng.* 506: 012061. https://doi.org/10.1088/1757-899X/506/1/012061.

21 Das, P.K., Nag, D., Debnath, S., and Nayak, L.K. (2010). Machinery for extraction and traditional spinning of plant fibres.

22 Hu, T.Q. (2008). *Characterization of Lignocellulosic Materials*. Oxford, UK: Blackwell Publishing Ltd. https://doi.org/10.1002/9781444305425.

23 Nikmatin, S., Irmansyah, I., Hermawan, B. et al. (2022). Oil palm empty fruit bunches as raw material of dissolving pulp for viscose rayon fiber in making textile products. *Polymers* 14: 3208. https://doi.org/10.3390/polym14153208.

24 Asim, M., Abdan, K., Jawaid, M. et al. (2015). A review on pineapple leaves fibre and its composites. *Int. J. Polym. Sci.* 2015: 1–16. https://doi.org/10.1155/2015/950567.

25 Jalalah, M., Khaliq, Z., Ali, Z. et al. (2022). Preliminary studies on conversion of sugarcane bagasse into sustainable fibers for apparel textiles. *Sustainability* 14: 16450. https://doi.org/10.3390/su142416450.

26 Ratna, A.S., Ghosh, A., and Mukhopadhyay, S. (2022). Advances and prospects of corn husk as a sustainable material in composites and other technical applications. *J. Cleaner Prod.* 371: 133563. https://doi.org/10.1016/j.jclepro.2022.133563.

27 Lopattananon, N., Payae, Y., and Seadan, M. (2008). Influence of fiber modification on interfacial adhesion and mechanical properties of pineapple leaf fiber-epoxy composites. *J. Appl. Polym. Sci.* 110: 433–443. https://doi.org/10.1002/app.28496.

28 Arib, R.M.N., Sapuan, S.M., Ahmad, M.M.H.M. et al. (2006). Mechanical properties of pineapple leaf fibre reinforced polypropylene composites. *Mater. Des.* 27: 391–396. https://doi.org/10.1016/j.matdes.2004.11.009.

29 Mohamed, A.R., Sapuan, S.M., and Khalina, A. (2010). Selected properties of hand-laid and compression molded vinyl ester and pineapple leaf fiber (PALF)-reinforced vinyl Ester composites. *Int. J. Mech. Mater. Eng.* 5: 68–73.

30 Ichazo, M., Albano, C., González, J. et al. (2001). Polypropylene/wood flour composites: treatments and properties. *Compos. Struct.* 54: 207–214. https://doi.org/10.1016/S0263-8223(01)00089-7.

31 Kamthan, R. and Tiwari, I. (2017). Agricultural wastes-potential substrates for mushroom cultivation. *Eur. J. Exp. Biol.* 7. https://doi.org/10.21767/2248-9215.100031.

32 Goswami, S.B., Mondal, R., and Mandi, S.K. (2020). Crop residue management options in rice–rice system: a review. *Arch. Agron. Soil Sci.* 66: 1218–1234. https://doi.org/10.1080/03650340.2019.1661994.

33 Kwon, H. and Kim, B.S. (2004). Bag cultivation. In: *Mushroom Growers' Handbook 1: Oyster Mushroom Cultivation*, 151–164. MushWorld.

34 Ahmad Zakil, F., Xuan, L.H., Zaman, N. et al. (2022). Growth performance and mineral analysis of *Pleurotus ostreatus* from various agricultural wastes mixed with rubber tree sawdust in Malaysia. *Bioresour. Technol. Rep.* 17: 100873. https://doi.org/10.1016/j.biteb.2021.100873.

35 Baktemur, G., Çelik, Z.D., Kara, E., and Taşkın, H. (2020). The effect of different agricultural wastes on aroma composition of Shiitake (*Lentinula edodes* (Berk.) Pegler) mushroom. *Turk. J. Agric. Food Sci. Technol.* 8: 1540–1547. https://doi.org/10.24925/turjaf.v8i7.1540-1547.3415.

36 Prado, F.C., Vandenberghe, L.P.S., Woiciechowski, A.L. et al. (2005). Citric acid production by solid-state fermentation on a semi-pilot scale using different percentages of treated cassava bagasse. *Braz. J. Chem. Eng.* 22: 547–555. https://doi.org/10.1590/S0104-66322005000400007.

37 Kumar, A., Gautam, A., and Dutt, D. (2016). Biotechnological transformation of lignocellulosic biomass in to industrial products: an overview. *Adv. Biosci. Biotechnol.* 07: 149–168. https://doi.org/10.4236/abb.2016.73014.

38 Kumar, A. and Jain, V.K. (2008). Solid state fermentation studies of citric acid production. *Afr. J. Biotechnol.* 7: 644–650.

39 Dhillon, G.S., Brar, S.K., Kaur, S., and Verma, M. (2013). Bioproduction and extraction optimization of citric acid from *Aspergillus niger* by rotating drum type solid-state bioreactor. *Ind. Crops Prod.* 41: 78–84. https://doi.org/10.1016/j.indcrop.2012.04.001.

40 Bari, M.N., Alam, M.Z., Muyibi, S.A. et al. (2009). Improvement of production of citric acid from oil palm empty fruit bunches: optimization of media by statistical experimental designs. *Bioresour. Technol.* 100: 3113–3120. https://doi.org/10.1016/j.biortech.2009.01.005.

41 Liu, R., Liang, L., Li, F. et al. (2013). Efficient succinic acid production from lignocellulosic biomass by simultaneous utilization of glucose and xylose in engineered *Escherichia coli*. *Bioresour. Technol.* 149: 84–91. https://doi.org/10.1016/j.biortech.2013.09.052.

42 Bao, H., Liu, R., Liang, L. et al. (2014). Succinic acid production from hemicellulose hydrolysate by an *Escherichia coli* mutant obtained by atmospheric and room temperature plasma and adaptive evolution. *Enzyme Microb. Technol.* 66: 10–15. https://doi.org/10.1016/j.enzmictec.2014.04.017.

43 Chugh, P., Soni, R., and Soni, S.K. (2016). Deoiled rice bran: a substrate for co-production of a consortium of hydrolytic enzymes by *Aspergillus niger* P-19. *Waste Biomass Valorization* 7: 513–525. https://doi.org/10.1007/s12649-015-9477-x.

44 Ferreira da Silva, I., Rodrigues da Luz, J.M., Oliveira, S.F. et al. (2019). High-yield cellulase and LiP production after SSF of agricultural wastes by *Pleurotus ostreatus* using different surfactants. *Biocatal. Agric. Biotechnol.* 22: 101428. https://doi.org/10.1016/j.bcab.2019.101428.

45 Dhillon, G.S., Oberoi, H.S., Kaur, S. et al. (2011). Value-addition of agricultural wastes for augmented cellulase and xylanase production through solid-state tray fermentation employing mixed-culture of fungi. *Ind. Crops Prod.* 34: 1160–1167. https://doi.org/10.1016/j.indcrop.2011.04.001.

46 USEPA (2001). *The Micronutrient Fertilizer Industry: From Industrial By-product to Beneficial Use*. Washington: BiblioGov.

47 Hua, Y., Heal, K.V., and Friesl-Hanl, W. (2017). The use of red mud as an immobiliser for metal/metalloid-contaminated soil: a review. *J. Hazard. Mater.* 325: 17–30. https://doi.org/10.1016/j.jhazmat.2016.11.073.

48 Madison. (2003). Guidance for the beneficial use of industrial byproducts under ch. nr. [Internet]. Wisconsin DNR. https://dnr.wi.gov/files/PDF/pubs/wa/WA1769.pdf (accessed 16 March 2023).

2

Antioxidants from Agricultural Wastes and their Potential Applications

Mohd A. Jenol[1], Yazmin Hussin[2], Pei H. Chu[1], Suraini Abd-Aziz[1], and Noorjahan B. Alitheen[2]

[1] *Universiti Putra Malaysia, Department of Bioprocess Technology, Faculty of Biotechnology and Biomolecular Sciences, Serdang, 43400, Selangor, Malaysia*
[2] *Universiti Putra Malaysia, Department of Cell and Molecular Biology, Faculty of Biotechnology and Biomolecular Sciences, Serdang, 43400, Selangor, Malaysia*

2.1 Introduction to Antioxidants and their Usages

Antioxidant molecules can scavenge free radicals and prevent lipids and proteins from oxidizing. The presence of antioxidants in our body helps stabilize and prevent free radical damage by providing electrons to the damaged cells, converting free radicals into waste by-products, which will then be excreted from the body [1]. The human body readily produces endogenous antioxidants, categorized into enzymatic and nonenzymatic antioxidants. However, exogenous antioxidants are often suggested to be incorporated through diet or consuming supplements [2]. An example of an enzymatic antioxidant playing an essential role in the biological system is superoxide dismutase (SOD), which catalyzes the disproportionation reaction of hydrogen peroxide and molecular oxygen from the reactive superoxide anion. Other examples include catalase, peroxiredoxins, glutathione peroxidase, and reductase, which all neutralize hydrogen peroxide into water and oxygen molecules [3]. On the other hand, nonenzymatic proteins, which are also categorized as preventive antioxidants found in the blood plasma, include ferritin, transferrin, albumin, and metallothionein, which function by preventing the formation of new reactive oxygen species (ROS) by binding transition metal ions [3].

Normal cell metabolism in living organisms commonly leads to the production of ROS, which are often developed by the accumulation of the standard use of oxygen by certain cell-mediated immune functions and respirations by the main organelles in the cells, such as mitochondria, chloroplasts, and peroxisomes [2]. The ROS can be categorized into free radicals and nonradicals. Free radicals are unstable molecules with unpaired electrons. They are by-products produced during

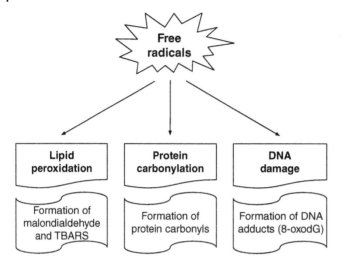

Figure 2.1 Schematic diagram of several types of free radicals. Source: Law et al. [5]/MPDI/Licensed under CC BY 4.0.

normal cell metabolism and have functional roles in cell signaling, cell death, ion transportation, and more [4]. However, ROS can be beneficial or detrimental to the living organism depending on their concentration. Examples of free radicals are hydroxyl radicals, superoxide anion, and nitric oxide. Free radicals can also be form because of environmental factors like pollution, nicotine or drug consumption, and high temperature. Figure 2.1 is a schematic diagram illustrating the detrimental effects of free radicals on biomolecules.

Oxidative stress can be described as the imbalance in the equilibrium between the formation of ROS and antioxidant protection [5]. Several diseases like Alzheimer's and Parkinson's diseases, hypertension, diabetes, cardiovascular diseases, and autoimmune disorders have been linked to oxidative stress, which could result in genetic instability and modification in DNA replication [6].

Antioxidant enzymes such as catalase and SOD act as the primary line of defense against oxidative stress caused by ROS, while non-enzymatic antioxidants such as uric acid and glutathione act as the second line of defense. ROS is constantly generated at low cell concentrations via electron-transport chain during aerobic respiration. Nevertheless, any excessively increased ROS concentrations to the point that the endogenous antioxidants in the body cannot neutralize them may lead to oxidative stress that could destroy cellular components like the membrane lipids, proteins, and DNA, thus affecting cellular homeostasis [7].

There are several mechanisms of action for antioxidants to maintain oxidative homeostasis, as shown in Figure 2.2. This includes directly scavenging free radicals or other ROS formed and inhibiting ROS formation at the cellular level by modulating antioxidant, prooxidant enzymes, and transcription factors [7]. Besides that, the formation of bioactive metabolites, which can remove or repair damages and modifications caused by ROS also affects the oxidative homeostasis [7].

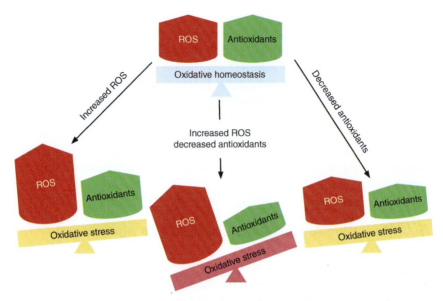

Figure 2.2 The schematic diagram of oxidative stress depicting the causal effect of increased reactive oxygen species (ROS) formation, decreased antioxidant capabilities, or the combination of both on oxidative homeostasis. Source: Li et al. [8]/Reactive Oxygen Species.

2.2 Sources of Antioxidants

The most abundant external sources of antioxidants can be found in fresh fruits and vegetables, spices, whole grains, nuts, certain meats, poultry, and fish. Figure 2.3 illustrates the antioxidant-rich sources and physiological effects of the antioxidants.

Vibrant-colored fruits like citruses and berries are great sources of antioxidant potential due to their phytochemical content, namely flavonoids, tannins, stilbenoids, phenolic acids, and lignans. The Goji berries are an excellent source of macronutrients and micronutrients, which contain high polysaccharides, carotenoids, and phenolics [9]. Herbs and traditional plant-based medicines are also rich source of antioxidants, which could contribute to their medicinal qualities.

Another interesting and widely consumed beverage in the daily diet which possesses substantial antioxidant potential is coffee. Coffee has an abundance of a

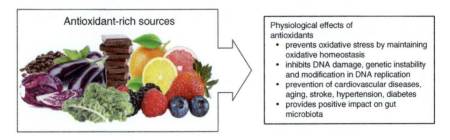

Figure 2.3 The antioxidant-rich sources and physiological effects of the antioxidants.

group of aromatic acids called hydroxycinnamic acids. Types of hydroxycinnamic acids like caffeic, coumaric, and ferulic, and several bioactive compounds such as caffeine, kahweol, nicotinic acid, and tocopherol have been reported to contribute to its antioxidant ability [10]. Celik and Gökmen [10] also found that the different brewing preparations of coffee and infused dark chocolate affected the interaction between the antioxidants in the brews. Dark chocolate is known for its health benefits due to its flavonoid-rich content compared to fruits, vegetables, and berries.

Antioxidants can also be found in various microalgae sources, such as brown seaweed, red seaweed, and green seaweed, which are rich in water-soluble vitamins, phlorotannin, carotenoids, fucoxanthin, and other bioactive compounds that contribute to their antioxidant capabilities [11].

2.3 Alternative Antioxidants Sources from Agricultural Wastes

Agricultural wastes are the organic by-products discarded from processes in agricultural-based industries. These agricultural wastes include residues from the field, such as stems, stalks, and leaves, and wastes from operations, such as husks, seeds, roots, bagasse, and molasses. Agricultural-based industrial wastes such as peels, nut oil cakes, coconut oil cakes, and soybean oil cakes are also produced in huge amounts due to the size of the demand for processed food fulfilled by the industry. It has been reported that fruit and vegetable waste contain very large amounts of bioactive molecules. The fruit peels and seeds have been reported to contain higher phenolic content than the edible fruit pulp by up to 15% for some fruits [12]. Through various waste valorization methods, converting waste materials into beneficial products, agricultural waste could be utilized more efficiently by extracting the available antioxidant molecules.

Antioxidants are substances, which can be found in plants and certain foods and have physiological effects on living organisms, which can be categorized into four common groups: polyphenols, alkaloids, terpenes, and saponins, with polyphenols being the most abundant secondary metabolites that can be found in plants. In horticultural crops, fruits and vegetables are the most widely consumed – either cooked or uncooked – making them the largest contributor to peel waste. The fruits and vegetables-based industries produce between 25% and 30% of the total product, commonly made up of peels, pomace, rind, and seeds [13]. Several examples of agricultural waste, which is reported to have potential bioactive compounds with antioxidative activity are shown in Table 2.1.

2.4 Extraction of Antioxidants from Selected Agricultural Waste

Extraction is the first step for isolating bioactive compounds from plants and agricultural wastes. Maceration is a traditional technique for the extraction of

Table 2.1 List of several agricultural wastes and the bioactive compounds with the reported antioxidant-related effects.

Agricultural waste	Bioactive compounds	Reported antioxidant-related effects	References
Pomegranate	Anthocyanin, quercetin, punicalagin, ellagic acid, gallic acid, and punicic acid from peel waste, and punicic acid, linoleic acid, and oleic acid from seed.	Has high phenolic, flavonoid, and flavonol content, utilized as nutraceutical, has great skincare benefits	[12–14]
Coconut	Various phenolic compounds	Efficient in remediating water pollutants	[15]
Citruses	Hesperidin, nobiletin, coumarin, D-limonene	Soothe sore muscles and indigestion, possess antidiabetic effects and antioxidant effects with comparable reduction power to Trolox	[16, 17, 20]
Brazillian savanna	Polyphenols and alkaloids from fruit peel	Antioxidant effects, promising antiproliferative and anti-migrative effects against human cancer liver cells	[18]
Lannea humilis (Oliv.)	Lupeol and camperstrol	Possess radical scavenging potential comparable to ascorbic acid	[19]

antioxidants. As an alternative to conventional extraction, the applications of green extraction techniques have been receiving great attention due to their vast range of advantages. Green extraction involves the use of modern green techniques with alternative green solvents in a circular economy (CE) concept. These extraction methods include pressurized liquid extraction (PLE), microwave-assisted, ultrasound, and supercritical fluid extraction (SFE). A comparison between the conventional and green extraction techniques applied in the antioxidants extraction is listed in Table 2.2. The extraction conditions and performance of these extraction techniques applied to sugar beet leave and feijoa leave are shown in Table 2.3.

2.4.1 Maceration

Maceration (Figure 2.4) is one of the conventional techniques for extracting phenolic compounds [21]. It is traditionally used to extract antioxidants from solid samples. Maceration extraction of antioxidants involves using organic solvents or water as the liquid phase based on solid–liquid separation [22]. Methanol, ethanol, acetone, and water are commonly used solvents for extraction. Other than being cheap and easily accessible, ethanol is effective in extracting various polyphenols,

Table 2.2 Comparison of conventional and green extraction techniques for the extraction of antioxidants.

Extraction technique	Advantages	Disadvantages	References
Maceration	• Adequate and subsequent recovery of antioxidants from various plant materials • The simplest method for solid-liquid extraction	• Time-consuming • Usage of high volume of toxic solvents, such as methanol	[21]
Pressurized Liquid Extraction (PLE)	• Fast extraction requires less raw material • Small volume of solvent needed • Lower energy consumption and low cost of extraction • High product recovery	• May lead to the degradation of thermolabile compounds • Involves complex sample matrices • Expensive equipment is required	[24, 25]
Microwave Assisted Extraction (MAE)	• Easy, low risks, and less expensive • Rapid, quick heating for extraction • Less solvent consumption can be used with or without solvent • More efficient and homogeneous	• A cleanup step is needed • May cause the oxidation of lipids in the matrix and interfere with the final product	[22, 23, 25]
Ultrasounds Assisted Extraction (UAE)	• Simple, rapid, and low overall cost • Low energy consumption and low maintenance • Preserves and increases biological activity of extracts • Feasible, economically profitable on a large scale • Improve products quality and safety	• High pressure or shear force could break down polysaccharide chains • Limited only to small-dimension reactors • May involve deleterious effects on the target compounds	[22, 23, 25–27]
Supercritical Fluid Extraction (SFE)	• Cost-effective, high yield, and environmental friendly • Suitable for large-scale recovery • Nontoxic, nonflammable, and high selectivity • Minimal alteration of active ingredients • Less deterioration of thermally labile compounds	• High capital equipment costs • Mechanical stirrers are not possible • Complex method development, and lower sample throughput, may lead to poor efficiency • Drying of the samples is needed before SFE • Difficult to extract polar compounds	[23, 25, 28, 29]

Table 2.3 Antioxidant extracts obtained from agricultural wastes using various extraction techniques.

Material	Extraction techniques	Extraction conditions	Findings	References
Sugar beet leaves	Maceration	1 : 10 (m/v) solid-to-liquid ratio; 50% ethanol; 24 hr; room temperature; 150 rpm	Total Phenolics Content (TPC) of < 0.9826 g GAE/100 g DW	[27]
	MAE	1 : 10 (m/v) solid-to-liquid ratio; 70% ethanol; 10 min; 600 W	TPC of 1.717 g GAE/100 g DW	
	UAE	1 : 10 (m/v) solid-to-liquid ratio; 30/50% ethanol; 30 min; 50/70 °C, 40 kHz, 60 W/l	TPC of 1.6191 g GAE/100 g DW	
	PLE	1 : 20 (m/v) solid-to-liquid ratio; 50% ethanol; 10 min; 150 °C; dynamic	TPC of 1.3173 g GAE/100 g DW	
	Subcritical water extraction	1 : 20 (m/v) solid-to-liquid ratio; water; 10 min; dynamic	TPC of 0.9826 g GAE/100 g DW	
Feijoa leaf extracts	SFE	CO_2/15% ethanol-water (50% v/v); 210 min; 55 °C; 30 MPa	TPC of 162 mg GAE/g	[29]
	PLE	Ethanol-water; 50 min; 80 °C; 10 MPa; dynamic	TPC of 166 mg GAE/g	
	UAE	Ethanol; 10 min; 20 kHz; 250 W	TPC of 92 mg GAE/g	

which makes it frequently used as an extraction solvent. However, there is no standard solvent for extracting phenolic compounds from plants or agricultural wastes [22].

Maceration extraction is time-consuming; however, it is adequate for extracting antioxidants from plant materials. During maceration, a longer time is needed to allow the solvents to penetrate and break the cell walls of the plants. This will eventually dissolve the analytes efficiently in the solvents [21]. In order to increase the efficiency of maceration extraction, optimization of various parameters should be conducted. These include the types of solvent, time and temperature for the extraction, sample-to-solvent ratio, agitation, and separation method.

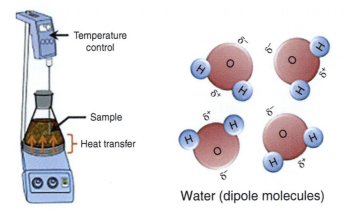

Figure 2.4 Experimental setup for conventional extraction of high value-added molecules from plant matrices used at laboratory scale. Source: Barba et al. [23] with permission.

2.4.2 Pressurized Liquid Extraction

PLE (Figure 2.5) is an advanced green extraction technique, which is extensively applied in the valorization of agricultural wastes for the recovery of bioactive compounds [27]. PLE is also referred to as pressurized fluid extraction (PFE), pressurized hot-solvent extraction (PHSE), or accelerated solvent extraction (ASE). It is also termed subcritical water extraction (SWE), superheated water extraction (SHWE), or pressurized hot-water extraction (PHWE) when water is used as the extraction solvent [23]. In PLE, high pressure and temperature are applied to maintain the solvent in a liquid state. The increase in its penetration into the matrix favors the extraction of the compounds of interest [22].

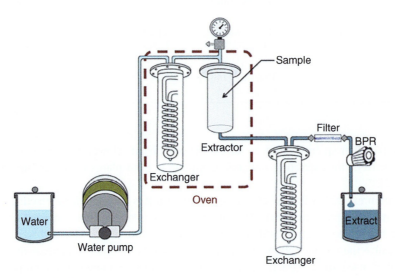

Figure 2.5 Schematic representation of pressurized hot water extraction equipment. BPR: backpressure regulator. Source: Barba et al. [23]/With permission from Elsevier.

At a determined temperature, the pressurized solvent is pumped into an extraction vessel, which consists of the sample matrix [25]. An oven controls the extraction temperature with different valves and restrictors for pressure control. Temperature is the most critical factor in PLE, making it a highly efficient and relatively selective technique for extracting antioxidants [27]. Solvent type, extraction time, pressure, and particle size can also affect extraction efficiency and selectivity [24].

The use of high temperature during the extraction improves its efficiency due to the disruption of analyte–sample matrix interactions caused by van der Waals forces, hydrogen bonding, and dipole attraction. London dispersion forces (induced dipole–induced dipole forces) become favorable over the hydrogen bonding in the subcritical zone, reducing the solvent's polarity. By lowering the activation energy for desorption, thermal energy will also decrease the cohesive (molecule–molecule) interactions between solute molecules and adhesive forces between dissimilar (solute and matrix) molecules [24]. These properties allow the solvent to reach deeper areas and increase the surface contact, thus enhancing the mass-transfer rate of the molecule in the solvent and thereby facilitating the extraction rate.

The elevated pressure during the extraction will increase the solvent's diffusivity and facilitate solvent penetration within the matrix particles to contact the analytes. The pressure exerted on the matrix during extraction also leads to disruption, thereby enhancing the mass transfer of the analyte from sample to solvent. Air bubbles within the matrix will hinder the solvent from reaching the analyte. Therefore, by applying high pressure, the problem related to air bubbles can be controlled, thus increasing the solubility of the analyte and desorption kinetics of the sample matrix. During PLE, a dual mixture of solvents is also possible. In this case, one of the solvents can enhance the analyte solubility, whereas the other can improve the analyte desorption [24].

2.4.3 Microwave-assisted Extraction

Microwave-assisted extraction (MAE) (Figure 2.6) is a green extraction method, which employs microwave energy, and it is based on the direct impact on polar compounds [28]. It consists of an electric and magnetic field oscillating perpendicularly, with nonionizing electromagnetic waves in a frequency range between 300 MHz and 300 GHz. The energy is absorbed and transferred as heat to the surrounding molecules following ionic conduction and dipole rotation [23, 28].

Heat generation using microwave energy directly affects the molecules through ionic conduction and dipole rotation. In ionic conduction, the ions undergo electrophoretic migration in the electromagnetic field. The flow of ions in the sample trying to align themselves in phase with the electromagnetic field will produce resistance and friction caused by intermolecular forces, thereby generating heat. Dipole rotation involves the realignment of dipoles with the applied field. Heat is generated due to forced molecular movement when the dipoles align and randomize at the frequency commonly used in commercial systems. In most applications, ionic conduction and dipole rotation occur simultaneously [28].

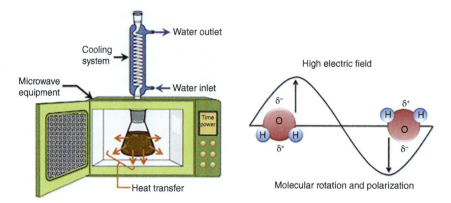

Figure 2.6 Microwave-assisted extraction equipment used at laboratory scale showing the molecular rotation mechanism. Source: Barba et al. [23]/With permission from Elsevier.

There are three steps in the MAE process, which include the separation of solutes from active sites of the sample matrix at the elevated temperature and pressure, followed by diffusion of the solvent across the sample matrix, and lastly, the release of solutes from the sample matrix to solvent [23]. A synergistic combination of heat and mass gradients working in the same direction can accelerate the process and result in a high extraction yield in MAE [25]. The heating process in MAE depends on the microwave absorbing capacity of the solvent, in which the solvent with a strong dipolar moment has a higher capacity, such as water and alcohol. This results in higher adaptation of MAE to extract polar compounds. Cell swelling may also occur due to increased pressure, which forces the cells to split. Additionally, the pressure in the biological material during MAE will alter the physical properties of the tissues, allowing better penetration of the extraction solvent by increasing the porosity of the biological matrix [28].

2.4.4 Ultrasounds-assisted Extraction

Ultrasound-assisted extraction (UAE) (Figure 2.7) is based on the use of ultrasounds with a frequency above the audible range, which is up to 20 kHz [28]. A vibrating body, the ultrasound output, allows waves to transfer energy to other neighboring particles by vibrating the surrounding medium. In UAE, the waves cause cavitation within the solvent, which produces cavitation bubbles when ultrasound is applied [22]. A local increase in pressure and temperature due to cavitation will promote the diffusivity, solubility, and transport of solute molecules, leading to the bubble's collapse and increasing the contact surface area, thus allowing better penetration of solvents into plant cell walls and eventually increasing the extraction of antioxidants.

The mechanism of UAE involves diffusion through cell walls and washing out of cell content when the walls break. The creation, expansion, and implosive collapse of microbubbles is the sonication principle, called acoustic cavitation. It is formed when bubbles form and collapse in a liquid medium due to alternating pressure changes [25, 28]. The extraction efficiency in UAE is influenced by factors, such

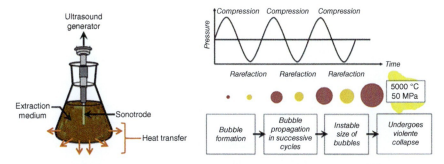

Figure 2.7 Ultrasound-assisted extraction principle and cavitation phenomenon. Source: Barba et al. [23]/With permission from Elsevier.

as frequency, ultrasonic power, ultrasonication time, temperature, solvent, and the solvent-to-solid ratio [23].

High temperature during UAW improves the solubility of the analytes in the solvent and increases the diffusion coefficient of the compounds. Other than enhanced material porosity, higher solvation and mass transfer and increased temperature also enhance the extraction yield by reducing the surface tension and viscosity of the extracts. However, as the temperature rises, the yield of antioxidants could decrease due to the decomposition of volatile bioactive compounds in the extracts. In addition, ultrasonication power also influences the yield of antioxidants in the UAE. Formation and collapse of more bubbles may take place at high ultrasonication power. The power increase may also increase the yield of antioxidants due to larger amplitudes of ultrasound waves that travel through the extracting solution. However, ultrasonication at excessively high power will generate high amounts of free radicals, which may degrade the antioxidant ingredients in the extract, thus reducing the antioxidant properties [26]. Furthermore, the choice of solvent used can also affect the efficiency of antioxidant extraction. The type of solvent, amount of solvent and concentration, and the ratio of solvent and solute are the factors, which contribute to the transmission of ultrasonic energy.

2.4.5 Supercritical Fluid Extraction

SFE is a solid–fluid extraction technique utilizing fluid in the supercritical state to solvate the solid matrix for the solute extraction within the matrix, as shown in Figure 2.8 [28]. SFE is based on using solvents maintained at pressure and temperature above their critical point [22]. At this phase, there is no liquid–gas phase boundary and surface tension. Therefore, it can retain both gas and liquid properties [28]. The solubilizing power is higher than gases due to the intermediate density of the supercritical fluids between gas and liquid. It can also penetrate porous solid materials more effectively compared to liquid solvents, thus leading to higher extraction yield [26]. These properties, which allow changing pressure and temperature values that may directly influence the density, make it a very selective technique for a target compound.

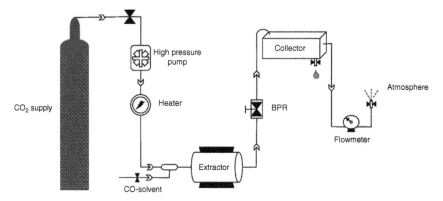

Figure 2.8 A flow diagram of the supercritical fluid extraction. Source: Ahangari et al. [30]/With permission from Elsevier.

Carbon dioxide is the most commonly used solvent in SFE due to its various advantages, such as mild critical conditions, nontoxic, nonflammable, nonexplosive, and easily available at a reasonable cost. It also allows solvent removal for solvent-free extracts through easy evaporation into the atmosphere. However, due to its low polarity, carbon dioxide has limited ability to dissolve highly polar compounds even at high densities [25]. For the extraction of polar polyphenols, carbon dioxide alone may not be able to extract them due to its nonpolar characteristic. Therefore, adding organic cosolvents at small proportions can help improve the solvating power and enhance the extraction efficiency. These cosolvents have polarity higher than carbon dioxide and are also called modifiers, such as ethanol, methanol, and acetone [23].

2.5 Potential Applications of Antioxidants Extracted from Selected Agricultural Wastes

Due to its significant value in living organisms, intensive efforts have been dedicated to identifying the origin of natural antioxidants. The World Health Organization (WHO) has recognized the relevance of antioxidants due to the escalating trend in dietary consumption of antioxidants worldwide. Therefore, in general, the potential of antioxidants covers various applications, especially the industries related to food and animal feed additives and pharmaceutics. This section focuses on the importance of antioxidants derived from selected agricultural wastes for several applications, including foods, cosmetics, and therapeutics.

2.5.1 Food

The food industry is one of the key players that can fully utilize the significance of antioxidants. This situation is due to the fact that the biggest challenge faced by the food industry is the impetuous spoilage of food, resulting in massive food

Table 2.4 Antioxidants obtained from several agricultural wastes and applications in the food industry.

Agricultural waste	Application in food industry	Finding(s)	Reference(s)
Agricultural residues	Functional food	• The antioxidant properties were found higher in coffee cherry husk as compared to other type of agricultural residues • The maximum yield of extraction was obtained using 50% methanol	[31]
Mixture of sugarcane straw (SS) and coffee husk (CH)	Functional ingredients and prebiotic	• The highest scavenging ability of xylooligosaccharides (XOS) of SS and CH reached 71% and 78%, respectively • XOS promotes the growth of probiotics, such as *Bifidobacterium longum* and *Lactobacillus paracasei*.	[32]
Phenolic extract from olive leaf	Food formulation	• Addition of olive leaf phenolic (OLP) extract in full-fat mayonnaise has improved the dispersion degree and physical properties of the food product	[33]
Grape seed and chestnut extract	Food enhancer	• Natural antioxidant from grape seed is found to be efficient as compared to synthetic antioxidant (BHT) • The oxidation of sausage was significantly reduced with the addition of grape seed extract	[34]

wastage that leads to a huge loss of money. Food spoilage is mainly caused by three major situations, including self-decomposition, oxidation, and growth of microorganisms. Therefore, the alternative solutions offered by compounds with antioxidant properties are deemed to impact the food industry positively. The discoveries of natural antioxidants in agricultural wastes have prompted research for a deeper understanding of how antioxidants could be utilized to solve the challenges faced by the food industry. The shift toward the utilization of agricultural wastes has elevated the novelty of antioxidant applications.

Antioxidants such as carotenoids, polyphenols, vitamin C, and tocopherols (vitamin E) are a vital component of food industry to improve flavor, aroma, and color as well as to inhibit oxidation. These compounds can be extracted from various natural resources as well as agricultural wastes. Table 2.4 summarizes several agricultural wastes used for producing antioxidant compounds and their potential applications, especially in the food industry.

The utilization of agricultural wastes in the production of antioxidant compounds is deemed a promising approach for several reasons, including (i) enhancement of the value-added of the waste generated, as well as (ii) providential production of antioxidant compounds. For instance, the utilization of several agricultural residues, including corn husk, coffee cherry husk, peanut husk, sugarcane bagasse, rice bran, olive leaf, and wheat have been reported to possess antioxidant properties and deemed as an alternative functional foods [31]. It was also shown that coffee cherry husk extract had higher antioxidant properties, in terms of total flavonoids, phenols, and tannins content, with the value of 10.15 g TFC/100 g of extract, 85.5 g GAE/100 g of extract, and 72.6 g TAE/g of extract, respectively, as compared to other agricultural residues tested. The antioxidant content of these agricultural residues is observed to be higher when 50% methanol extraction process was used as compared to water and ethanol extraction. These potential findings are beneficial in food industry as these compounds extracted from agricultural residues are potentially be utilized in the development of functional food with possibly help in some disease prevention.

Phenolic compounds, which are natural antioxidants are found in olive leaf extract and were reported to have antioxidant and antimicrobial properties [31]. Flamminii et al. [33] reported that the enrichment of mayonnaise with encapsulated phenolic-rich olive leaf extract was efficient in improving the physical properties, albeit further study was required to improve the sensory profile. Nevertheless, olive leaf extract derived from olive oil chain waste and by-products can be considered as a beneficial multifunctional ingredient. Furthermore, the antioxidant compounds from food by-products benefit animal health when utilized as animal feed due to its nutritional properties due to their nutritional properties when utilized as animal feed [35]. Fruits and vegetable wastes have antioxidant potency and can be potentially utilized as a source of natural antioxidants. Agro-wastes such as apple by-products, tomato wastes, and juice waste are potential sources of antioxidant compounds. They have been used in the poultry industry as animal feed, while mango by-product was used as fish feed. Other wastes, such as citrus and olive by-products, rich sources of antioxidants, have been used as alternative feeds for ruminants. Correddu et al. [36] reported that ingesting agro-food residues rich in phenolic compounds could decrease the methane and urea emissions in ruminants, thus avoiding environmental issues.

All in all, the implication of antioxidant compounds in the food industry is relevant and beneficial in every perspective, including the shelf life of the food as well as the health of the consumers. Leyva–Porras et al. [37] summarize that the antioxidant compounds have contributed to vital improvement in antioxidant activities and food preservation. Therefore, intensive effort should be put into expanding the discovery of antioxidant compounds, especially from agricultural wastes, and their full potential.

2.5.2 Cosmetics

The formulation of products related to the cosmetic industry requires important features, including attenuation of free radicals, skin aging as well as oxidative stress, and utilization of antioxidants is deemed to meet the needs. The mechanisms of action by the antioxidants are chelating agents, oxygen scavengers, reducing agents, synergistic agents, and skin protection. As a consequence, antioxidants became vital elements in cosmetic products to enhance their effectiveness and stability as well as a favorable ingredient in the cosmetics formulation for consumers [38], leading to the increment in the exploration of alternative sources of antioxidants. Hence, antioxidants from agricultural wastes can be considered an ideal option for a cost-effective substitute in cosmetics formulation.

In the development of skin care products, the incorporation of vitamins and antioxidants is of great interest due to their effectiveness. It is a well-known fact that each vitamin, including vitamin A, B, C, E, and K has their own significant benefits in cosmetics product formulation. In addition, flavonoids are also deemed as important ingredients in cosmetic products. Over the years, the effort toward the identification of potential agricultural wastes to be the source of flavonoids has increased. As demonstrated by Guerrero–Castillo et al. [39], *Pouteria lucuma* seeds, agricultural wastes generated from commercialized pulp and flour contain 19 flavonoid compounds with the total flavonoid content of 5.99 μmol Q/g dry weight. The results obtained suggest that this extract can be used as a nutritional supplement in the cosmetic formulations and this can benefit the cosmetic industry.

In addition, Soto et al. [38] have demonstrated the extraction of antioxidants from agricultural by-products, including grape pomace and *Pinus pinaster* wood chips. The antioxidant property obtained from grape pomace extract is 52% Trolox activity, one of the highest among the four tested samples. The results concluded as the valorization of these agricultural wastes attributed with appropriate characteristics enabled incorporation into cosmetic products. To the best of our knowledge, the current applications of antioxidant extract in cosmetics mainly focus on natural resources instead of agricultural wastes. An intensive focus should be given into the utilization of agricultural wastes with various advanced technologies available to utilizing agricultural wastes with various advanced technologies. These include oil palm biomass, lemongrass, pineapple wastes, and other types of biomass, which are abundantly available.

2.5.3 Therapeutics

The application of antioxidants in therapeutic industries has been well established. Humans are constantly exposed to free radicals resulting from electromagnetic radiation. This situation is due to the surrounding environment, either natural or man-made causes. These will eventually lead to human oxidative stress caused by the continuous generation of oxygen free radicals. As explained by Uttara et al. [40], antioxidant defense mechanisms act in several ways in humans, including (i) counteracting the oxidative damage from oxygen free radicals; (ii) assisting in modulating the ROS; and supporting by antioxidant enzymes.

Agricultural wastes contain various bioactive components that can be utilized in various industry types; one is the antioxidant. Several studies have investigated numerous potential alternatives for antioxidant compounds from many agricultural waste sources. Khalaf et al. [41] reviewed that the selected native seeds of tropical fruits, including *durian*, jackfruit, mango, papaya, pumpkin, and *rambutan*, contained various bioactive compounds with a wide range of antioxidant properties, which are also beneficial for therapeutic applications. This is supported by another study that demonstrated the potential of papaya extract for antiparasitic activity against several Trypanosoma species in animals and humans [42].

Other than that, Sinsinwar et al. [43] demonstrated the utilization of coconut (*Cocos nucifera*) shell as an antibacterial agent against human pathogens. These results revealed that the silver nanoparticles (AgNPs) synthesized by coconut shell extract are effective against several human pathogens, including *Staphylococcus aureus*, *Escherichia coli*, *Salmonella typhimurium*, and *Listeria monocytogenes*. Besides, Fazio et al. [44] also proved that pomegranate peel extracts possessed potent cytotoxic activity against human breast cancers and HeLa cancer cells. It was proved that the pomegranate peel extracts possess potent cytotoxic activity against human breast cancers and HeLa cancer cells.

Agricultural wastes hold remarkable potential as a source of bioactive compounds that play a vital role in therapeutics. Therefore, continuous efforts are required to investigate the effectiveness of antioxidants from agricultural wastes as a therapeutic agent against various diseases.

2.6 Future Direction of Antioxidants from Agriculture Wastes

There was practically no waste in the earlier time since all the by-products were reused as nutrients and recycled back into the soil. Ever since the Green Revolution in the 1950s, the transformation of modern agriculture has focused on meeting global demand, thus increasing profitability. In order to meet the global demand, the agricultural sector is expected to grow exponentially in sustainable ways since the world population is expected to reach 10 billion by 2050 [45]. This situation has led to the overexploitation of natural resources, thus resulting in an excessive amount of waste generation. This is mainly due to the unwanted by-products produced, which are not further utilized and result in the uncontrollable accumulation, ultimately, environmental pollution problem. Technology and knowledge advancements have afforded these wastes with high potential to be transformed into value-added products. This helps in waste management issues and benefits economic growth, especially for underdeveloped and developing countries.

The CE concept has played a vital role in the modern Industrial Revolution era. This concept implies the 6Rs: reduce, reuse, recycle, redesign, remanufacture, and recover. Globalization has transformed the traditional agricultural sector into modern or industrial agriculture. It is vital to keep it sustained and circular, although it is challenging. As Jimenez–Lopez et al. [46] mentioned CE and agriculture have always been synergistic and thus support each other. Despite all the advantages of the implication of CE in the agricultural sector, unfortunately, practicalities are still low. The European Union provided the countermeasure for this matter by providing research and launching initiatives [47]. However, the awareness of each individual in the sector is needed to support this matter and thus successfully transform the agriculture industry.

Bioactive compounds are essential and nonessential compounds that appear in most natural resources. Examples of these compounds are vitamins, micronutrients, and others. For decades, the availability of these compounds has been extensively explored, especially for their antioxidant properties. This is due to the limitless advantages of these compounds in various applications, including food, therapeutic, and cosmetic industries. One of the main sources of bioactive compounds is agricultural waste. In 2021 alone, more than 16,000 articles have been published on this topic. The trend is increasing for the articles published up to Oct 2022; there are more than 13,000 articles. This situation indicates the potential and importance of agricultural waste in bioactive compound production, especially antioxidants. One of the key pathways of antioxidants from agricultural wastes is commercialization and final consumer acceptance. Numerous studies have focused on the benefits and green technology of extracted antioxidants from agricultural wastes, yet it is still considered as an infant sector in giant commercialization industries. The adaptation of agricultural wastes as an alternative source of antioxidants contributed to several advantages, including low-price products and the sustainability of natural wastes. This also can create more job opportunities, resulting in total national income growth.

2.7 Conclusions

Innovation using agricultural wastes as raw materials to extract antioxidants is a principal strategy to overcome undesired environmental impacts from the accumulation of waste. Agricultural wastes are rich in bioactive compounds such as antioxidants that can be valued in different applications such as food and animal feed additives, cosmetics, and therapeutics. Optimization of extraction techniques is critical for the recovery and maximization of bioactive compound yields. Subsequently, the recovered antioxidants can be explored and reutilized to produce high-value products to achieve circular and sustainable bio-economy and environmental conservation. Further research and biological strategies are needed to improve the extraction efficiency of raw material wastes to increase the yields of bioactive compounds with high antioxidant capacity in an environmentally friendly and profitable way. The ongoing investigation should also be conducted to exploit and uncover various uses of antioxidants derived from agricultural wastes for value creation and their effectiveness in final application in the food and pharmaceutical industries.

References

1 Rahman, M.M., Reza, A.S.M.A., Khan, M.A. et al. (2021). Unfolding the apoptotic mechanism of antioxidant enriched-leaves of *Tabebuia pallida* (lindl.) miers in EAC cells and mouse model. *J. Ethnopharmacol.* 278: 114297.

2 Mut-Salud, N., Álvarez, P.J., Garrido, J.M. et al. (2016). Antioxidant intake and antitumor therapy: toward nutritional recommendations for optimal results. *Oxid. Med. Cell Longev.* 2016.

3 Mirończuk-Chodakowska, I., Witkowska, A.M., and Zujko, M.E. (2018). Endogenous non-enzymatic antioxidants in the human body. *Adv. Med. Dent. Sci.* 63 (1): 68–78.

4 Lü, J., Lin, P.H., Yao, Q., and Chen, C. (2010). Chemical and molecular mechanisms of antioxidants: experimental approaches and model systems. *J. Cell. Mol. Med.* 14 (4): 840–860.

5 Law, B.M.H., Waye, M.M.Y., So, W.K.W., and Chair, S.Y. (2017). Hypotheses on the potential of rice bran intake to prevent gastrointestinal cancer through the modulation of oxidative stress. *Int. J. Mol. Sci.* 18 (7): 1352.

6 Adisakwattana, S., Pasukamonset, P., and Chusak, C. (2020). *Clitoria ternatea* beverages and antioxidant usage. In: *Pathology*, 189–196. Elsevier.

7 Xu, I.M.J., Lai, R.K.H., Lin, S.H. et al. (2016). Transketolase counteracts oxidative stress to drive cancer development. *Proc. Natl. Acad. Sci. U.S.A* 113 (6): E725–E734.

8 Li, R., Jia, Z., and Trush, M.A. (2016). Defining ROS in biology and medicine. *React. Oxyg. Species (Apex, NC).* 1 (1): 9.

9 Ma, Z.F., Zhang, H., Teh, S.S. et al. (2019). Goji berries as a potential natural antioxidant medicine: an insight into their molecular mechanisms of action. *Oxid. Med. Cell Longev.* 2019.

10 Çelik, E.E. and Gökmen, V. (2018). A study on interactions between the insoluble fractions of different coffee infusions and major cocoa free antioxidants and different coffee infusions and dark chocolate. *Food Chem.* 255: 8–14.

11 Cömert, E.D. and Gökmen, V. (2018). Evolution of food antioxidants as a core topic of food science for a century. *Food Res. Int.* 105: 76–93.

12 Grisales-Mejía JF, Torres-Castañeda H, Andrade-Mahecha MM, Martínez-Correa HA. Green extraction methods for recovery of antioxidant compounds from epicarp, seed, and seed tegument of avocado var. Hass (*Persea americana* Mill.) *Int. J. Food Sci.* 2022;2022.

13 Charalampia, D. and Koutelidakis, A.E. (2017). From pomegranate processing by-products to innovative value added functional ingredients and bio-based products with several applications in food sector. *BAOJ Biotech.* 3 (025): 210.

14 Fellah, B., Bannour, M., Rocchetti, G. et al. (2018). Phenolic profiling and antioxidant capacity in flowers, leaves and peels of Tunisian cultivars of *Punica granatum* L. *J. Food Sci. Technol.* 55 (9): 3606–3615.

15 Stavrinou, A., Aggelopoulos, C.A., and Tsakiroglou, C.D. (2018). Exploring the adsorption mechanisms of cationic and anionic dyes onto agricultural waste peels of banana, cucumber and potato: adsorption kinetics and equilibrium isotherms as a tool. *J. Environ. Chem. Eng.* 6 (6): 6958–6970.

16 Fresh FAOCF (2017). Processed statistical bulletin 2016. FAO Rome, Italy

17 Morrow, N.M., Burke, A.C., Samsoondar, J.P. et al. (2020). The citrus flavonoid nobiletin confers protection from metabolic dysregulation in high-fat-fed mice independent of AMPK [S]. *J. Lipid Res.* 61 (3): 387–402.

18 Justino, A.B., Barbosa, M.F., Neves, T.V. et al. (2020). Stephalagine, an aporphine alkaloid from *Annona crassiflora* fruit peel, induces antinociceptive effects by TRPA1 and TRPV1 channels modulation in mice. *Bioorg. Chem.* 96: 103562.

19 Achika, J.I., Ayo, R.G., Habila, J.D., and Oyewale, A.O. (2020). Terpenes with antimicrobial and antioxidant activities from Lannea humilis (Oliv.). *Sci. Afr.* 10: e00552.

20 Shah, B.B. and Mehta, A.A. (2018). In vitro evaluation of antioxidant activity of D-Limonene. *Asian J. Pharm. Pharmacol.* 4: 883–887.

21 Zannou, O., Pashazadeh, H., Ibrahim, S.A. et al. (2022). Green and highly extraction of phenolic compounds and antioxidant capacity from kinkeliba (*Combretum micranthum* G. Don) by natural deep eutectic solvents (NADESs) using maceration, ultrasound-assisted extraction and homogenate-assisted extraction. *Arabian J. Chem.* 15 (5): 103752.

22 Cacique, A.P., Barbosa, É.S., de Pinho, G.P., and Silvério, F.O. (2020). Maceration extraction conditions for determining the phenolic compounds and the antioxidant activity of *Catharanthus roseus* (L.) g. don. *Sci. Agrotechnol.* 44: e017420.

23 Barba, F.J., Zhu, Z., Koubaa, M., and Sant'Ana AS, Orlien V. (2016). Green alternative methods for the extraction of antioxidant bioactive compounds from winery wastes and by-products: a review. *Trends Food Sci. Technol.* 49: 96–109.

24 Mustafa, A. and Turner, C. (2011). Pressurized liquid extraction as a green approach in food and herbal plants extraction: a review. *Anal. Chim. Acta* 703 (1): 8–18.

25 Chemat, F., Abert Vian, M., and Zill-E-Huma (2009). Microwave assisted – separations: green chemistry in action. *Green Chem. Res. Trends* 33–62.

26 Esclapez, M.D., García-Pérez, J.V., Mulet, A., and Cárcel, J.A. (2011). Ultrasound-assisted extraction of natural products. *Food Eng. Rev.* 3 (2): 108–120.

27 Maravić, N., Teslić, N., Nikolić, D. et al. (2022). From agricultural waste to antioxidant-rich extracts: green techniques in extraction of polyphenols from sugar beet leaves. *Sustain. Chem. Pharm.* 28 (June): 1–12.

28 Amran, M.A., Palaniveloo, K., Fauzi, R. et al. (2021). Value-added metabolites from agricultural waste and application of green extraction techniques. *Sustainability* 13 (20): 11432.

29 Santos, P.H., Kammers, J.C., Silva, A.P. et al. (2020). Antioxidant and antibacterial compounds from feijoa leaf extracts obtained by pressurized liquid extraction and supercritical fluid extraction. *Food Chem.* 2021 (344): 128620.

30 Ahangari, H., King, J.W., Ehsani, A., and Yousefi, M. (2021). Supercritical fluid extraction of seed oils – a short review of current trends. *Trends Food Sci. Technol.* 111 (February): 249–260.

31 Vijayalaxmi, S., Jayalakshmi, S.K., and Sreeramulu, K. (2015). Polyphenols from different agricultural residues: extraction, identification and their antioxidant properties. *J. Food Sci. Technol.* 52 (5): 2761–2769.

32 Ávila, P.F., Martins, M., and Goldbeck, R. (2021). Enzymatic production of xylooligosaccharides from alkali-solubilized arabinoxylan from sugarcane straw and coffee husk. *Bioenergy Res.* 14 (3): 739–751.

33 Flamminii, F., Di Mattia, C.D., Sacchetti, G. et al. (2020). Physical and sensory properties of mayonnaise enriched with encapsulated olive leaf phenolic extracts. *Foods* 9 (8): 997.

34 Lorenzo, J.M., González-Rodríguez, R.M., Sánchez, M. et al. (2013). Effects of natural (grape seed and chestnut extract) and synthetic antioxidants (buthylatedhydroxytoluene, BHT) on the physical, chemical, microbiological and sensory characteristics of dry cured sausage "chorizo.". *Food Res. Int.* 54 (1): 611–620.

35 de la Luz, C.-G.M., del Carmen, V.-A.M., Leyva-Jiménez, F.J. et al. (2020). Revalorization of bioactive compounds from tropical fruit by-products and industrial applications by means of sustainable approaches. *Food Res. Int.* 138: 109786.

36 Correddu, F., Fancello, F., Chessa, L. et al. (2019). Effects of supplementation with exhausted myrtle berries on rumen function of dairy sheep. *Small Rum Res.* 170: 51–61.

37 Leyva-Porras, C., Román-Aguirre, M., Cruz-Alcantar, P. et al. (2021). Application of antioxidants as an alternative improving of shelf life in foods. *Polysaccharides* 2 (3): 594–607.

38 Soto, M.L., Parada, M., Falqué, E., and Domínguez, H. (2018). Personal-care products formulated with natural antioxidant extracts. *Cosmetics* 5 (1): 13.

39 Guerrero-Castillo, P., Reyes, S., Robles, J. et al. (2019). Biological activity and chemical characterization of *Pouteria lucuma* seeds: A possible use of an agricultural waste. *Waste Manage. (Oxford)* 88: 319–327. https://doi.org/10.1016/j.wasman.2019.03.055.

40 Uttara, B., Singh, A.V., Zamboni, P., and Mahajan, R. (2009). Oxidative stress and neurodegenerative diseases: a review of upstream and downstream antioxidant therapeutic options. *Curr. Neuropharmacol.* 7 (1): 65–74.

41 Khalaf, A.A., Desa, S., and Baharum, S.N.B. Overview of selected native seeds in agricultural wastes and its properties. *Medico-Legal Update* 19 (2): 324–330.

42 Maisarah, A.M., Amira, N.B., Asmah, R., and Fauziah, O. (2013). Antioxidant analysis of different parts of Carica papaya. *Int. Food Res. J.* 20 (3): 1043.

43 Sinsinwar, S., Sarkar, M.K., Suriya, K.R. et al. (2018). Use of agricultural waste (coconut shell) for the synthesis of silver nanoparticles and evaluation of their antibacterial activity against selected human pathogens. *Microb. Pathog.* 124: 30–37.

44 Fazio, A., Iacopetta, D., La Torre, C. et al. (2018). Finding solutions for agricultural wastes: antioxidant and antitumor properties of pomegranate Akko peel extracts and β-glucan recovery. *Food Funct.* 9 (12): 6618–6631.

45 Ramankutty, N., Mehrabi, Z., Waha, K. et al. (2018). Trends in global agricultural land use: implications for environmental health and food security. *Annu. Rev. Plant Biol.* 69 (1): 789–815.

46 Jimenez-Lopez, C., Fraga-Corral, M., Carpena, M. et al. (2020). Agriculture waste valorisation as a source of antioxidant phenolic compounds within a circular and sustainable bioeconomy. *Food Funct.* 11 (6): 4853–4877.

47 European Environment Agency. (2015). Closing the loop - An EU action plan for the *Circular* Economy. https://www.eea.europa.eu/policy-documents/com-2015-0614-final.

3

Lemongrass Oleoresin in Food Flavoring

Madihah Md Salleh[1], Shankar Ramanathan[2], and Rohaya Mohd Noor[3]

[1] *Universiti Teknologi Malaysia, Department of Biosciences, Faculty of Science, 81310, Skudai, Johor Bahru, Malaysia*
[2] *Bioprout (M) Sdn Bhd, No. 10 Jalan Perniagaan Setia 6, Taman Perniagaan Setia, Johor Bahru, 81100, Malaysia*
[3] *ALS Technichem (M) Sdn Bhd, No.21, Jalan Astaka U8/84, Bukit Jelutong, Shah Alam, 40150, Selangor, Malaysia*

3.1 Introduction

Lemongrass plant, or the scientific name known as *Cymbopogon citratus*, is mainly found in tropical and subtropical parts of the world. It is of great interest in commercially valuable essential oils and is widely used in functional food and also in traditional medicines [1, 2]. Lemongrass has three main components, which are essential oil, phytoconstituents, and oleoresins. Lemongrass has a typical lemon-like odor, which is mainly due to the presence of citral components. Citral is a combination of two stereoisomeric, acyclic monoterpene aldehydes which are known as geranial and neral [46]. The trans-isomer of geranial is predominant to the cis-isomer neral. Lemongrass essential oil has been used since ancient times in traditional medicine as a remedy to improve blood circulation, stabilize menstrual cycles, promote digestion, and increase immunity. Nowadays, it has been widely used to produce perfumes, flavors, detergents, and in the pharmaceutical industry.

Besides essential oil, one of the high-potential phytoconstituents used in industries is oleoresins. Oleoresins consist of a natural mixture of essential oils and resin that can be extracted from plant or natural sources. Oleoresin is also known as a natural flavor ingredient that can be prepared by extracting lemongrass with a non-potable volatile solvent. There are various types of extraction methods that can be applied for oleoresin extraction. Combination of physical and chemical methods using solvent has been widely used for oleoresin extraction. The typical solvents used for the extraction of oleoresin are acetone, CO_2, ethyl acetate, ethylene dichloride, methanol, and methylene chloride. The chemical composition of the oleoresins of *C. citratus* varies according to various factors, such as geographical origin, farming practices, plant age, photoperiod, harvest period, genetic differences, and extraction methods [3–5].

Chemical Substitutes from Agricultural and Industrial By-Products: Bioconversion, Bioprocessing, and Biorefining, First Edition. Edited by Suraini Abd-Aziz, Misri Gozan, Mohamad Faizal Ibrahim, and Lai-Yee Phang.
© 2024 WILEY-VCH GmbH. Published 2024 by WILEY-VCH GmbH.

Oleoresins are widely used in the food industry and beverages, as well as restraint. They serve as a natural spice that adds flavor and color, or as a natural antioxidant [6]. High concentrations of oleoresin usually need to be diluted before application in the final use in industry. Oleoresins are usually blended with a dry crystalline ingredient, such as salt, sugar, or dextrose, or they are liquid blended with other vegetable oils. For water-based applications in beverages or pickling, oleoresins can be blended with emulsifiers to make them water dispersible. There are several advantages of using spice oleoresins rather than ground spices. They are natural, cost-effective, clean (no microbiological growth), and have a long shelf life [4]. The application of oleoresins in concentrated form makes them easier for storage and transportation. Lemongrass oleoresins show a wide spectrum of biological activities, such as high antibacterial and remarkable antifungal activities, that make them a potential food preservative [7].

3.2 Types of Lemongrass and Their Components

Lemongrass is a fragrant type of plant with high economic value and potential. It produces high amount of essential oil with significant economic value because of its various cookery, pharmaceutical, and aesthetic applications [30]. It is easily obtained all over the world and is well known for some broad therapeutic applications in the traditional medical field, such as Ayurvedic, in countries such as India. Genus *Cymbopogon* is largely grown in regions of Asia, Africa, and America with tropical climate. The genus *Cymbopogon* includes more than 144 species and is well known for its high content of essential oils [8]. Around 60 types of different species of lemongrass are found in various countries in Asia and Africa [9]. Table 3.1 list the differences in lemongrass species leaves found in different countries in the world.

This majestic plant with a lemon aroma came from the Poaceae family, which has beyond 9000 species and 635 genera. Data show that of 140 plant species that were planted, 45 were found in India, 6 in Australia, 52 in Africa, around 6 in South America, 4 in Europe, 2 in North America, and the rest in South Asia. The two most widely cultivated varieties of lemongrass, *Cymbopogon exuosus* and *C. citratus*, are well-known because they contain important essential oils in diverse places in various countries. Lemongrass is known by different names in different countries, such as West Indian lemongrass or lemongrass, verveine des indes in French, zacate de limón in Spanish, and xiang mao in Chinese. In Portuguese, on the other hand, it is known as capim-santo. Overall around 28 indigenous names were recorded in different parts of the world [6, 10]. Commonly, lemongrass is tall in size, has a harsh texture, and has a tingling lemon taste.

It has many firm stems ascending from a small rhizomatous root and can grow up to 1.5 m tall. Flowers infrequently bloom on it. The general characteristics of the plant are blue–green leaves that are erect, linear in shape, and produce a lemon-like flavor when massacred. *C. citratus* is considered to be originated from Malaysia and is widely grown in different parts of America and certain parts of

Table 3.1 Different types of lemongrass species in various countries.

Type of lemongrass	Native	Constituents of oleoresin (%)	References
Cymbopogon citratus	Indonesia, Malaysia, and Brunei	Geranialdehyde (71.30) β-myrcene (19.03) 4,5-Epoxycarene (2.80) β-linalool (1.71)	[3]
Cymbopogon martinii inflorescence	India, Vietnam, and Myanmar	Geraniol (66.90) Trans, cis-farnesol (9.43) Geranyl acetate (7.50)	[5]
Cymbopogon schoenanthus	Africa and Iran	Peritone (46.8) 4-Carene (16) D-limonene (3.2) Elemol (8.0)	[11]
Cymbopogon flexuosus	India	Isointermedeol (24.97) Geraniol (20.08) Geranyl ethanoate (12.20) α-bisabolol (8.42) Limonene (3.5) Borneol (1.90)	[9]
Cymbopogon winterianus	Indonesia	Rhodinal (2.2–55.4%) Geraniol (14.2–53.0%) Dihydrogeraniol (8.2–16.4%) Isopulegol (0.3–12.6%) Elemol (0.8–8.2%) Limonene (0.2–5.0%)	[4]

Africa, Southeast Asia, and the Indian Ocean Islands, on a commercial scale and in gardens [12]. Citral in lemongrass is a blend of two stereoisomeric monoterpene aldehydes; the trans-isomer geranial is a major compound to the cis-isomer neral, a type of oleoresin. Lemongrass essential oil has been vastly exploited since early times in traditional medicine as a tonic to upgrade health or enhance immunity. It is also used in daily products such as perfumes, flavors, detergents, and pharmaceuticals. The substance constituents of the essential oil of *C. citratus* differ by the location, age of the plant, farming practices, cultivating time, photoperiod, genetic characteristics, and extraction technique.

The crucial composition of the essential oil obtained has commonly been examined for its active compounds, such as esters, alcohols, aldehydes, and terpenes. The extracted oil, which contains various oleoresins, shows a broad scale of biological activities. The essential oil is also proven to show very strong antibacterial and antifungal properties, which makes it a potential food preservative. Dried and powdered lemongrass is widely used as a food taste enhancer [13]. It is normally used in hot drinks, such as teas, broths, and various curries. It may also be used as a key ingredient in cooking different types of meat, such as beef and chicken, and even seafood. Various studies show that the addition of lemongrass in food can help to protect the stomach from various infections, help in digestion and excretion, and prevent ulcers. It is also well-known for treating nausea or constipation [14]. Therefore, many countries around the world use it as an important medicinal herb [3]. In Egypt, dried leaves of lemongrass are boiled and the liquid obtained is taken as a diuretic and renal antispasmodic, whereas in Indonesia and Malaysia, the whole plant is boiled and the liquid obtained is taken orally to improve blood flow in the pelvic area and uterus [14].

3.3 Potential Chemical Substitutes from Lemongrass

3.3.1 Essential Oil

Essential oils are hydrophobic liquids comprising aromatic and volatile compounds that are extracted and synthesized by natural reactions from potentially different parts of the plants. They are a complex mixture of secondary metabolites, often terpenoids, which play an important role in plant defense systems and have powerful antimicrobial properties. The major constituents extracted from the essential oil of lemongrass include citrala (41.82%), myrcene (12.75%), geranyl acetate (3.00%), methyl heptanene (2.62%), geraniol (1.85%), β-elemene (1.33%), elemol (1.2%), citronellyl acetate (0.96%), citronellal (0.73%), β-caryophylleneoxide (0.61%), dipentene (0.23%), delta-3-catrene (0.16%), β-caryophyllene (0.18%), citral b (0.18%), α-pinene (0.13%), β-phellandrene (0.07%), and β-cymene (0.2%) [15]. However, the variations in the chemical composition of lemongrass essential oil were influenced by several factors, such as geographical origin, farming practices, plant age, photoperiod, harvest period, genetic differences, and extraction methods.

Citral represents the dominant component of lemongrass essential oil and is a general combination of two geometric isomers, which are E-isomer and Z-isomer. The E-isomer is referred to as geranial or citral A, and the Z-isomer is referred to as neral or citral B. Geranial composition (0.99–48.14%) dominates over neral (0–38.32%). The quality of lemongrass essential oil is generally evaluated by the citral content with at least 75% of its value to be considered as a quality product. The presence of citral imparts a lemony scent, but its use in perfume formulations is still limited due to its great tendency to polymerize, oxidize, and discolor. Despite all these disadvantages, however, citral is used in many formulations due to its powerful ability to cause a variety of interesting reactions. The quality and quantity of lemongrass essential oil strongly depend on the time of harvesting the plant, since the composition and content of the essential oil are closely related to the stage of development of the whole plant, organs, and plant cells [16].

Phenolics, hydrocarbon terpenes, aldehydes, ketones, ethers, epoxides, organic sulfur, benzoids, terpenoids, and many other compounds dominate the chemical composition of the essential oil that has antimicrobial properties against a variety of diseases [17]. Lemongrass essential oil has considerable commercial importance because it is used in the manufacturing of perfumes, flavors, cosmetics, detergents, and pharmaceuticals. As a culinary flavoring agent, lemongrass essential oil is added to alcoholic and soft drinks, frozen milk desserts, sweet baked goods, candies, jellies and puddings, meat and meat products, and fats and oils. It may also enhance the taste of fish, wines, and sauces. Biological studies have shown that various chemical compounds of the essential oil have antimicrobial, antioxidant, antiparasitic, insecticidal, and insect-repellent properties. Antimicrobial and antioxidant properties of essential oil have become an important element in preservation in the food industry. Due to the growing consumer demand for natural resources of food, essential oil has the potential to be an alternative to synthetic compounds [13].

3.3.2 Phytoconstituents

Plant products contain a variety of compounds, including phenolics and flavonoids. These polyphenols are a large group of natural compounds with multiple biological activities and applications in a variety of fields due to their antioxidant, antitumor, antiviral, antidiarrheal, anti-inflammatory, antibiotic, antifilarial, and allelopathic properties. Due to the presence of antioxidant properties in natural polyphenols, their applications in several fields have been recognized, such as in chicken patties, herbal cookies, meat products, chicken sausages, and cosmetics field.

Phenolic acid compounds in lemongrass extract were dominated by caffeic acid and syringic acid, with gallic acid, dihydrobenzoic acid, catechin, vanillic acid, epicatechin, *p*-coumaric acid, *trans*-ferulic acid, quercetin, and *trans*-cinnamic acid present in trace amount [18]. Differences in phenolic compound profile and content may be explained by two factors, such as region and the harvest season. Similarly, the differences in profile and content may be the primary factors influencing the antioxidant activities of lemongrass extract.

3.3.3 Oleoresins

Oleoresins are natural mixtures of oils and resins extracted from a variety of plants and contain varying concentrations of essential oil [19]. The whole natural spices or herb products obtained from the extraction process represent two types of components, which are volatile and nonvolatile. Oleoresins produce an instant taste and aroma, which meet the element of food specifications, and contribute to an excellent economy and sterilization procedure during the manufacturing production process. When spice oleoresins are used in food flavoring instead of raw spice, they release a better active ingredient, which attributes to a key flavor in food production.

Lemongrass oil has been the traditional source of citral. This oil has been used as a raw material for ionone and methyl ionone. There are no specific restrictions on the use of lemongrass oil in flavors and fragrances. Citral, the major component of lemongrass essential oil, is widely used in line-scented soaps, aftershaves, perfumes, detergents, cosmetics, candles, and aromatherapy. Lemongrass oil has no adverse effects on blood, liver and kidney function, protein, carbohydrate, and lipid metabolism of rats.

3.4 Characteristics and Properties of Oleoresin

Extracted in a concentrated form, oleoresins are important natural chemicals from various herbaceous plants. They are a combination of essential oils and resin, which comprise volatile and non-volatile component. Their structure is similar to resins that are extracted in a semisolid compound with a strong aroma and are composed of essential or fatty oils that give the food its aromatic taste. Oleoresins from Cymbopogon species or well known as lemongrass, which originates from countries in tropical regions, are being extracted for various uses in many countries [20]. Research shows that the extraction of lemongrass leaves resulted in 17.3% of oleoresin. The oleoresin content in varieties of lemongrass ranged from 16% to 23% (Table 3.1).

Lemongrass-extracted oleoresin is found to have a lemon aroma mixed in dark green-colored liquid with a sticky characteristic. Its lemon-like properties make it a suitable food addictive [7]. The lemon-like aroma and taste may be due to its major content, which is citral [20]. This property makes it a good additive in beauty products, food flavoring, cleansing agents, and even in pharmaceutical products. Although many extraction methods were used to get high-quality oleoresin, steam distillation, however, was found to be the best method for optimum extraction. The types of chemical constitution extracted from lemongrass depend on various factors, such as the time of harvesting, age of the plant, farming method, and genetic properties [4, 5].

As research shows that oleoresin has antimicrobial properties, it does not have microorganisms or enzymes in the extract. This also contributes to longer shelf life [7]. Suitable extraction method is important to prevent the destruction of oleoresin. Besides having unique flavor properties, some oleoresin exhibits the ability to dissolve in fats and oils, which could not be found in essential oils. However, oleoresin has a milder flavor compared to essential oil.

3.5 Lemongrass Oleoresin Composition and Function

Lemongrass oleoresin is a dark green viscous liquid with a distinct lemon aroma and flavor and is primarily used as a flavoring agent in various industries (Table 3.2). The lemon prefix in lemongrass specifies the characteristic lemon-like odor caused by the presence of 75% of citral content, which is a mixture of two isomeric acyclic monoterpene aldehydes: geranial (trans) and neral (cis). The citral in lemongrass essential oil has been used to treat various diseases and health issues, such as oily skin, acne, flatulence, scabies, and excess sweating, additionally, citral also act as antimicrobial, antiviral, and antifungal, as well as its application as flavouring agent. Used medicinally, citral also has been found to act as an anti-inflammatory compound that has the potential to reduce chronic inflammation and relieve pain, such as arthritis and cardiovascular disease [24].

The major drive for lemongrass cultivation is its oil, which has wide applications in culinary flavoring. It is used in most major categories of food, including alcoholic

Table 3.2 The composition and function of lemongrass oleoresin.

Composition of Oleoresin	Function	References
Citral (geranial and neral)	• Antimicrobial efficiency against pathogenic bacteria and fungi • Treatment of skin conditions, such as oily skin and acne, as well as health conditions, such as flatulence and scabies • Reduce inflammation pain • Applied in perfumery, cosmetics, and pharmaceutical to control pathogens	[21, 22]
Myrcene	• It has antibacterial and antifungal properties, analgesic and antispasmodic effects on uterine and intestinal tissue, antiparasitic properties, and central nervous system activity	[26]
Citronellol	• Act as an anti-inflammatory agent by speeding up the healing process of wound • Function as an antifungal against a strain of *Aspergillus niger* that causes lung and sinus infections • Used as an insect repellent • Lift mood or battle fatigue, which can give calmness to the brain	[27]
Linalool	• Balancing emotional stress and soothing a busy mind • Has inflammatory activity, including inflamed joints, respiratory, bee sting, or skin problems like psoriasis and dermatitis	[28, 29]
Limonene	• Reduce the inflammation effects • Applied in aromatherapy as an antistress and antianxiety agent • Act as an antioxidant by reducing the cell damage caused by unstable free radicals • Inhibit the growth of skin tumors to prevent inflammation and oxidative stress	[30, 31]
Geranyl acetate	• Used as perfumes for soaps and creams as well as a flavoring ingredient	[32]

and nonalcoholic beverages, frozen dairy desserts, pastries, gelatin and puddings, meat and meat products, and fats and oils. It is used to isolate citral for vitamin A and many other aromatic chemicals. The oil has a very good smell of natural citral and can be used in citrus fragrances. It can be used as a unique flavor for green tea, has a very good aroma and therapeutic properties, has good medicinal properties, and is a popular ingredient in aromatherapy. The oil has extensive potential as a bactericide, insect repellent, and in medicine. Its mosquito-repellent activity lasts two to three hours, showing significant anti-feeding and larvicidal activity against *Helicoverpa armigera*.

The chemical composition of the essential oil of lemongrass varies depending on its geographical origin and types of compounds, such as hydrocarbon terpenes, alcohols, ketones, esters, and aldehydes [25]. Aside from citral, lemongrass oil contains geraniol, geranyl acetate, and monoterpene olefins, such as myrcene. As a result of its monoterpene and myrcene content, lemongrass essential oil possesses the ability to treat bacterial and fungal infections, analgesic and antispasmodic effects on urine and intestinal tissue, antiparasitic properties, activity on the central nervous system, and anxiolytic and sedative effects [26].

Lemongrass oil effectively works as a panacea against bacteria, flu, and colds because it contains stimulating agents, tonics, aromas, diuretics, and antispasmodic. Aside from improving digestion, nausea, and menstruation problems, it is also used to treat headaches, muscle cramps, spasms, and rheumatism. Lemongrass oil is among the most popular essential oils used in aromatherapy for relieving stress, anxiety, and depression. It is used as a natural remedy to heal wounds and prevent several types of infections, including skin infections, pneumonia, blood infections, and severe intestinal infections. Lemongrass oil was also reported as an effective antifungal property against four types of fungi that causes various types of disease, such as ringworm and jock itch [33].

3.6 Extraction Technique of Lemongrass Oleoresin

The research development on extraction techniques of lemongrass oleoresins has been studied by various researchers. Although most of the researchers are focusing on the extraction of essential oil from lemongrass, there are several methods used for the extraction methods of oleoresin (Table 3.3). The extraction of oleoresin from lemongrass (*Cymbopogon flexuosus*) has been standardized, based on the part of lemongrass, preprocessing requirements, type of solvent, and the time of extraction [34].

The extraction methods need to be optimized to get efficient parameters for the extraction of certain oleoresin or bioactive compounds from lemongrass. The optimization of extraction can be conducted using one factor at a time (OFAT) or statistical methods design. The OFAT methods involve a lot of experiments and analysis. Only one factor can be changed at one time for this type of experiment. This method is not economic and needs a longer time to identify the optimized condition. One of the most common statistical methods based on experimental designs for optimization is response surface methodology (RSM). This method involves the evaluation effects of multiple factors and their interactions with one or more response variables, which reflect interactions of more than two factors at one time [35, 36]. RSM has been used by many researchers as an effective tool to optimize the extraction processes of oleoresin [7, 18, 35]. Statistical analysis, such as Central Composite Design (CCD), provides more efficient information in the shortest time with the minimum number of experiments and also analyses (Table 3.3). This method is more economical and time-saving as compared to OFAT.

3.6.1 Chemical Extraction

Chemical extraction involves the use of food-grade solvents, such as hexane and ethanol. These solvents were used for the isolation of essential oils and oleoresin

Table 3.3 Extraction methods of lemongrass oleoresin.

Methods	Optimization	Factors influencing oleoresin extractions	Products (%)	References
Molecular sieves carbon dioxide extraction	OFAT	• Temperature 293–313 K • Pressure 100–200 bar	• 85% essential oil with neral and geranial	[36]
Chemical hot extraction	OFAT and CCD	Soxtec system • Methanol 19.7% (w/v) • 100 g dried fine powder lemongrass leaves • 10 min rinsing • 75 min boiling	• 92% of oleoresin	[7]
Conventional hydrodistillation (HD)	OFAT	• Plant material ratio of 8 : 1, • 180 min of extraction	• 85.15 citral • 49.46 geranial • 35.67% neral • Minor mycerine	[34]
Microwave-assisted hydrodistillation (MAHD)	OFAT	• Plant material ratio of 8:1 • Microwave power of 250 W • 90 minutes extraction time	• 86.48% citral • 50.81% geranial • 35.69% neral • Minor amount of myrcene • Minor amount of linalool, geranic acid, and citronellol	[34]
Decoction extraction procedure	CCD	• Solid/liquid ratio (5 g/100 ml), • Temperature (93.8 °C), • Time (1.3 min)	• Antioxidant activity (DPPH) 71.98 ± 0.33 mg GAE/100 ml extract • Total polyphenol content (TPC) 80.63 ± 0.49 mg TE/100 ml extra	[18]
Pressurized liquid extraction (PLE)	CCD	• Temperature 167 °C, • Pressure 1203 psi • Static time of 20.43 min	• 72% oleoresin consisted of geranial and geraniol	[35]
Supercritical fluid extraction (SFE) with CO_2	OFAT	• Lemongrass leaves 0.7% (w/v) • Pressure 1700 psi • Temperature 50°C.	• Essential oil consisted of citral, geranial, and neral	[18]
Vacuum fractional distillation	OFAT	• Column height of 400 mm, • Power input of 165 W • Pressure of 15 mm Hg	• 92% of citral	[24]

Note:
OFAT: One Factor at a Time.
CCD: Central Composite Design.

from plant material. This method is suitable for extraction of low yield of essential oil from plant material. Furthermore, this method is also suitable for samples that do not tolerate pressure and steam distillation. This method is more favorable for the production of high-quality fragrance as compared to another type of distillation method. During this process, the nonvolatile compound from plant materials, such as waxes and pigments, will be extracted and removed through other processes [7].

The chemical extraction method involves the treatment of plant materials with solvent to produce a waxy aromatic compound known as "concentrate." Additional concentrated substances with alcohol will release the oil particles. The design of the flow process for the chemical extraction concept for oleoresin is presented in Figure 3.1. The application of the aforementioned chemicals in this process will remain in the oil. This oil can be applied for various purposes in the industry, such as in the perfume industry and also for aromatherapy purposes.

The extraction of lemongrass oleoresin was conducted using a different type of solvent. This process can be divided into two methods, which are known as maceration and percolation. The maceration process involves macerating the plant material in a suitable solvent, such as hexane, followed by filtering and concentrating the oleoresin extract. This method involved the use of a cold solvent to reduce the chances of the decomposition process. Cold extraction gave a great benefit for the preservation of fresh aroma and flavor. However, this method normally takes a long time and consumes a high amount of solvent [38].

During the solvent extraction process, a solvent (usually n-hexane) is added to the plant material for the purpose to dissolve the essential oil. After the filtration process, the solution is concentrated by using a distillation process. At this stage, the substance containing resin (resinoid) or a combination of wax and essential

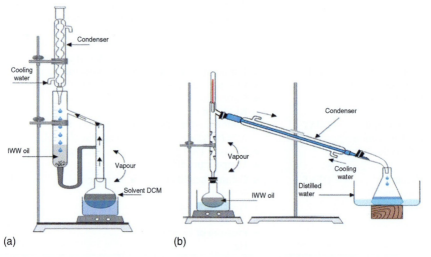

Figure 3.1 Design of flow process of solvent extraction for oleoresin. (a) Laboratory assembly used for the continuous liquid–liquid extraction with dichloromethane (CLLEDCM) and (b) high-power fractional distillation (HPFD). Source: Mendoza et al. [37]/MDPI/CC BY 4.0.

oil will remain. Although this method is quite efficient and relatively simple for lemongrass essential oil extraction, the application of a high amount of solvent and low yield of oleoresin make this system look unsatisfactory for reproducibility [36]. After the extraction process, the sample will be concentrated by using an evaporation process for the reduction of the volatile compounds. However, at this stage, contamination of the essential oil with solvent residues may still occur. One of the equipment that has been used for the extraction of lemongrass essential oil is the soxhlet apparatus [7, 38].

During the soxhlet extraction process, the plant material will have direct exposure continuously with the reusing liquid phase, to increase extraction efficiency. This method is comparatively most significant as compared to other conventional methods of oleoresin extraction. However, the drawback of soxhlet extraction is a long heating period at a high temperature, which usually involves the application of high temperatures close to the boiling point of the solvent. This condition may lead to the thermal degradation of fragile compounds. The application of both solvent extraction by maceration and Soxhlet extraction need to have optimized condition. A requirement of the correct choice of solvent is very essential to obtain a good extraction yield, as well as to prevent the loss of volatile compounds. Currently, both dry and fresh lemongrass leaves were used for essential oil using solvent extraction. Application of hexane yields essential oil up to 1.5%. Integration of sonication and hexane extraction manage to produce comparable contents of the main compounds to steam distillation [7].

In the percolation technique, solvent percolation through a column of the material is faster and less amount of solvent is used. Although this technique is quick and uses less solvent, it contributes to high decomposition due to the heating process. Soxhlet extraction is a form of continuous percolation with an additional fresh solvent and uses special glassware. The plant material is separated from the extract by encasing it in a paper known as thimble. When this part is full, the solvent in the thimble is transferred to the main vessel containing the extract and the process will continue. This technique let the fresh solvent continuously extracts the bioactive compound efficiently with a combination of a minimum amount of solvent and a heating process. However, heating is again another disadvantage of this process.

Besides hexane, other solvents such as methanol also show promising results for oleoresin extraction. A combination of 300 ml methanol with 75 min boiling, 10 min rinsing, and one-time washing is the most efficient protocol that can produce up to 92% of oleoresin present in 100 g lemongrass leaf powder [7].

3.6.2 Steam Distillation

Steam distillation is one of the most popular methods used for the extraction and isolation of essential oils from plants for application in natural products. During this process, the steam will vaporize the plant material's volatile compounds, which eventually go through a condensation and collection process [38].

The steam distillation comprises a large container which is made of stainless steel. This part contains the plant material with the steam added to it (Figure 3.2).

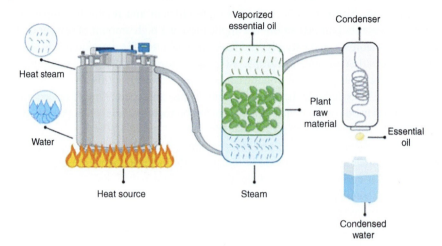

Figure 3.2 Design of flow process of steam distillation for oleoresin extraction. Source: Machado et al. [23]/MDPI/CC BY 4.0.

The steam will be injected through the plant material containing the desired oils and release the plant's aromatic molecules into vapor. The vaporized plant compounds will be transferred to the condensation flask, which is known as the condenser. Two separate pipes enter the condenser. The hot and cold water channels are used to control the temperature of the condenser. There are two separate pipes for the hot water to exit and for cold water to enter the condenser [24]. The differences in the temperature in the condenser will produce vapor cool back into liquid form. At this stage, the aromatic liquid by-product drops from the condenser and collects inside a receptacle underneath it, which is known as a separator. At this stage, the water and oil do not mix and the essential oil will float on top of the water. This situation will make the collection of essential oil easier. There are some cases where certain essential oils, such as clove essential oil, that are heavier than water are found at the bottom of the separator.

3.6.3 Pressurized Liquid Extraction (PLE)

PLE is a process that integrates pool temperature and pressure with liquid solvents to get to accomplish rapid and efficient extraction of analytes from several matrices. PLE technique is a more recent extraction technique proposed to obtain bioactive compounds. It uses less solvent, in a shorter period, is an automated system, and can retain the sample in an oxygen and light-free environment, which is in contrast to the traditional organic solvent extraction concept [7]. PLE design was based on the application of conventional solvent set at controlled temperatures and pressures and was well established for extraction of valuable compounds from natural sources. A major benefit of PLE above conventional solvent extraction methods conducted at atmospheric pressure is that pressurized solvents remain in the liquid state, even above their normal atmospheric pressure boiling points allowing high temperatures.

PLE is used for the extraction of compounds present in vegetable oil, such as lipids and fatty acids, owing to its short time extraction. It also can be used in the extraction of minor components, such as anthocyanins, carotenoids, phenols, sterols, phospholipids, tocopherols, free fatty acids, and glycerides, because these compounds may be degraded during long extractions at higher temperatures [34, 37]. Moreover, PLE has been shown to present a safe and rapid technique for extracting antioxidant compounds from plants. This technique involves a combination of parameters such as temperature and pressure with liquid solvents to achieve fast, efficient, and reliable extraction. Approximately, 72% of oleoresin, which consisted of geranial and geraniol, has been extracted under an optimized temperature of 167 °C, pressure 1203 psi, and static time of 20.43 min from lemongrass [35].

PLE was considered a consolidated high-throughput and green extraction technique for the sustainable extraction of bioactive compounds from natural sources and for the determination of a broad variety of analytes of interest in food and environmental samples. Although PLE has a lot of advantages as compared to conventional methods, it is not widely used because the cost is quite high compared to conventional methods.

3.7 Application of Lemongrass Oleoresin as Food Flavoring

On a dry weight basis, around 1–2% essential oil is found in lemongrass (*C. citratus*). High content of oleoresins was obtained based on its characterization analysis, such as citral and geranial. This oleoresin is used as raw material in the synthesis of vitamins, such as vitamin A and β-carotene [34]. Various parts of plants are used to make herbs and spices that give food flavor and aroma. The herb has a variety of medicinal properties, such as antimicrobial, antidiabetic, antioxidant, and anti-inflammatory properties. As a result, herbs and spices can deliver nutritional and medicinal benefits to baked goods. Consumers are increasingly seeking healthier, more nutritious, and safer food products in recent years, which has increased their anxiety about food quality and safety. Research on the latest food technology reveals that lemongrass oleoresins can be applied in food for increasing food preservation qualities during production, transportation, and storage. Due to this, new substitutes have been studied, such as essential oils and oleoresins. The extract containing the oleoresins from the leaves of *C. citratus* is a fragrant essential oil rich in a wide range of oleoresins that is well analyzed and studied for its application in food preservatives [39].

The essential oil with the oleoresins is highly biosynthesized primarily in the leaves of *C. citratus* [42]. Food and Drug Administration ascertained that oleoresins from lemongrass extraction have proven to be safe and can be used as a food flavoring and addictive. The tangy taste of lemongrass makes it suitable to be used in alcoholic and nonalcoholic drinks, dairy food products, candies, puddings, baked foods meat-based products, and even fats and oils. It is also known to enhance food flavors, such as wine, various sauces, and certain fish types. Many types of research have shown that oleoresins found in the extraction of lemongrass exhibit strong antiparasitic, antimicrobial [41], and insecticidal properties [42]. Owing to its natural antioxidant and antimicrobial properties, it can be exploited in food preservation as a substitute for synthetic chemical-based food preservatives, which are lately less acceptable by consumers [38]. Table 3.4 list various functions and applications of lemongrass oleoresins in the food industry.

Table 3.4 Applications of lemongrass oleoresins.

Applications	Types of oleoresin	Functions	References
Natural preservative			
Food preservative	Geranial, myrcene, and neral	Possess antimicrobial and antifungal effects toward Gram-positive and *Candida* sp.	[43, 46]
Food preservative	Citral and geraniol	Possess excellent antifungal and antibacterial properties	[13]
Food preservative	Citral, geraniol, citronellol, elemol, is geranial, citronellal, and geranyl acetate	Antifungal and antibacterial	[44]
Food preservative	Citral	Against *Escherichia coli*, *Salmonella enterica*, and *Staphylococcus aureus*	[45]
Food preservative	α-citral and β-citral	Microencapsulated lemongrass essential oil to Coalho cheese	[48]
Health food			
Various essential vitamins	Citral	Vitamin A, vitamin B1, B2, B3, B5 and B6, vitamin C, and folate	[8]
Flavor			
Matcha green lemongrass tea	Citral	Extends food shelf life and gives citral flavor	[45]
Cake flavoring (citral)	Citral	Natural flavoring	[33]
Fish, sauces, and wines	—	Improve flavoring	[47]
Spice	Citral	Natural spice	[48]
Fragrant agent	Citral	Lemony fragrance	[48]
Natural antioxidant			
Cakes	—	Preventing oxidative loss of shelf-life	[49]
Wine		Antioxidant	[13]

3.8 Oleoresin Prospect

Oleoresins, which are important plant extracts of resin and essential oil, can be extracted from various plants, such as tomatoes or lemongrass [3, 27, 49]. Oleoresins are known to exhibit various aromas and flavors depending on the type of plant. This substance constitutes volatile and nonvolatile compounds that can be obtained by using solvent extraction [7]. The whole spice is less shelf-stable compared to oleoresins extracted and oleoresins extraction can be formulated to suit the customer's requirement. Oleoresins are widely used as a flavor in food products, the pharmaceutical industry, insect repellent, and food and beverages industries.

Oleoresins provide various benefits compared to whole spices when used as flavoring additives. For example, oleoresins give food formulation with lower price options and a strong spiciness aroma as compared to whole spices [1, 2, 28]. The current trend shows that oleoresin's suppliers and manufacturers from developing countries can benefit from the continuous growth of European natural food colors and flavors.

Global market analysis shows that oleoresins will reach estimation market value of US$ 1.7 billion by 2025 compared to a market value of US$ 1.2 billion in 2019. The increment of compounded was calculated to grow 6.0% by annual growth rate (CAGR) from 2019 to 2025, which shows high demand for this compound in the industry. The increment in natural flavor additives in food products and a growing number of quick-service restaurants contribute to an increment in the need for oleoresins in the food industry. The continuous growth of the oleoresin market may be due to the current trend where people prefer and are well aware of the benefits of natural phytomedicines and plant extracts compared to chemically synthesized flavors [30, 44–46].

Oleoresins also give a great benefit as base flavors for several drinks. The rapid growth of the beverage industry also leads to increased demand for oleoresins. With the availability of a wide range of oleoresins for drinks, bakeries and processed food manufacturers can use combination flavors in foods and drinks to make them more appealing [1, 2, 43]. This will, in turn, be projected to drive the growth of the oleoresin product in the food and beverages industries. In addition, the growing usage of health supplements and rising activities of R&D in the research sector have also contributed to the rising demand for oleoresins in various industries. Besides that, the increment in preference of users for clean-label personal care products also gives a solid rise in the oleoresin market [3, 8, 43].

3.9 Conclusions

Lemongrass contains various types of bioactive compounds that can be used in different industries and sectors, such as food product and beverages industry, pharmaceutical industry, medical sector, and cosmetics. One of the bioactive compounds that attract a lot of interest is oleoresin. The component of oleoresin in lemongrass depends on the species of lemongrass. With the advance in research development, oleoresin can be extracted and customized for various applications. The trend of oleoresin demand growing tremendously is due to its versatile criteria as a food preservative, a natural antioxidant, a flavor, an insect repellent, and a rich source of essential vitamins. Lemongrass oleoresins extracted from the Cymbopogon species grow in the tropical and subtropical regions of the world. Various technology has been used for the extraction of oleoresin from lemongrass. A combination of physicochemical extraction techniques contributes to high standard and purity of oleoresin. The interaction of several parameters during the extraction process needs to be optimized to improve the quality and quantity of oleoresins. Extraction of green oleoresin using various methods and technology increased the standard purity of the oleoresin. High purity of oleoresin is more shelf-stable and can be custom-made to suit the requirements of the industries. The researchers and industry have to work together on the efficiency of extraction techniques and the application of oleoresins. In the future, with the increased demands from industries, oleoresin has a high prospect to be market, which will contribute to income generation for developing countries.

References

1 Abdulazeez, M.A., Abdullahi, A.S., and James, B.D. (2016). Lemongrass (*Cymbopogon* spp.) oils. In: *Essential Oils in Food Preservation, Flavor and Safety*, 509–516. Elsevier https://linkinghub.elsevier.com/retrieve/pii/B9780124166417000584.

2 Anggraeni, N.I., Hidayat, I.W., Rachman, S.D., and Ersanda. (2018). Bioactivity of essential oil from lemongrass (*Cymbopogon citratus* Stapf) as antioxidant agent. In: *AIP Conference Proceedings 1927*, 030007. AIP Publishing http://aip.scitation.org/doi/abs/10.1063/1.5021200.

3 Avoseh, O., Oyedeji, O., Rungqu, P. et al. (2015). *Cymbopogon* species; ethnopharmacology, phytochemistry and the pharmacological importance. *Molecules* 20 (5): 7438–7453. http://www.mdpi.com/1420-3049/20/5/7438.

4 Haerussana, A.N.E.M. and Chairunnisa, H.F. (2022). Essential oil constituents and pharmacognostic evaluation of Java Citronella (*Cymbopogon winterianus*) stem from Bandung, West Java, Indonesia. *Open Access Maced J. Med. Sci.* 10 (A): 1338–1346. https://oamjms.eu/index.php/mjms/article/view/9546.

5 Kaur, G., Arya, S.K., Singh, B. et al. (2019). Transcriptome analysis of the palmarosa *Cymbopogon martinii* inflorescence with emphasis on genes involved in essential oil biosynthesis. *Ind. Crops Prod.* 140: 111602. https://linkinghub.elsevier.com/retrieve/pii/S0926669019306120.

6 Haque, A.N.M.A., Remadevi, R., and Naebe, M. (2018). Lemongrass (*Cymbopogon*): a review on its structure, properties, applications and recent developments. *Cellulose* 25 (10): 5455–5477. http://link.springer.com/10.1007/s10570-018-1965-2.

7 Irfan, S., Zahra, S.M., Murtaza, M.A. et al. (2022). Characterization of lemongrass oleoresins. In: *Handbook of Oleoresins*, 261–283. CRC Press.

8 Soorya, C., Balamurugan, S., Basha, A.N. et al. (2021). Profile of bioactive phyto-compounds in essential oil of *Cymbopogon martinii* from Palani Hills, Western Ghats, INDIA. *J. Drug Deliv. Ther.* 11 (4): 60–65. http://jddtonline.info/index.php/jddt/article/view/4887.

9 Sharma, P.R., Mondhe, D.M., Muthiah, S. et al. (2009). Anticancer activity of an essential oil from *Cymbopogon flexuosus*. *Chem. Biol. Interact* 179 (2–3): 160–168. https://linkinghub.elsevier.com/retrieve/pii/S0009279708006613.

10 Naik, M.I., Fomda, B.A., Jaykumar, E., and Bhat, J.A. (2010). Antibacterial activity of lemongrass (*Cymbopogon citratus*) oil against some selected pathogenic bacterias. *Asian Pac. J. Trop. Med.* 3 (7): 535–538. http://linkinghub.elsevier.com/retrieve/pii/S1995764510601290.

11 Eloh, K., Kpegba, K., Sasanelli, N. et al. (2020). Nematicidal activity of some essential plant oils from tropical West Africa. *Int. J. Pest. Manag.* 66 (2): 131–141. https://www.tandfonline.com/doi/full/10.1080/09670874.2019.1576950.

12 Mitra, A. and Zaman, S. (2020). Biodiversity and its conservation. In: *Environmental Science – A Ground Zero Observation on the Indian Subcontinent*, 143–214. Cham: Springer International Publishing http://link.springer.com/10.1007/978-3-030-49131-4_6.

13 Majewska, E., Kozłowska, M., Gruczyńska-Sękowska, E. et al. (2019). Lemongrass (*Cymbopogon citratus*) essential oil: extraction, composition, bioactivity and uses for food preservation – a review. *Polish J. Food Nutr. Sci.* 69 (4): 327–341. http://www.journalssystem.com/pjfns/Lemongrass-Cymbopogon-citratus-essential-oil-extraction-composition-bioactivity-and,113152,0,2.html.

14 Carbajal, D., Casaco, A., Arruzazabala, L. et al. (1989). Pharmacological study of *Cymbopogon citratus* leaves. *J. Ethnopharmacol.* 25 (1): 103–107. https://linkinghub.elsevier.com/retrieve/pii/0378874189900494.

15 Singh, N., Ratnapandian, S., and Sheikh, J. (2021). Durable multifunctional finishing of cotton using -cyclodextrin-grafted chitosan and lemongrass (*Cymbopogon citratus*) oil. *Cell. Chem. Technol.* 55 (1–2): 177–184. https://www.cellulosechemtechnol.ro/pdf/CCT1-2(2021)/p.177-184.pdf.

16 Verma, R.K., Verma, R.S., Chauhan, A., and Bisht, A. (2015). Evaluation of essential oil yield and chemical composition of eight lemongrass (*Cymbopogon* spp.) cultivars under Himalayan region. *J. Essent. Oil Res.* 27 (3): 197–203. http://www.tandfonline.com/doi/full/10.1080/10412905.2015.1014936.

17 Wani, A.R., Yadav, K., Khursheed, A., and Rather, M.A. (2021). An updated and comprehensive review of the antiviral potential of essential oils and their chemical constituents with special focus on their mechanism of action against various influenza and coronaviruses. *Microb. Pathog.* 152: 104620. https://linkinghub.elsevier.com/retrieve/pii/S0882401020309864.

18 Muala, W.C.B., Desobgo, Z.S.C., and Jong, N.E. (2021). Optimization of extraction conditions of phenolic compounds from *Cymbopogon citratus* and evaluation of phenolics and aroma profiles of extract. *Heliyon* 7 (4): e06744. https://linkinghub.elsevier.com/retrieve/pii/S2405844021008471.

19 Dussault, D., Vu, K.D., and Lacroix, M. (2014). In vitro evaluation of antimicrobial activities of various commercial essential oils, oleoresin and pure compounds against food pathogens and application in ham. *Meat. Sci.* 96 (1): 514–520. https://linkinghub.elsevier.com/retrieve/pii/S0309174013004993.

20 Lulekal, E., Tesfaye, S., Gebrechristos, S. et al. (2019). Phytochemical analysis and evaluation of skin irritation, acute and sub-acute toxicity of *Cymbopogon citratus* essential oil in mice and rabbits. *Toxicol. Rep.* 6: 1289–1294. https://linkinghub.elsevier.com/retrieve/pii/S2214750019302525.

21 Li M, Liu B, Bernigaud C, Fischer K, Guillot J, Fang F. Lemongrass (*Cymbopogon citratus*) oil: a promising miticidal and ovicidal agent against *Sarcoptes scabiei*. Taylan Ozkan A, editor. *PLoS Negl.Trop. Dis.* 2020;14(4):e0008225. https://dx.plos.org/10.1371/journal.pntd.0008225

22 Viktorová, J., Stupák, M., Řehořová, K. et al. (2020). Lemon grass essential oil does not modulate cancer cells multidrug resistance by citral—its dominant and strongly antimicrobial compound. *Foods* 9 (5): 585. https://www.mdpi.com/2304-8158/9/5/585.

23 Machado, C.A., Oliveira, F.O., de Andrade, M.A. et al. (2022). Steam distillation for essential oil extraction: an evaluation of technological advances based on an analysis of patent documents. *Sustainability* 14: 7119. https://doi.org/10.3390/su14127119.

24 (a) Do, D.N., Nguyen, D.P., Phung, V.D. et al. (2021). Fractionating of lemongrass (*Cymbopogon citratus*) essential oil by vacuum fractional distillation. *Processes* 9 (4): 593. https://www.mdpi.com/2227-9717/9/4/593. (b) Shah, G., Shri, R., Panchal, V. et al. (2011). Scientific basis for the therapeutic use of *Cymbopogon citratus*, stapf (Lemon grass). *J. Adv. Pharm. Technol. Res.* 2 (1): 3. http://www.japtr.org/text.asp?2011/2/1/3/79796.

25 Saada, N.S., Abdel-Maksoud, G., Abd El-Aziz, M.S., and Youssef, A.M. (2020). Evaluation and utilization of lemongrass oil nanoemulsion for disinfection of documentary heritage based on parchment. *Biocatal. Agric. Biotechnol.* 29: 101839. https://linkinghub.elsevier.com/retrieve/pii/S1878818120315814.

26 Baldacchino, F., Tramut, C., Salem, A. et al. (2013). The repellency of lemongrass oil against stable flies, tested using video tracking. *Parasite* (20): 21. http://www.parasite-journal.org/10.1051/parasite/2013021.

27 An, Q., Ren, J.N., Li, X. et al. (2021). Recent updates on bioactive properties of linalool. *Food Funct.* 12 (21): 10370–10389. http://xlink.rsc.org/?DOI=D1FO02120F.

28 Peana, A.T., D'Aquila, P.S., Panin, F. et al. (2002). Anti-inflammatory activity of linalool and linalyl acetate constituents of essential oils. *Phytomedicine* 9 (8): 721–726. https://linkinghub.elsevier.com/retrieve/pii/S0944711304701804.

29 Meenapriya, M. and Jothi, P. (2017). Effect of lemongrass oil on rheumatoid arthritis. *Int. J. Curr. Adv. Res.* 2694–2696. http://journalijcar.org/issues/effect-lemongrass-oil-rheumatoid-arthritis.

30 Mukhtar, Y.M., Adu-Frimpong, M., Xu, X., and Yu, J. (2018). Biochemical significance of limonene and its metabolites: future prospects for designing and developing highly potent anticancer drugs. *Biosci. Rep.* 38 (6): https://portlandpress.com/biroscirep/article/38/6/BSR20181253/98155/Biochemical-significance-of-limonene-and-its.

31 Kulkarni, R.N., Mallavarapu, G.R., Baskaran, K. et al. (1997). Essential oil composition of a citronella-like variant of lemongrass. *J. Essent. Oil Res.* 9 (4): 393–395. http://www.tandfonline.com/doi/abs/10.1080/10412905.1997.9700738.

32 Dao, T.P., Do, H.T., Khoi, L.Q. et al. (2020). Evaluation of physico-chemical properties of lemongrass (*Cymbopogon citratus* L.) essential oil grown in tien giang province, Vietnam. *Asian J. Chem.* 32 (5): 1248–1250. http://www.asianjournalofchemistry.co.in/user/journal/viewarticle.aspx?ArticleID=32_5_41.

33 Wagh, A.M., Jaiswal, S.G., and Bornare, D.T. (2021). A review: extraction of essential oil from lemon grass as a preservative for animal products. *Pharma. Innov.* 10 (10): 1562–1567. https://www.thepharmajournal.com/archives/?year=2021&vol=10&issue=10&ArticleId=8423.

34 Ain, N.A.H., Zaibunnisa, A.H., Zahrah, H.M.S., and Norashikin, S. (2013). An experimental design approach for the extraction of lemongrass (*Cymbopogon citratus*) oleoresin using pressurised liquid extraction (PLE). *Int. Food Res. J.* 20 (1): 451.

35 Paviani, L., Pergher, S.B.C., and Dariva, C. (2006). Application of molecular sieves in the fractionation of lemongrass oil from high-pressure carbon dioxide extraction. *Braz. J. Chem. Eng.* 23 (2): 219–225. http://www.scielo.br/scielo.php?script=sci_arttext&pid=S0104-66322006000200009&lng=en&tlng=en.

36 Schaneberg, B.T. and Khan, I.A. (2002). Comparison of extraction methods for marker compounds in the essential oil of lemon grass by GC. *J. Agric. Food. Chem.* 50 (6): 1345–1349. https://pubs.acs.org/doi/10.1021/jf011078h.

37 Mendoza, S.M.V., Moreno, E.A., Fajardo, C.A.G., and Medina, R.F. (2019). Liquid–liquid continuous extraction and fractional distillation for the removal of organic compounds from the wastewater of the oil industry. *Water* 11: 1452. https://doi.org/10.3390/w11071452.

38 (a) Ribeiro-Santos, R., Andrade, M., de Melo, N.R., and Sanches-Silva, A. (2017). Use of essential oils in active food packaging: recent advances and future trends. *Trends Food Sci. Technol.* 61: 132–140. https://linkinghub.elsevier.com/retrieve/pii/S0924224416303521. (b) d'Ávila, J.V., Martinazzo, A.P., dos Santos, F.S. et al. (2016). Essential oil production of lemongrass (*Cymbopogon citratus*) under organic compost containing sewage sludge. *Rev. Bras. Eng. Agrícola e Ambient* 20 (9): 811–816. http://www.scielo.br/scielo.php?script=sci_arttext&pid=S1415-43662016000900811&lng=en&tlng=en.

39 Kpoviessi, S., Bero, J., Agbani, P. et al. (2014). Chemical composition, cytotoxicity and in vitro antitrypanosomal and antiplasmodial activity of the essential oils of four *Cymbopogon* species from Benin. *J. Ethnopharmacol.* 151 (1): 652–659. https://linkinghub.elsevier.com/retrieve/pii/S0378874113008210.

40 Brügger, B.P., Martínez, L.C., Plata-Rueda, A. et al. (2019). Bioactivity of the *Cymbopogon citratus* (Poaceae) essential oil and its terpenoid constituents on the predatory bug, *Podisus nigrispinus* (Heteroptera: Pentatomidae). *Sci. Rep.* 9 (1): 8358. http://www.nature.com/articles/s41598-019-44709-y.

41 Boukhatem, M.N., Kameli, A., Ferhat, M.A. et al. (2014). The food preservative potential of essential oils: is lemongrass the answer? *J. Verbraucherschutz*

Lebensmittelsicherh. 9 (1): 13–21. http://link.springer.com/10.1007/s00003-013-0852-x.

42 Mukarram, M., Choudhary, S., Khan, M.A. et al. (2021). Lemongrass essential oil components with antimicrobial and anticancer activities. *Antioxidants* 11 (1): 20. https://www.mdpi.com/2076-3921/11/1/20.

43 Rodrigues, L., Coelho, E., Madeira, R. et al. (2022). Food ingredients derived from lemongrass byproduct hydrodistillation: essential oil, hydrolate, and decoction. *Molecules* 27 (8): 2493. https://www.mdpi.com/1420-3049/27/8/2493.

44 de Melo, A.M., Turola Barbi, R.C., de WFC, S. et al. (2020). Microencapsulated lemongrass (*Cymbopogon flexuosus*) essential oil: a new source of natural additive applied to Coalho cheese. *J. Food Process Preserv.* 44 (10): https://onlinelibrary.wiley.com/doi/10.1111/jfpp.14783.

45 Skaria, B.P., Joy, P.P., Mathew, S., and Mathew, G. (2006). Lemongrass. In: *Handbook of Herbs and Spices*, 400–419. Elsevier https://linkinghub.elsevier.com/retrieve/pii/B9781845690175500245.

46 Liao, P.C., Yang, T.S., Chou, J.C. et al. (2015). Anti-inflammatory activity of neral and geranial isolated from fruits of *Litsea cubeba* Lour. *J. Funct. Foods* 19: 248–258. https://linkinghub.elsevier.com/retrieve/pii/S175646461500451X.

47 Gaba, J., Bhardwaj, G., and Sharma, A. (2020). Lemongrass. In: *Antioxidants in Vegetables and Nuts – Properties and Health Benefits*, 75–103. Singapore: Springer Singapore http://link.springer.com/10.1007/978-981-15-7470-2_4.

48 Antonious, G.F. (2008). *Presence of Zingiberene and Curcumene in Wild Tomato Leaves. Tomatoes and Tomato Products*, 193–214. CRC Press.

49 Amer, T.A.M. (2018). Effect of lemon and orange oils on shelf life of cake. *Middle East J. Appl. Sci.* 14: 1364–1374.

4

Nanocarbon Material and Chemicals from Seaweed for Energy Storage Components

Tirto Prakoso[1], Hary Devianto[1], Heri Rustamaji[2], Praswasti PDK Wulan[3], and Misri Gozan[3]

[1] *Bandung Institute of Technology (ITB), Faculty of Industrial Technology, Department of Bioenergy Engineering and Chemurgy, Jl. Ganesa No. 10, Bandung, 40132, Indonesia*
[2] *Universitas Lampung, Department of Chemical Engineering, Bandar, Lampung, 351145, Indonesia*
[3] *Universitas Indonesia, Faculty of Engineering, Chemical Engineering, Depok, 16424, Indonesia*

4.1 Introduction

Algae are photosynthetic aquatic organisms, including seaweeds (i.e. marine macroalgae) and microalgae. Algae play an essential role in marine environments by providing the source of energy for all nautical living things in the food chain. They also offer numerous ecological opportunities and environmental impacts, such as contamination prevention, carbon capture or sequestration, acidification restoration, biodiversity availability, and coastal security, among other things [1]. As opposed to terrestrial plants widely cultivated on land for biofuel, algae can grow without using farmland acreage. Numerous plants and animals can survive in salt or brackish water, eluding conflict between freshwater and land demand for nourishment production [2].

Algae, particularly seaweeds, play a significant role in global aquaculture. Algae cultivation accounted for almost 30% of the 120 million metric tonnes of worldwide aquatic manufacturing in 2019, with red macroalgae (Rhodophyta) and brown macroalgae (Phaeophyceae) ranking second and third, respectively [3]. However, seaweeds are not well recognized in several areas since their development is mainly focused on Eastern and Southeastern Asia. On the consumer side, even macroalgae have become broadly and commonly ingested human foods in Eastern Asia, they remain mainly segmented or innovative foods in the rest of the world. They were mainly consumed by relatively few people for varying reasons, such as dietary uses, on several coastlines as traditional dishes, e.g. plant-based diets for animal welfare). Other benefits of seaweeds involve biofertilizers, bio-packaging, cosmetic products, medical products, nutraceuticals, food ingredients, animal feeds, pharmaceuticals, cosmeceuticals, and biofuel, among others [4]. However, only the seaweed industry and academic researchers are aware of their participation in these products [5].

Chemical Substitutes from Agricultural and Industrial By-Products: Bioconversion, Bioprocessing, and Biorefining, First Edition. Edited by Suraini Abd-Aziz, Misri Gozan, Mohamad Faizal Ibrahim, and Lai-Yee Phang.
© 2024 WILEY-VCH GmbH. Published 2024 by WILEY-VCH GmbH.

Aquaculture was the primary industry supporting global seaweed manufacturing and accounted for 97% (34.7 million tonnes) of world seaweed production in 2019 from cultivation, and the rest is wild seaweed (1.1 million tonnes). In 2019, Asia contributed 97.4% of global seaweed production (99.1% from cultivation), and seven of the world's top 10 major seaweed producers were from Eastern or

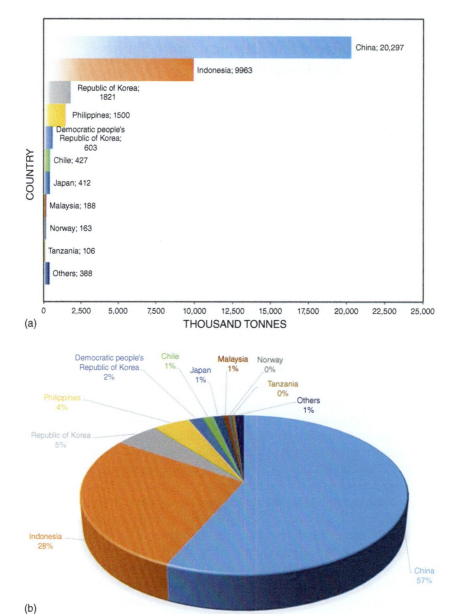

Figure 4.1 Seaweed aquaculture production per country (a) in wet weight tonnes and (b) in global percentage share in 2019. Source: Adapted from Cai et al. [1].

Southeastern Asia. In 2019, the output of seaweed in the America and Europe was 1.4% and 0.8% of the total, respectively. In these two regions, the wild collection was the primary fount of seaweed production, with inculcation counted for only 4.7% and 3.9% of the world's seaweed production, respectively. Conversely, in Africa (81.3%) and Oceania, seaweed development was mainly based on cultivation (85.3%), despite contributing only 0.4% and 0.05% to global seaweed production, respectively. Figure 4.1 shows seaweed production by country in tonnage capacity and percentage contribution [1, 6].

In general, seaweeds can be categorized into three taxonomic groups: 2000 species of brown seaweed fall under the Phaeophyceae family (*Saccharina japonica*, *Undaria pinnatifid, and Sargassum fusiforme*); red seaweeds with over 7200 species under Rhodophyta (*Porphyra*, *Eucheuma*, *Kappaphycus alvarezii*, and *Gracilaria* spp.); and green seaweeds more than 1800 macroalgae species are classified as Chlorophyta (*Enteromorpha* sp., *Monostroma* sp., *Caulerpa* sp., and *Codium* sp.) [7].

The leading producers of seaweed are China and Indonesia with combined production of more than 23 million metric tonnes in 2019. The majority of the red algae from the genera *Gracilaria*, *Pyropia*, and kelp from the genera *Saccharina japonica* and *Undaria pinnatifida*, are produced in China. On the other hand, *Kappaphycus* and *Eucheuma* are primarily produced in Indonesia. *Saccharina*, *Undaria*, *Porphyra*, *Eucheuma/Kappaphycus*, and *Gracilaria* are the top five genera, accounting for approximately 98% of the world's cultivars of seaweed production. Furthermore, Chile, China, and Norway lead the exploitation of the wild stocks of seaweeds, of which kelps are the most sought-after [1, 8]. Figure 4.2 shows the percentage of each species produced from seaweed aquaculture and country in 2019.

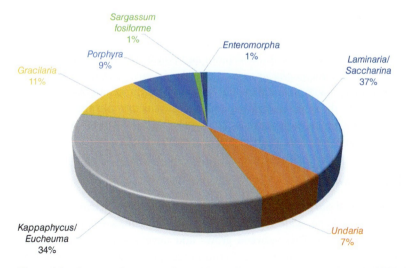

Figure 4.2 Percent of each species produced from seaweed aquaculture in 2019. Source: Adapted from Cai et al. [1].

4.2 Source of Seaweed

Based on their pigmentation, seaweeds are divided into three broad categories: red, brown, and green. Botanists classify these groups as Phaeophyceae, Rhodophyceae, and Chlorophyceae, in that order. Seaweeds are also known as macroalgae. This differentiates them from microalgae (Cyanophyceae), which are microscopic in size, frequently single cells, and are renowned for blooming and contaminating rivers and waterways [4].

4.2.1 Red Seaweed

Two warm-water genera (*Kappaphycus/Eucheuma and Gracilaria*) and one cold-water genus are the main targets for red seaweed cultivation (*Porphyra*, also known as nori). Twenty-three nations or territories, including nine Asian nations, four Eastern African nations or zones, four Pacific Island states, and six Latin American and Caribbean nations, produced 11.6 million tonnes of *Kappaphycus/Eucheuma* (33.6% of all seaweeds) in 2019 [7].

The weather in Indonesia is wet and tropical, with two seasons (dry and rainy) and temperatures between 21 and 33 °C. The tropical seaweeds, *Eucheuma* and *Kappaphycus*, which can also withstand the atmospheric temperature of 22–33 °C, depend on this climate to grow. However, the excessive seasonal variability necessitates the use of temporary regional calendars for healthy growth and preventing disease [9]. Fundamental knowledge of the transmission, growing circumstances, biological variation, and major threats to the *Eucheuma denticulatum*, *Kappaphycus striatus*, and *K. alvarezii* seaweeds, is required for all sectors of the supply chain, in particular those operating on the Indonesia seaweed distribution network for good environmental techniques.

The three seaweeds under consideration are native to the Philippines, Malaysia, and Indonesia, except for *K. alvarezii* and *E. denticulatum*, both endemic to the coral reefs. All three species were presented to other regions of southeast Asia, the Pacific Islands, the western Indian Ocean, and Latin and South America for commercial planting and research. *Kappaphycus striatus* was also invented in Indonesia. Due to distinctions in morphological features and the kind of carrageenan obtained from these two seaweeds, the researchers grouped *Kappaphycus striatum* and *K. alvarezii* as they belo to the genus *Kappaphycus*. The most valuable form of carrageenan, kappa, comes from the algac *K. striatum* and *K. alvarezii* [10].

4.2.2 Brown Seaweed

The type of seaweed *Sargassum* sp. is a relatively large brown marine plant with a flattened thallus shape, wide leaf shape, oval like a lush sword, and air-filled bubbles called bladders. This seaweed grows and develops on complex objects such as dead rocks, but it is also often found floating in water carried away [11]. In addition, *Sargassum* sp. also has finger-like branches and is a brown plant, relatively small in size, grows and thrives on a solid base substrate. Part of the plant resembles a bilaterally or radially symmetrical bush and is equipped with producing components [7].

Sargassum sp. contains the most compounds in the form of alginate, in addition to other chemical compounds that are relatively small in number, including laminarin, cellulose, fucoidan, mannitol, and other bioactive compounds. Besides, these algae contain fat, protein, crude fiber, and antibacterial and mineral substances (trace elements). Alginate is a salt form of alginic acid in the form of membrane mucilage. The food, fabrics, cosmetics, and pharmaceutical industries use alginic acid extensively. The properties of alginates involve their capacity to create gels, raise viscosity, and maintain stability at a range of pH of 3.8–5 [12].

4.2.3 Green Seaweed

About 0.05% of all seaweeds were grown globally for green seaweed in 2019, in contrast to the rapid growth in brown and red seaweed cultivation in 2019, with 6321 tonnes. *Monostroma nitidum* also referred to as green laver, had the world's highest planting in 2019. Green seaweed can be used in salad dressings and other foods as a sea vegetable. According to their common or marketing names, *Caulerpa lentillifera*, also called sea grape or green caviar, and *M. nitidum*, also known as green laver, are both regarded as delicacies. In addition to their use as food for animals, biofertilizers and biostimulants, pharmaceuticals, cosmetics, and waste treatment, green seaweeds also have a variety of other uses [1, 4]. Figure 4.3 shows several types of seaweed species.

Figure 4.3 Some kind of seaweed genus photograph. (a) *Glacilaria* sp., (b) *Euchema cottoni*, (c) *Eucheuma Spinosum*, (d) *Kappaphycus* sp., (e) *Sargassum* sp., and (f) *Undaria pinnatifida*. Source: Ferdouse et al. [7] and Waters et al. [9]. Reproduced with permission of Taylor & Francis.

4.3 Potential Material Substitute from Seaweed

Conceivable biomass algal production per area is frequently higher than terrestrial plants. Brown seaweeds, for instance, produced 13.1 kg of dry weight per square meter per year under cultivated conditions, as opposed to 10 kg from sugarcane [13]. As a prospective sunlight-driven cell processing plants for converting carbon dioxide into biofuels and chemical feedstocks, seaweed is thought to rank among the most important viable options for second-generation biofuels in the future. However, neither micro nor macroalgae have been successfully used as a source of commercial-scale fuel, even with their evident prospects [2].

The low energy return on investment (EROI) values calculated for microalgae compared to those for petroleum and diesel are part of the issue. Numerous research projects are now being undertaken to address this, considering the full range of products made from microalgae and biofuel in so-called biorefineries. In contrast, there has not been much research on macroalgae going on that focuses on creating or developing fuel feedstocks. However, in wet tonnage terms, the utilization of macroalgal feedstocks for nonfuel purposes is 100 times more widespread than microalga [2].

The application of marine macroalgae as a promising raw material for creating renewable fuels has a few different benefits. Without agricultural inputs like arable land, freshwater fertilizers, the capacity for effective depolymerization, etc. algae can produce much food. The development of macroalgae as a biofuel feedstock is the focus of an international research effort that aims to make the necessary technological advancements in biomass conversion process technology and planting technology [14].

A readily available and inexpensive natural source of carbon is found in seaweed. Seaweeds have a very diverse chemical makeup that varies depending on the species and, occasionally the season. They typically include ash, carbonate, proteins, lipids, and fiber. Because of this versatility, simple pyrolysis can produce carbon materials from seaweed with varying physicochemical properties. Seaweed has additional lignocellulose content. In addition, seaweed also contains fixed carbon, moisture ash, and volatile matter, which vary according to their species [14]. Table 4.1 presents the results of the proximate analysis on the types of seaweed. The type of grass suitable for carbon materials must meet a high fixed carbon content.

The source content of carbon in biomass can be converted into nanocarbon materials, which include activated carbon, graphite, fullerene, graphene, and carbon nanotubes. Nanocarbon materials are nanoscale structures made of carbon atoms with distinct properties not discovered in bulk carbon materials, such as high electrical conductivity, thermal conductivity, good mechanical properties, and elevated surface area.

4.3.1 Activated Carbon from Seaweed

Activated carbon, one of the most significant types of carbon, has a large number of micropores and a high surface area. With a specific surface area of up to 3000 m^2/g,

Table 4.1 Proximate analysis result of species seaweed biomass.

Seaweed	Fixed carbon (wt%)	Moisture (wt%)	VM (wt%)	Ash (wt%)	References
C. pilulifera[a]	18.4	10.5	32.2	38.6	[15]
P. yezoensis[a]	22.1	9.2	36.8	31.3	[15]
Sagarsum sp.[a]	9.34	9.34	44.5	36.82	[16]
Sargassum spp.[a]	25.1	9.8	43.9	21.2	[17]
Eucheuma cottonii[a]	27.82	11.2	47.4	13.58	[18]
Saccharina latissima[b]	13.3	—	75.8	10.9	[19]
Fucus serratus[b]	11.8	—	70.8	17.4	[19]
Red seaweed[b]	12.44	—	63.9	24.28	[20]
Kappaphycus alvarezii[b]	0.4	—	61.3	38.3	[21]
Eucheuma denticulatum[b]	23.9	—	61.5	14.6	[21]
Sargassum sp.[a]	9.6	14.4	61.4	14.6	[22]
Undaria pinntifida[a]	13.3	9.9	49.4	27.4	[23]
Laminaria ochroleuca[a]	15.7	9.6	47.8	26.9	[23]

VM, volatile matter.
a) As received basis.
b) Dry basis.

activated carbon is very efficient at expelling inorganic contaminants from water, such as heavy metals. Since it can take part in chemical reactions and serves as a support for catalysis, activated carbon is also referred to as "active carbon." Activated carbons have recently been used in energy conversion and storage [24, 25].

In general, porous carbon was distinguished by its physicochemical characteristics, such as a sizable surface area, a wide range of pore sizes, a relatively low density, etc. The term "activated carbon" describes carbon materials that have had their surfaces activated or had their internal structures altered through surface modification, metal or oxide deposition, etc., for specific applications. All types of activated carbon are porous carbons. Nevertheless, not all porous carbons are activated carbons. Even though activated carbons are essentially microporous materials, porous carbons have a wide range of pore sizes [25].

Activated carbon materials were produced involving three common processes. The process includes the pretreatment of biomass raw materials, carbonization, and activation. During the pretreatment process, the majority of the nutrients and solvable impurities are eliminated. The deashing process is a commonly used treatment method that aims to reduce the mineral content in the ash, usually aiming to increase the recovery of activated carbon. Organic lignin and cellulose can be transformed into carbonaceous materials through the carbonization process. Additionally, water, nutrients, oxygen, hydrogen, sulfur, and other elements are reduced by carbonization. Heating temperatures above 400 °C during the carbonization process could cause carbon loss. The particles in raw materials become dehydrated as the temperature rises. The oxygen from the raw materials was released as H_2O, CO, CO_2, etc. Such reaction by-products promote additional activation reactions [26]. Usually, carbonization is followed by the activation procedure. Physical activation and chemical activation are two methods for activating carbonaceous materials (biochars). A vacuum or inert gas atmosphere was used to activate the raw material during the physical activation process at temperatures between 600 and 1000 °C. Chemical activating agents are infused into raw materials during the chemical activation process. They are heated in an inert gas, and the activation and carbonization coincide [26].

Activated carbon can be synthesized from seaweed using various methods. Several studies have been synthesized in different ways, including one-step pyrolysis with flowing nitrogen gas [24, 27–31]; a combination of pyrolysis, pre-carbonization, pyrolysis activation [25, 32, 33], and a combination of hydrothermal carbonization (HTC) and pyrolysis activation [18, 34, 35]. The seaweed HTC stage generally produces hydrochar and liquid products, which can then be processed into crude biochemicals or biofuels [19, 21, 22]. Meanwhile, the hydrochar product is activated by pyrolysis to produce carbon material with a higher specific surface area. Physical characterization of SEM and porosity of activated carbon from *Sargassum horneri* made by hydrothermal and pyrolysis is shown in Figure 4.4.

Prakoso et al. [36] studied activated carbon's properties from *Sargassum* spp. seaweed hydrothermal process using KOH, $ZnCl_2$, and $CaCl_2$ as activating agents and CO_2. An investigation on the effect of different operational conditions on the characteristics of activated carbon was also carried out. They were followed by the interaction between the type of activator and the ratio of activators to biomass that significantly affect the result of activated carbon. The kind of activator and the interaction between the type of activator and the activator to raw material ratio have a significant impact on activated carbon yield. The highest product of activated carbon was obtained at 27.5% for the $CaCl_2$ activating agent sample at 250 °C. The specific surface area (S_{BET}) of activated carbon was significantly affected by the type of activator, with the highest specific surface area for various activators being 1552, 1368, and 1799 m²/g, respectively, for the activating agent of $ZnCl_2$, $CaCl_2$, and KOH at 250 °C. The SAC product has a mesoporous structure with a pore size distribution between 2.16 and 18 nm.

The research findings show that the HTC process by applying an activator followed by the pyrolysis activation process can be a promising method to produce high-quality activated carbon. Table 4.2 shows that different types of seaweed and synthesis methods result in other porosity properties. However, it can be said that seaweed is a potential raw material for activated carbon that can be applied to various regions.

4.3.2 Graphene from Seaweed

Graphene is a hexagonal or honeycomb-shaped two-dimensional (2D) material with a thickness equivalent to an atom's diameter made of sp^2-hybridized carbon atoms. Also, every carbon atom in graphene is covalent bonded to each other in the identical

4.3 Potential Material Substitute from Seaweed

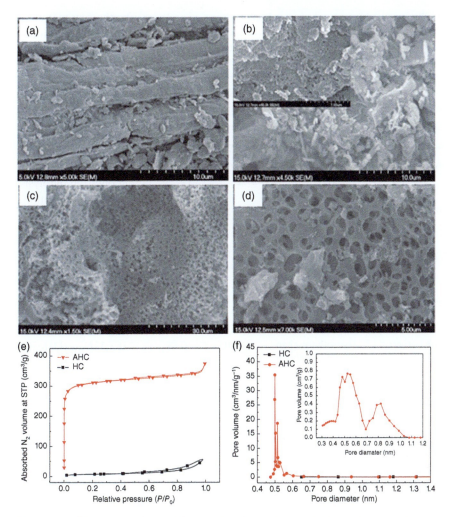

Figure 4.4 SEM images of the (a) *S. horneri*, (b) hydrochar, (c, d) KOH-activated carbon (AHC). Source: (a–d) from Zeng et al. [35]/Reproduced with permission from Hindawi/CC BY 4.0. (e) N_2 adsorption–desorption isotherms, and (f) pore size distributions.

plane and van der Waals forces hold the monolayer graphene sheets together [37]. The exceptional electronic, mechanical, chemical, thermal, and optical properties of single-atom-thick graphene make it a very attractive object for fundamental research and practical use. It is the lightest electronic conductor of all materials for a given surface area and could be manufactured cheaply and in significant quantities derived from natural graphite resources. This opens the way to its prospective application in a wide range of fields for better power production (solar cells), enhanced energy storage (supercapacitors, batteries, and fuel cells), sensors, biomedical, and even cutting-edge membrane substances for separations [38].

Biomass substances have already been investigated as potential raw materials for making graphene. Biomass materials have the prospects to substitute fossil

Table 4.2 The seaweed-based activated carbon's surface area and pore volume.

Alga type	Synthesis method	Activating agent	V_p (cm^3/g)	D_p (nm)	S_{BET} (m^2/g)	References
Lessonia nigrescens	Pyrolysis at 750 °C	Physical	0.29	0.8	1082	[24]
Lessonia nigrescens	Pyrolysis at 600 °C	Physical	0.29	<2	746	[27]
Turbinaria turbinata	Pyrolysis at 800 °C	Physical	0.71	4.4	812	[28]
Sargassum	Pyrolysis at 800 °C	Physical	0.25	2.8	361.1	[31]
Red Seaweed	Pre-carbonization and pyrolysis at 900 °C	Physical	0.15	2.1	3418	[25]
Enteromorpha prolifera	Pre-carbonization and pyrolysis at 800 °C	Physical	1.73	2.06	3345	[33]
Ascophyllum nodosum	Pre-carbonization and pyrolysis at 700 °C	KOH	—	2.3	1493	[32]
Spirulina platensis	Hydrothermal and pyrolysis activation at 900 °C	NH$_3$	1.16	2.1	1610.3	[34]
Sargassum horneri	Hydrothermal and pyrolysis activation at 600 °C	KOH	0.58	2.16	1221	[35]
Sargassum sp.	Hydrothermal and pyrolysis activation at 800 °C	KOH	0.86	2.72	1269	[36]
		ZnCl$_2$	0.93	2.81	1552	
		CaCl$_2$	1.51	10.19	594	

fuels and mined graphite in the production of graphene due to the elevated carbon centile (45–50 wt%) [39]. Furthermore, biomass materials are inexpensive and widely available, allowing for significant cost reductions in graphene. Graphene has been successfully synthesized from many types of biomass precursors, including cotton, corncobs, and camphor leaves. Biomass materials are frequently used as scaffolds and biocatalysts. Future research must identify biomass sources and methods that produce high-quality graphene at affordable prices and with a low carbon impact [40].

Using biomass resources and waste substances as carbon credits is a suitable technique that will result in expensive, large-scale graphene production. Moreover, using a catalyst to produce massive volumes of graphene is difficult. Furthermore, while using plants and waste as a carbon source, the absence of a catalyst may result in problems like low quality and a high number of defects. Much research has been done on the thermochemical conversion of biomass. Several methods have been constructed, as well as HTC, liquefaction, gasification, and pyrolysis. By employing these methods, biomass has been transformed into various products [38].

Carbonization and graphitization are frequently required steps in extracting graphene from biomass. Biomass materials are typically composed of carbon, hydrogen, and oxygen, and this biomass was first dehydrated, then graphitized at high temperatures to create graphene. Biomass-based graphene is frequently made up of aligned nanographene domains, in contrast to the ideal 2D layered graphene sheets, resulting in irregular shapes, special functional groups, and impressive properties [41]. These carbons derived from biomass are frequently known as graphene-like substances. Numerous techniques modify the starting biomass materials, such as chemical activation, thermal pretreatment, and atom doping. The amount of layers commonly determines whether carbons derived from biomass are mainly graphite (>10 layers), multilayer graphene (2–10 layers), or single-layer graphene. When biomass is applied as the initial material to make graphene, exfoliation is often necessary after carbonization and graphitization to create single or few-layer graphene. Many techniques have been researched for exfoliating graphitized biomass into graphenes, such as the Hummers technique and shear mixing.

Several researchers study the use of seaweed for graphene synthesis. Sharma et al. studied the partially reduced graphene oxide (prGO) made with a seaweed sap that was taken from the red seaweed *K. alvarezii* (Figure 4.5a,b). The seaweed sap is a fungicide splatter used to stimulate plant growth. The solution contained similar organic and inorganic micro and macronutrients, including salts of potassium, magnesium, iron, zinc, and other metals, as well as plant-growing controls, flavonols, choline, and glycine betaine. Different mixtures of the liquid completely void of plant growth regulators (PGRs) and flavonols, in addition to all organic substances, had been created to ascertain the precise concentration of the sap accountable for the reduction of graphene oxide (GO). It was found that *K. alvarezii* and organic matter both underwent a partial reduction of GO, whereas flavonols did not. It was determined that the reduction was caused by the synergistic interaction between the transition metals and flavonols in the sap. A plant foliar spray could be made from the retrieved fluid acquired after the isolation of prGO because its chemical makeup was similar to that of the original fluid [42].

Hummers and Offeman's method produces synthetic GO with FT-IR spectral bands that are in good agreement with the oxide (Figure 4.5c,d). Pure GO's powder X-ray diffraction (PXRD) pattern revealed a distinctive peak at $2° = 11.58°$ with an interlayer spacing of 7.65°, indicating the existence of oxygenated functional group following graphite oxidation. Original graphite peaks include (002) at $2° = 26.53°$ and (010) at $2° = 42.42°$. With a much lower interlayer spacing of 3.35°, the X-ray diffraction (XRD) pattern of GO treated with flavonols, and organic matter showed a bright peak getting closer to (002) at $2° = 20.73°$ (Figure 4.5e). Providing additional evidence of GO was reduced, the micro-Raman spectra of the GO samples were confirmed. Pure GO displayed a D/G intensity ratio of 0.89, while pristine graphite displayed a grade of 0.45. The ratio increased to 0.98 when GO was treated with *Kappaphycus* sap (Figure 4.5f). Since the sap that was partially free of PGRs and flavonols was unable to reduce GO, it is obvious from the analyses that pure *Kappaphycus* sap clean from all organic substances could do so. Therefore, the PGRs or flavonols must have been crucial to the reduction process [42].

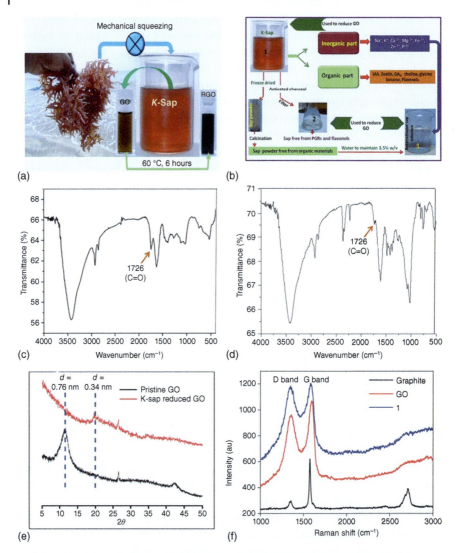

Figure 4.5 (a) Scheme reduction of graphene oxide in the existence of seaweed sap from *Kappaphycus alvarezii*, (b) preparation of various *Kappaphycus* sap compositions. Source (a, b) from Sharma et al. [42]/Reproduced with permission from Royal Society of Chemistry. (c) FT-IR spectrum of graphene oxide, (d) GO processed with K-sap (1), (e) Powder XRD trend of pristine GO (black) and reduced GO by 1 (red), and (f) Raman spectrum of graphite (black), GO (red), and GO processed with pure K-sap (1) (blue). Source: Sharma et al. [42]. Reproduced with permission of Royal Society of Chemistry.

Mondal et al. [43] reported a simple technique for the mass processing of Fe_3O_4/Fe–GN (Fe_3O_4/Fe-doped graphene nanosheets) from bountiful macroalgae biomass. The raw material used to create graphene nanosheets was the granules left over after extracting the liquid juice from fresh *Sargassum tenerrimum*. A deep eutectic solvent (DES) created by the complex formation of choline chloride and iron chloride (ChCl–FeCl$_3$) was used as a framework and catalyst for the production of graphene nanosheets. The creation of Fe$_3$O$_4$/Fe–GN with a moderate surface area (225 m^2/g) and electrical conductivity (2385 mS/m) resulted from the pyrolysis of alga particle and DES at 710–910 °C in a 95.5% N$_2$ and 4.5% H$_2$ atmospheric. As shown in Figure 4.6a,b, all the pyrolysis products exhibited magnetic properties. The FT-IR spectrum of the three products is depicted in Figure 4.6c. Figure 4.6d shows the presence of the Fe$_3$O$_4$ phase in SAR-700 powder XRD (JCPDS file no: 00-003-0863). Raman spectroscopy was utilized to examine graphene materials. The characteristic peaks in the Raman spectrum at 1350 cm^{-1} (D-band), 1570 cm^{-1} (G-band), and 2700 cm^{-1} (2D-band) are typical graphene signatures (Figure 4.6e). Figure 4.6f informs the atomic percent of C (52.84%), O (36.46%), N (1.7%), Fe (7.94%), and scrimpy of S (0.93%) in the XPS survey of functionalized graphene (SAR-700). Figure 4.6g,h shows high-resolution XPS images of graphitic nitrogen of graphene sheets interacting with Fe atoms. *Sargassum tenerrimum* functionalized graphene sheets as a long-term alternative to valuable metal-based oxygen reduction reactions (ORR) catalysts [43].

Utilizing discarded seaweed biomass as new precursors, Liu et al. [44] created a nanoscale Kirkendall effect-aided procedure for the straightforward and expandable production of three-dimensional (3D) Fe$_2$O$_3$ hollow nanoparticles (HNPs)/graphene aerogel (Fe$_2$O$_3$-HNPs/N-GAs). On 3D graphene aerogel, Fe$_2$O$_3$ HNPs with an ordinary wall thickness of about 6 nm are dispersed and serve as spacers to facilitate the splitting of nearby graphene nanosheets. The hollow structure of the active Fe$_2$O$_3$ NPs and the 3D graphene aerogel framework's unique system outperform all previously reported Fe$_2$O$_3$/graphene hybrid electrodes, which is why they are so effective. The current work is a critical stage forward into excellent performance control 3D graphene-based NPs aerogels for optimizing lithium storage while opening up new avenues for primary theoretical and industrial uses [34].

Fe$_2$O$_3$-HNPs/N-GAs-x composites' XRD analysis is depicted in Figure 4.7a, revealing that the final products are α-Fe$_2$O$_3$ (JCPDS No. 33-0664). The fact that the graphitic diffraction peak at 20°–30° could not be found indicates that Fe$_2$O$_3$-HNPs were effectively fastened to the graphene sheets, preventing the aggregation of graphene layers. The α-Fe$_2$O$_3$ distinctive peaks and the D and G peaks of graphene can be seen in the Raman spectra of Fe$_2$O$_3$-HNPs/N-GAs-x (Figure 4.7b). Fe$_2$O$_3$-HNPs/N-GAs-10 TGA and DTA analyses show that a noticeable endothermic peak seems at 460 °C, which relates to the disintegration (Figure 4.7c). The C1s, O1s, N1s, and Fe 2p core levels are visible in the XPS full-scan spectrum (Figure 4.7d). Fe$_2$O$_3$ has typical distinctive peaks in the XPS survey spectrum at 711 and 725 eV, which correspond to Fe2p$_{3/2}$ and Fe2p$_{1/2}$, respectively. Figure 4.7e illustrates the prevalent Fe2p$_{3/2}$ and Fe2p$_{1/2}$ peaks with two satellite peaks on their high binding-energy side (8 eV), which suggests the characteristic peaks of Fe$_2$O$_3$. Figure 4.7f depicts three peaks from the high-resolution N1s spectrum – pyridinic-N (398.3 eV), pyrrolic-N (400.3 eV), and graphitic-N – can be used to fit the data well (402.7 eV) [34].

Gupta et al. [45] investigated the synergistic effects of combining polyaniline (PANI) and seaweed derived graphene (SDG) produced through a straightforward

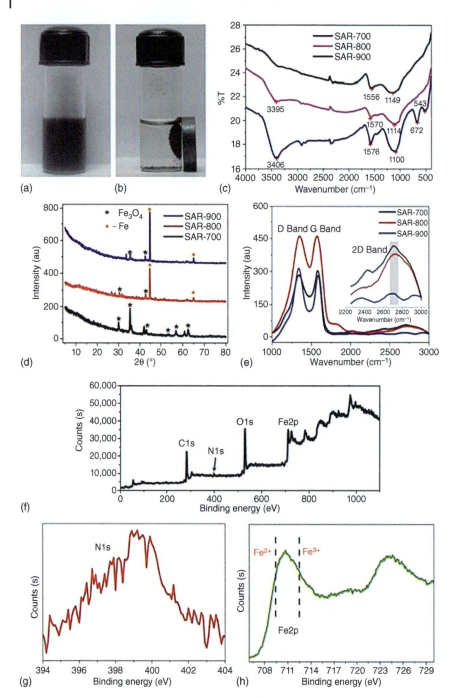

4.3 Potential Material Substitute from Seaweed | 73

Figure 4.6 (a) Granules of *Sargassum* combined with DES were pyrolyzed at 700 °C to produce an aqueous dispersion of GNs, (b) their magnetic characteristic in the liquid mixture. Source: (a, b) Mondal et al. [43]/Reproduced with permission from Royal Society of Chemistry. (c) Graphene sheet FT-IR spectrum to verify iron oxides doping, (d) graphene sheet's X-ray diffraction (XRD) patterns after being pyrolyzed at various temperatures, (e) Raman spectra of graphene sheet's qualitative properties, (f) XPS scan survey curve with highlights for natural dopants, and (g, h) graph of the N1s interaction with the Fe active sites in high-resolution XPS. Source: Mondal et al. [43]. Reproduced with permission of Royal Society of Chemistry.

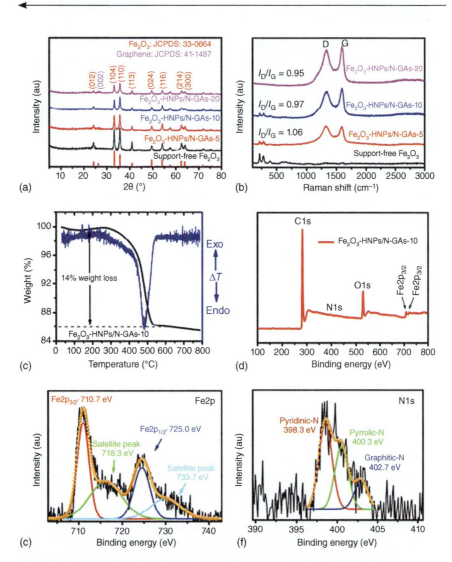

Figure 4.7 Characterization of the Fe_2O_3-HNPs/N-GAs structure. (a) XRD patterns, (b) Fe_2O_3-HNPs/N-GAs-*x* and support-free Fe_2O_3 Raman spectra, (c) Fe_2O_3-HNPs/N-GAs-10 thermal analysis using DTA and TG, (d) High-resolution XPS spectra of Fe_2O_3-HNPs/N-GAs-10 and XPS survey spectra of these materials, (e) Fe2p and (f) N1s peak. Source: Liu et al. [44]. Reproduced with permission of American Chemical Society.

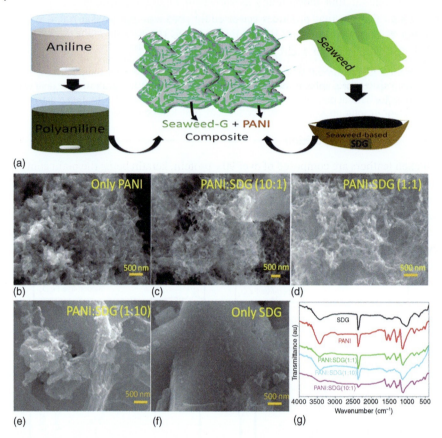

Figure 4.8 (a) Schematics of PANI-coated seaweed graphene's manufacturing processes, (b–f) SEM images of various PANI/SDG compositions (g), and FTIR spectra indicate the presence of PANI on the product. Source: (b–f) from Gupta et al. [45]/Reproduced with permission from Springer Nature.

aqueous synthetic process for energy storage applications. By in situ polymerizing aniline monomer (PANI:SDG), PANI nanofiber-coated seaweed-derived graphene nanocomposites were produced. As shown in Figure 4.8a, an easily scaled-up synthetic method developed nanocomposites with enhanced electrical conductivity (>70 mS/cm) and thermal stability. Figure 4.8b–f, where SDG is present at comparable compositions, displays a nanoflake-shaped substrate covered in PANI nanofibers. The electropolymerized PANI on SDG is visible in the FT-IR spectra of Figure 4.8g. Investigation indicates that formations of the PANI-SDGs offer comparable structural shapes due to synergistic surface interactions. Pure PANI nanofibers' XRD pattern (inset of Figure 4.9a) reveals the typical peaks at 2° = 15.3°, 20.4°, and 26.32°, which correspond to PANI's (011), (020), and (200) semicrystalline planes. This value demonstrates that aniline has successfully been polymerized into PANI. The PANI-Raman SDG's spectrum displays all of the PANI's distinctive peaks as well as extra D and G bands that are indicative of SDG's inclusion in the nanocomposites (Figure 4.9b). Additionally, nanocomposites maybe caused by

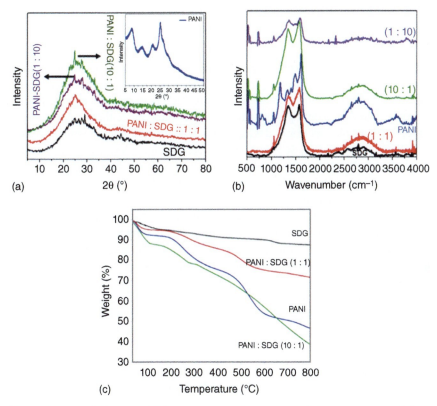

Figure 4.9 (a) XRD pattern, (b) Raman spectra of graphene (SDG) nanocomposite derived from PANI-seaweed with various compositions, (c) Solid mass weight loss measured thermogravimetrically for various compositions. Source: Gupta et al. [45]/Reproduced with permission of Springer Nature.

SDG's ordered crystal structure because the D band's intensity is lower than the G band's, indicating substantially less defective substance [45].

Thermogravimetric analysis was used to investigate the composition and thermal stability of SDG, PANI, and their nanocomposites (Figure 4.9c). As a result of the elimination of adsorbed water molecules, all samples displayed a slight weight loss during the initial processing stage up to 100 °C. The control SDG experienced a minor mass loss (7% w/w), while the control PANI experienced the most significant mass loss (55% w/w) up to 800 °C [45].

Although graphene produced from biomass is commonly very pure, its structural integrity varies depending on the feedstocks and manufacturing processes. It is necessary to optimize biomass materials and compatible methods for producing graphene. The most efficient biomass has been layer-based cellulose fiber, which frequently yields high-quality, few-layered graphene with a long edge length – noting the need for increased biomass yield to graphene conversion. According to reports, seaweed reduced graphene by 30%, while sacrificing 45% of its purity [43]. Optimization and new conversion methodologies are required to achieve high throughput, high yield, and high-quality graphene simultaneously.

4.4 Utilization of Seaweed-based Material for Energy Storage Component

Advanced energy storage technologies, such as rechargeable batteries and supercapacitors, are crucial to developing clean and sustainable energy sources like solar and wind. SCs on the other hand, are high-performance energy storage systems, which are becoming more and more necessary due to the quickly growing market for portable electronic devices. As an alternative energy storage source for rechargeable batteries, SCs with special benefits like elevated power density, extended cycle life (>100,000 cycles), and security will see increasing use in the future. Several top-notch review articles have covered the SC's device configuration and energy storage systems.

Biomass is a good source of functional carbon materials for applications like electrodes, separators, or binders in electrocapacitive energy storage. Recent studies have revealed that inexpensive carbon materials made from biomass that have electrocapacitive qualities are similar to industrial activated carbon. It can be made using direct activation or carbonization techniques. Several researchers have studied using seaweed as a component in electronic devices. Supercapacitor electrodes have been used with a carbon source derived from seaweed [24, 25, 27, 28, 32, 33, 46, 47], as an agar-based membrane for battery separators [48], and as alginic acid for electrode binder of lithium batteries [49].

4.4.1 Seaweed-derived Carbon Material for Supercapacitor Component

Bichat et al. [24] have characterized *Lessonia nigrescens* (LN) seaweed carbons-based electrode-supercapacitors in H_2SO_4, KOH, and Na_2SO_4 aqueous electrolytes. They informed the capacitance values obtained were 255, 201, and 125 F/g for H_2SO_4, KOH, and Na_2SO_4. The reliability potential window and capacitance values are both influenced by the properties of the electrode material and the electrolyte pH. For a specific electrolyte, the highest capacitance values and stability potential window are found in the carbon with the greatest oxygen content. Additionally, it has been demonstrated for the first time that a neutral electrolyte may contribute to pseudo-faradic reactions if the material has sufficient surface functionality. The most active groups seem to resemble quinones. Asymmetric carbon/carbon cell function up to 1.6 V in an aqueous electrolyte of Na_2SO_4.

Raymundo-Pinero et al. [27] investigated the effect of adding CNTs in LN-based carbon materials for supercapacitor electrodes. The result shows that in an aqueous electrolyte of $1 M H_2SO_4$, LN-CNT composites are more suitable than LN for designing high-power supercapacitors that can be charged and discharged while maintaining significant energy values. The supercapacitor with LN-10% CNT electrodes has an energy specific of about 8.8 Wh/kg and a power specific of 6.7 kW/kg at 10 seconds discharge time, which is a 20 and 29% improvement over the LN-based capacitor, respectively. Pintor et al. [28] study the effect of pyrolysis temperature on *Turbinaria turbinata* brown seaweed's carbon characteristics. In symmetric, electrochemical supercapacitors, carbon serves as an electrode material.

Cyclic voltammetry (CV), the galvanostatic charge/discharge (GCD) method, and electrochemical impedance spectroscopy (EIS) were used to characterize the electrochemical properties of carbon materials. Initial findings indicated that the sample prepared by pyrolysis at 800 °C produced the best behavior. The carbon had an average surface area of 812 m^2/g. Interesting results were obtained from electrochemical tests using an organic electrolyte of TEABF$_4$ with a capacitance of 74.7 F/g, the equivalent series resistance of 0.51 Ω/cm^2, and ionic resistance of 1.5 Ω/cm^2. These findings demonstrate the potential capacitive characteristic of seaweed-derived carbon and its use in electrochemical supercapacitors.

Wu et al. [33] investigated *Enteromorpha prolifera* macroalgal pollutants to produce nitrogen-rich carbons through carbonization and activation. Algae are an excellent option for nitrogen-rich carbons as adsorbents and electrode materials because they have surface areas that reach 3345 m^2/g, hierarchical pores, and nitrogen (1.6–3.8%) functionalities. By changing the activation environments, one can alter the pore structure and surface properties of carbons. One of the most significant findings concerning cutting-edge biomass-based carbons is that hydrogen consumption is substantial, reaching up to 7.02 wt% at −197 °C and 21 bar. Additionally, at 1 bar, an outstanding uptake of 2.71 wt% is also seen. According to the electrochemical test results, a capacitance of 441 F/g at 1 A/g in a 6 M KOH electrolyte was recorded for the most surface-area-rich carbon. As a result of its large surface area and stratified pores that provide numerous active sites to help electrolyte diffusion more effectively, the carbon with the most microporous surface area had proper cycle consistency with 86% beginning capacitance after 5000 cycles. After being constructed into a symmetric supercapacitor, in a 6 M KOH aqueous electrolyte, the electrochemical properties of the EP-based carbons were studied. Typical CV curves, GCD, EIS, and cycle numbers are depicted in Figure 4.10.

Carrageenan derived from red algae and Fe hydrogel was used by Li et al. [47] as a precursor for creating carbon material. The 3D stratified macro-meso-microporous sulfur-doped carbon aerogel (HPSCA), with controllable nanopores and a whopping surface area of up to 4038 m^2/g, is created by them. After carbonization, acid washing, and activation, the ultrahigh surface area of HPSCA is a result of the Fe^{3+} ion dispersions at the molecular level in carrageenan. It is a suitable nominee resource for lithium–sulfur (Li–S) batteries and double-layer supercapacitors (SCs) due to its appealing structural features. To create high-energy composite cathodes for Li–S batteries with an elevated specific capacity, cycle stability, and extended cycle life (400 cycles), because of its relatively sophisticated porous structure, the HPSCA can hold more sulfur (up to 80 wt%). Meanwhile, HPSCA can exhibit high specific capacitance of 336 and 218 F/g (1 A/g) in both the aqueous and organic electrolytes. High capacitance in aqueous and organic electrolytes of 205 F/g at 100 A/g and 175 F/g at 50 A/g, respectively, are retained while achieving superior rate performance.

Jiang et al. [25] investigated bleached red seaweed carbonized as a supercapacitor electrode. A carbonization temperature of 900 °C gives a high super capacitance (226.3 F/g) and good stability within 2400 cycles. Comparing these capacitance, pulp, or filter paper values, it is noticeably higher. Meanwhile, Perez-Scalzedo et al. [32] looked at chemically activated biocarbon (AKPH) made

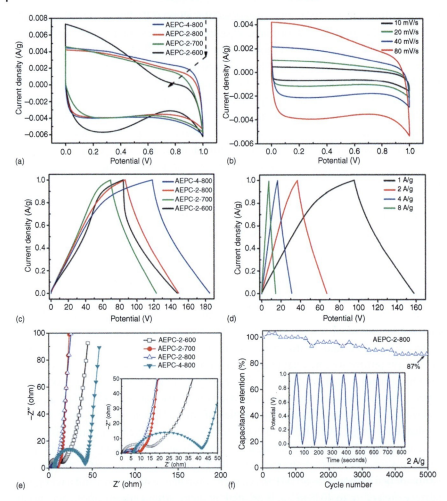

Figure 4.10 (a) Voltammogram of four supercapacitor products at 80 mV/s, (b) AEPC-2-800 voltammogram at varied scan rates, (c) GCD profile of four samples at 1 A/g, (d) AEPC-2-800 GCD profile at different current densities, (e) Four carbon samples Nyquist plots, (f) the cyclic performance of AEPC-2-800 at 2 A/g. Source: Wu et al. [33]/Royal Society of Chemistry/Licensed under CC BY 3.0.

from Ascophyllum nodosum with KOH as the electrode material for ORR and supercapacitors. With an onset voltage of 0.88 and 0.76 V versus RHE and a current density of 5.3 mA/cm^2, the electrochemical performance is good compared to commercial platinum. At 0.5 A/g, AKPH showed a capacitance of 208 F/g. With a retention capacity of 92.3%, this material is stable after 2500 cycles at 5 A/g. Seaweed is now a promising source of materials for energy conversion and storage applications thanks to its performance [32]. Figure 4.11 shows the fabricated supercapacitor characterization, including CV, GCD, EIS curve, and cycle stability. Table 4.3 summarizes the types of seaweed based-carbon material and the electrochemical characteristics of supercapacitors.

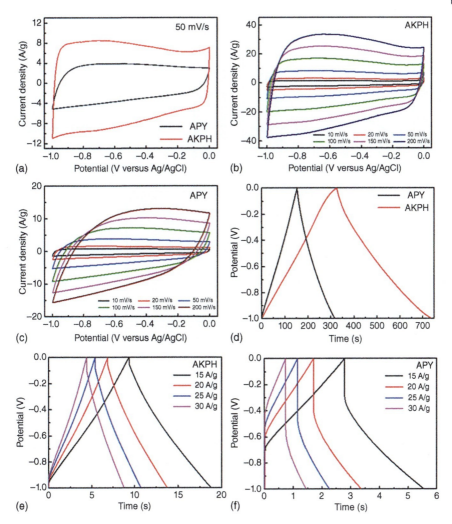

Figure 4.11 (a) CV for APY and AKPH in 1 M KOH at 50 mV/s, (b) AKPH CV at various scan rates, (c) APY CV at various scan rates, (d) GCD curves of APY and AKPH, (e) AKPH' GCD plots at various current densities, (f) APY' GCD plots at various current densities, (g) APY and AKPH Nyquist plots, (h) specific capacitance versus current density plot, and (i) AKPH cyclic performance. Source: Perez-Salcedo et al. [32]/Reproduced with permission of Springer Nature.

4.4.2 Seaweed-derived Chemical Materials for Battery Component

The chemical-derived seaweed can also be applied as another energy storage component. According to Kovalenko et al. [48], the high-modulus natural polysaccharide alginate obtained from brown algae produces an incredibly stable battery anode. Alginates, a significant part of brown algae and many aquatic microorganisms, differ from many polysaccharides typically found in terrestrial plants. They contain carboxylic groups in each of the polymer's monomeric units. A higher percentage of

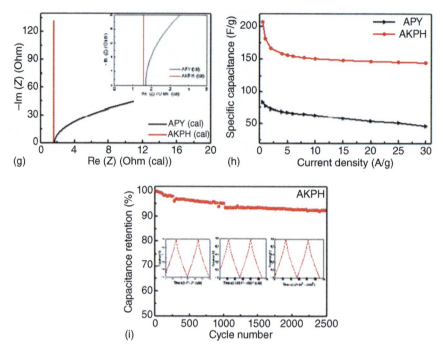

Figure 4.11 (Continued)

Table 4.3 Seaweed-based-carbon material and supercapacitor's electrochemical properties.

Seaweed type	Carbon product	Electrolyte	Capacitance (F/g)[a]	Cycle life (after cycle)	References
Lessonia nigrescens	AC	6 M KOH	149 (1 A/g)	89% (5,000 cycles)	[24]
Lessonia nigrescens	Carbon composite	1 M H_2SO_4	273 (1 A/g)	84% (15,000 cycles)	[27]
Turbinaria turbinata	AC	1 M NEt_4BF_4	74.5 (2 mV/s)	—	[28]
Red algae based-carrageenan	Doped carbon–Fe aerogel	6 M KOH 1 M $TEBF_4$	335 (1 A/g) 217 (1 A/g)	93% (10,000 cycles) 89.2% (5,000 cycles)	[47]
Enteromorpha prolifera	AC	6 M KOH	440 (1 A/g)	87% (5,000 cycles)	[33]
Ulva fasciata	Graphene-PANI	1 M H_2SO_4	320 (10 A/g)	90% (1,000 cycles)	[45]
Red algae	AC	5 M KCl	226.3 (2 A/g)	98% (2,400 cycles)	[25]

AC, activated carbon.
a) At current density or scan rate.

the carboxylic group in the binder should increase the number of potential binder–Si bonds and improve the stability of the Si electrode. A copolymer of b-D-mannuronic acid (M) and a-L-guluronic acid (G) residues forms alginate, also known as alginic acid. They demonstrated that combining Si nanopowder with alginate produces a stable battery anode with an eightfold increase in reversible capacity over the most advanced graphitic anodes. Na-alginate films displayed approximately 6.7 times greater rigidity in a dry condition than dry films of polyvinylidene difluoride (PVDF), according to atomic force microscopy indentation investigations. Li-ion battery (LIB) electrodes often contain PVDF, a typical commercial binder [47].

Alginate's rigidity did not significantly change when immersed in the electrolyte solution. At the same time, the PVDF sheets softened by roughly 50 times, according to Kovalenko et al. Ellipsometry studies show no noticeable swelling of thin (~70 nm) Na-alginate films in the electrolyte solvent vapors. Comparable-thickness PVDF films, however, fluctuate in thickness by about 20% and draw a sizable amount of carbonates from the moisture. Minor swelling of alginate suggests little interaction between the polymer and the electrolyte. This characteristic may prevent the electrolyte liquid from approaching the binder/Si interface too closely. Their similar behavior explains why Na-CMC binders with Si anodes perform so well [48]. Electrochemical analysis of the LIB show alginate binder has higher performance than PVDF and CMC, as depicted in Figure 4.12.

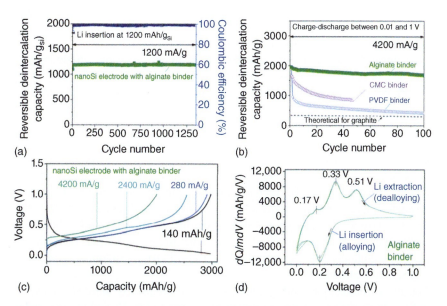

Figure 4.12 Electrochemical performance of alginate-based nano-Si electrodes. (a) Cycle number versus reversible Li-extraction capacity and CE of nano-Si electrodes, (b) Reversible Li-extraction capacity of nano-Si electrodes with alginate, CMC, and PVDF binders versus cycle number, (c) Galvanostatic discharge profiles of the nano-Si electrode at varied current densities, (d) Differential capacity curves of the nano-Si electrode in the potential window of 0–1 V. Source: Kovalenko et al. [48]/Reproduced with permission of American Association for the Advancement of Science.

Shin et al. [49] showed that agar, readily collected from brown algae, might be used for the usable binder and separator membrane of LIBs for another battery application. To control the agar polymer's phase separation performance in the nonsolvent-induced phase separation (NIPS) procedure, they added 3-glycidoxypropyl trimethoxysilane (GPTMS) as a surface modifier and created the porous separator membrane. This method made it possible to determine the ultimate morphologies of membranes for LIBs. They created a LIB separator membrane with suitable porosity, satisfactory electrolyte wettability, increased ionic conductivity, and increased thermal stability by accurately regulating the quantity of GPTMS modification. Additionally, the agar demonstrated advantageous properties like PF_5 stabilization and Mn^{2+} chelation. In high-energy-density batteries with an Mn-based cathode (LMO half-cell and LNMO/graphite entire cell), the advantages of the G-Agar separator membrane and agar binder make them compatible. The results in an exceptional cycle performance (84.1% after 100 cycles at 55 °C). The agar is practical for various energy storage devices due to its versatility and affordability [48].

4.5 Future Prospects and Challenges

A detailed explanation of the use of seaweed biomass-based materials as essential components in supercapacitors and batteries have been presented in this chapter. Problems and challenges must be resolved while producing carbon materials derived from helpful seaweed biomass in energy storage.

First, due to the wide range of seaweed biomass supplies available, a basic study on the impact of seaweed biomass contexture on the electrocapacitive property of the carbon produced is necessary. Second, the vast majority of the electrode materials made from seaweed biomass that has been identified as having excellent electrocapacitive performance were developed in laboratories. Scale-up research is necessary to create the process protocols for further technological advancement. Third, while the KOH-activation technique can produce biomass-derived hierarchical porous carbons with elevated specific surface areas, it has limited control over the pore geometry, size, and connection. Additionally, making graphitic carbons, which affect an electrolyte's electrical conductivity and surface wettability, is not a good idea when using the KOH-activation method. One may produce biomass-derived carbons with superior specific surface area, suitable pore structure, high electrical conductivity, and great wettability toward electrolytes by using KOH activation in conjunction with thermal treatment.

Fourth, carbonization and graphitization are frequently needed to extract graphene from seaweed. Seaweed is typically consist of C, H, and O, first undergoing dehydration and then graphitized at high temperatures to create graphene. Biomass-based graphene frequently consists of aligned nanographene domains rather than the ideal 2D layered graphene sheets, giving it irregular shapes, distinctive functional groups, and incredible characteristics. Fifth, by boosting electrode capacitance or expanding the operation voltage, supercapacitor's energy density can be improved. Future research will focus on increasing the capacitance by incorporating pseudocapacitive substances like heteroatoms, metal oxides, or conductive polymers with carbon derived from biomass. Another efficient way to increase energy and power density is to increase the potential window by utilizing electrolytes of ionic liquids, designing asymmetric cells, or creating hybrid capacitors. Finally, conducting in-depth techno-economic studies on the utilization of seaweed-derived based biomaterials and biochemicals for their use as energy storage components, supercapacitors, and batteries is important.

4.6 Conclusions

Seaweed biomass is an excellent source of usable carbon substances and chemicals for electrocapacitive energy storage utilization, such as electrodes, separators, and binders. Current studies have revealed that simple carbonization and activation methods can produce economical biomass-derived carbon materials. The electrocapacitive characteristics of this material are comparable to commercial activated carbon and graphene. It is possible to produce superior hierarchical porous carbons from biomass that perform as electrocapacitors using chemical activation techniques, such as potassium hydroxide, zinc chloride, and calcium chloride. Chemicals derived from seaweed, such as alginic acid, have also been shown to improve the performance of LIBs in their role as membrane separators and electrode binders. Utilization of seaweed-based materials in energy storage is potential and should not cause a conflict of interest with its use as a food ingredient but can increase a higher added value.

References

1 Cai, J., Lovatelli, A., Aguilar-Manjarrez, J. et al. (2021). Fisheries and seaweeds and microalgae: an overview for unlocking. *FAO Fisheries and Aquaculture Circular* 1229: 36p.
2 Milledge, J.J., Smith, B., Dyer, P.W., and Harvey, P. (2014). Macroalgae-derived biofuel: a review of methods of energy extraction from seaweed biomass. *Energies* 7: 7194–7222.
3 FAO (2021). FAO Global Fishery and Aquaculture ProductionStatistics (FishStatJ; March 2021; www.fao.org/fishery/statistics/software/fishstatj/en).
4 McHugh DJ (2003). Seaweeds uses as human foods. A Guide to the Seaweed Industry. p. 105. http://afrilib.odinafrica.org/handle/0/15549%5Cnhttp://www.fao.org/docrep/006/y4765e/y4765e00.htm (accessed 08 May 2023).
5 Campbell, I., Macleod, A., Sahlmann, C. et al. (2019). The environmental risks associated with the development of seaweed farming in Europe – prioritizing key knowledge gaps. *Front. Marine Sci.* 6: 00107.
6 FAO (2020). *World Fisheries and Aquaculture*, 1–11. Rome: FAO https://www.fao.org/3/ca9229en/online/ca9229en.html#chapter-1_1.
7 Ferdouse, F., Lovstad Hold, S., Smith, R. et al. (2018). The global status of seaweed production, trade and utilization. *FAO Globefish Res. Programme* 124: 120.

8 Buschmann, A.H., Camus, C., Infante, J. et al. (2017). Seaweed production: overview of the global state of exploitation, farming and emerging research activity. *Eur. J. Phycol.* 52 (4): 391–406. https://doi.org/10.1080/09670262.2017.1365175.

9 Waters, T., Jones, R., Theuerkauf, S. et al. (2019). Coastal conservation and sustainable livelihoods through seaweed aquaculture in Indonesia: a guide for buyers, conservation practitioners, and farmers. *Nat. Conserv.* V1: 1–47.

10 Rimmer, M.A., Larson, S., Lapong, I. et al. (2021). Seaweed aquaculture in Indonesia contributes to social and economic aspects of livelihoods and community wellbeing. *Sustainability* 13 (19): 1–22.

11 Yarish C, Valderrama D, Brummett R E, Radulovich R, (2016). Seaweed aquaculture for food security, income generation and environmental health in tropical developing countries. World Bank Group. *107147*.

12 Via, M., Abdillah, A.A., and Alamsjah, M.A. (2019). Physics and chemical characteristics of *Sargassum* sp. seaweed with addition of sodium alginate stabilizer to different concentrations. *IOP Conf. Ser.: Earth Environ. Sci.* 236 (1): 012128.

13 Leu, S. and Boussiba, S. (2014). Advances in the production of high-value products by microalgae. *Ind. Biotechnol.* 10 (3): 169–183.

14 Baghel, R.S., Trivedi, N., Gupta, V. et al. (2015). Biorefining of marine macroalgal biomass for production of biofuel and commodity chemicals. *Green Chem.* 17 (4): 2436–2443.

15 Li, D., Chen, L., Zhang, X. et al. (2011). Pyrolytic characteristics and kinetic studies of three kinds of red algae. *Biomass Bioenergy* 35 (5): 1765–1772. https://doi.org/10.1016/j.biombioe.2011.01.011.

16 Kim, S.S., Ly, H.V., Kim, J. et al. (2013). Thermogravimetric characteristics and pyrolysis kinetics of alga *Sagarssum* sp. biomass. *Bioresour. Technol.* 139: 242–248. https://doi.org/10.1016/j.biortech.2013.03.192.

17 Ali, I. and Bahadar, A. (2017). Red sea seaweed (*Sargassum* spp.) pyrolysis and its devolatilization kinetics. *Algal Res.* 21: 89–97. https://doi.org/10.1016/j.algal.2016.11.011.

18 Prakoso, T., Nurastuti, R., Hendriansyah, R. et al. (2018). Hydrothermal carbonization of seaweed for advanced biochar production. *MATEC Web Conf.* 156: 1–5.

19 Brown, A.E., Finnerty, G.L., Camargo-Valero, M.A., and Ross, A.B. (2020). Valorisation of macroalgae via the integration of hydrothermal carbonisation and anaerobic digestion. *Bioresour. Technol.* 312: 123539. https://doi.org/10.1016/j.biortech.2020.123539.

20 Patel, N., Acharya, B., and Basu, P. (2021). Hydrothermal carbonization (HTC) of seaweed (macroalgae) for producing hydrochar. *Energies* 14 (7): 1–16.

21 Nallasivam, J., Francis Prashanth, P., Harisankar, S. et al. (2022). Valorization of red macroalgae biomass via hydrothermal liquefaction using homogeneous catalysts. *Bioresour. Technol.* 346: 126515. https://doi.org/10.1016/j.biortech.2021.126515.

22 Rustamaji, H., Prakoso, T., Rizkiana, J. et al. (2022). Synthesis and characterization of hydrochar and bio-oil from hydrothermal carbonization of *Sargassum* sp. using choline chloride (ChCl) catalyst. *Int. J. Renewable Energy Dev.* 11 (2): 403–412.
23 Cassani, L., Lourenço-lopes, C., Barral-martinez, M. et al. (2022). Thermochemical characterization of eight seaweed species and evaluation of their potential use as an alternative for biofuel production and source of bioactive compounds. *Int. J. Mol. Sci.* 23 (4): 2355.
24 Bichat, M.P., Raymundo-Piñero, E., and Béguin, F. (2010). High voltage supercapacitor built with seaweed carbons in neutral aqueous electrolyte. *Carbon* 48 (15): 4351–4361.
25 Jiang, L., Han, S.O.K., Pirie, M. et al. (2021). Seaweed biomass waste-derived carbon as an electrode material for supercapacitor. *Energy Environ.* 32 (6): 1117–1129.
26 Inagaki, M. and Kang, F. (2014). Fundamental science of carbon materials. In: *Materials Science and Engineering of Carbon: Fundamentals*, 17–217. Butterworth-Heinemann.
27 Raymundo-Piñero, E., Cadek, M., Wachtler, M., and Béguin, F. (2011). Carbon nanotubes as nanotexturing agents for high power supercapacitors based on seaweed carbons. *ChemSusChem* 4 (7): 943–949.
28 Pintor, M.J., Jean-Marius, C., Jeanne-Rose, V. et al. (2013). Preparation of activated carbon from *Turbinaria turbinata* seaweeds and its use as supercapacitor electrode materials. *C.R. Chim.* 16 (1): 73–79.
29 Divya, P., Prithiba, A., and Rajalakshmi, R. (2019). Biomass derived functional carbon from *Sargassum wightii* seaweed for supercapacitors. *IOP Conf. Ser.: Mater. Sci. Eng.* 561 (1): 012078.
30 Terakado, O., Tanaka, F., and Tsunamori, Y. (2021). Preparation of activated carbon from holdfasts of kelp, large brown seaweed, *Saccharina japonica*. *Engineering* 13 (02): 71–81.
31 Yang, W. and Liu, Y. (2021). Removal of elemental mercury using seaweed biomass-based porous carbons prepared from microwave activation and H_2O_2 modification. *Energy Fuel* 35 (3): 2391–2401.
32 Perez-Salcedo, K.Y., Ruan, S., Su, J. et al. (2020). Seaweed-derived KOH activated biocarbon for electrocatalytic oxygen reduction and supercapacitor applications. *J. Porous Mater.* 27 (4): 959–969.
33 Wu, X., Tian, Z., Hu, L. et al. (2017). Macroalgae-derived nitrogen-doped hierarchical porous carbons with high performance for H_2 storage and supercapacitors. *RSC Adv.* 7 (52): 32795–32805. https://doi.org/10.1039/C7RA05355J.
34 Liu, F., Liu, L., Li, X. et al. (2016). Nitrogen self-doped carbon nanoparticles derived from spiral seaweeds for oxygen reduction reaction. *RSC Adv.* 6 (33): 27535–27541.
35 Zeng, G., Lou, S., Ying, H. et al. (2018). Preparation of microporous carbon from *Sargassum horneri* by hydrothermal carbonization and koh activation for CO_2 capture. *J. Chem.* 2018: 4319149.

36 Prakoso, T., Rustamaji, H., Yonathan, D. et al. (2022). The study of hydrothermal carbonization and activation factors' effect on mesoporous activated carbon production from *Sargassum* sp. using a multilevel factorial design. *Reaktor* 22 (2): 59–69.

37 Lee, X.J., Hiew, B.Y.Z., Lai, K.C. et al. (2019). Review on graphene and its derivatives: synthesis methods and potential industrial implementation. *J. Taiwan Inst. Chem. Eng.* 98: 163–180. https://doi.org/10.1016/j.jtice.2018.10.028.

38 Saha, J.K. and Dutta, A. (2022). A review of graphene: material synthesis from biomass sources. *Waste Biomass Valorization*. Springer Netherlands 13: 1385–1429. https://doi.org/10.1007/s12649-021-01577-w.

39 Ravi, S. and Vadukumpully, S. (2016). Sustainable carbon nanomaterials: recent advances and its applications in energy and environmental remediation. *J. Environ. Chem. Eng.* 4 (1): 835–856. https://doi.org/10.1016/j.jece.2015.11.026.

40 Zhou, Y., He, J., Chen, R., and Li, X. (2022). Recent advances in biomass-derived graphene and carbon nanotubes. *Mater. Today Sustain.* 18: 100138. https://doi.org/10.1016/j.mtsust.2022.100138.

41 Kong, X., Zhu, Y., Lei, H. et al. (2020). Synthesis of graphene-like carbon from biomass pyrolysis and its applications. *Chem. Eng. J.* 399: 125808. https://doi.org/10.1016/j.cej.2020.125808.

42 Sharma, M., Mondal, D., Das, A.K., and Prasad, K. (2014). Production of partially reduced graphene oxide nanosheets using a seaweed sap. *RSC Adv.* 4 (110): 64583–64588.

43 Mondal, D., Sharma, M., Wang, C.H. et al. (2016). Deep eutectic solvent promoted one step sustainable conversion of fresh seaweed biomass to functionalized graphene as a potential electrocatalyst. *Green Chem.* 18 (9): 2819–2826.

44 Liu, L., Yang, X., Lv, C. et al. (2016). Seaweed-derived route to Fe_2O_3 hollow nanoparticles/n-doped graphene aerogels with high lithium ion storage performance. *ACS Appl. Mater. Interfaces* 8 (11): 7047–7053.

45 Gupta, R., Vadodariya, N., Mahto, A. et al. (2018). Functionalized seaweed-derived graphene/polyaniline nanocomposite as efficient energy storage electrode. *J. Appl. Electrochem.* 48 (1): 37–48. https://doi.org/10.1007/s10800-017-1120-z.

46 Ruiz, V., Santamaría, R., Granda, M., and Blanco, C. (2009). Long-term cycling of carbon-based supercapacitors in aqueous media. *Electrochim. Acta* 54 (19): 4481–4486.

47 Li, D., Chang, G., Zong, L. et al. (2019). From double-helix structured seaweed to S-doped carbon aerogel with ultra-high surface area for energy storage. *Energy Storage Mater.* 17: 22–30. https://doi.org/10.1016/j.ensm.2018.08.004.

48 Kovalenko, I., Zdyrko, B., Magasinski, A. et al. (2011). A major constituent of brown algae for use in high-capacity Li-ion batteries. *Science* 334 (6052): 75–79.

49 Shin, M., Song, W.J., Han, J.G. et al. (2019). Metamorphosis of seaweeds into multitalented materials for energy storage applications. *Adv. Energy Mater.* 9 (19): 1–10.

5

Spent Mushroom Substrate as Alternative Source for the Production of Chemical Substitutes

Vikineswary Sabaratnam[1,5], Chia Wei Phan[2,5], Hariprasath Lakshmanan[3], and Jegadeesh Raman[4]

[1] *Universiti Malaya, Institute of Biological Sciences, Faculty of Science, Kuala Lumpur, 50603, Malaysia*
[2] *Universiti Malaya, Faculty of Pharmacy, Department of Pharmaceutical Life Sciences, Kuala Lumpur, 50603, Malaysia*
[3] *JSS Academy of Higher Education & Research, School of Life Sciences (Ooty Campus), Department of Biochemistry, Mysuru, 570004, India*
[4] *Rural Development Administration (RDA), Agricultural Microbiology Division Wanju-gun, 55365, Jeollabuk-do, Republic of Korea*
[5] *Mushroom Research Centre, Universiti Malaya, Kuala Lumpur, 50603, Malaysia*

5.1 Introduction

Mushroom cultivation has been explored by mankind since millennia ago and rapid development in mushroom growing techniques contributes to the industry's exponential rise. However, this great increase in mushroom production also comes with its challenges, one of them is the management of the spent mushroom substrate (SMS), which is the remains of the renewable agri-residues –specifically referring to the substrate waste produced after the mushroom cultivation process. According to Food and Agricultural Organization (FAO) in 2020, there was a 13.8-fold rise in the mushroom industry in 30 years from 1990 [1]. Due to the increasing demand and consumption of mushrooms as food sources, the mushroom industry is fast growing globally even though it was temporarily stagnated by the recent COVID-19 pandemic. Mushrooms are highly sought-after mainly for a nutrition source and also for their recently highlighted various medicinal properties. Since mushroom is a decomposer in nature, cultivating it can be perceived as a way of promoting circular agriculture, where waste from other agricultural sectors is used to grow commercial mushrooms, even better that they essentially do not need chemicals to grow well. Hence, mushroom cultivation, or fungi culture, is widely exploited for its sustainability since the negative impact of the said practice on the environment is low.

China, where the mushroom production technique is still preserved using a traditional yet novel way, is the biggest producer of mushrooms, accounting for around 93% of the world's mushroom production [1]. Shiitake (*Lentinula edodes*) is the most cultivated mushroom in China with 11.159 million kg produced in 2019 followed by

Chemical Substitutes from Agricultural and Industrial By-Products: Bioconversion, Bioprocessing, and Biorefining, First Edition. Edited by Suraini Abd-Aziz, Misri Gozan, Mohamad Faizal Ibrahim, and Lai-Yee Phang.
© 2024 WILEY-VCH GmbH. Published 2024 by WILEY-VCH GmbH.

wood ear mushroom (*Auricularia auricula*), and oyster mushroom (*Pleurotus ostreatus*), as reported by the China Edible Fungi Association (CEFA) [2]. Japan, another leading mushroom-producing country, mushroom production reached more than 450 million kg in 2019 with *L. edodes*, also leading in amount [2]. Other major mushroom producers include the USA, India, and the Netherlands. Figure 5.1 illustrates the amount of mushrooms produced by a few prominent producer countries in 2019. The steady growth of mushroom production and demand is also experienced by other countries, including Malaysia, wherein 2012, the total mushroom produced per day is 24.000 kg [3].

Growing commercial mushrooms can be done by cultivating them in polybags filled with formulated lignocellulosic substrates, which are rich in three main fibrous polysaccharides namely: lignin, cellulose, and hemicellulose, which are all the constituents of plant cell walls. The amount of each component, however, differs across different kinds of agri-residues used as substrates. Cellulose is a polymer of glucose monomers with $\beta(1 \rightarrow 4)$ linkages, hemicellulose is an amorphous polysaccharides formed by the combination of various hexoses and pentoses, and lignin is a polyphenolic polymer that is constituted by sinapyl alcohol (4-[(1*E*)-3-hydroxy-1-propen-1-yl]-2,6-dimethoxyphenol), coniferyl alcohol (4-[(1*E*)-3-hydroxy-1-propen-1-yl]-2-methoxyphenol), and *p*-coumaryl alcohol (4-[(1*E*)-1-propen-1-yl]phenol) [4, 5].

The substrate for mushroom cultivation, as said earlier, typically comes from the by-product of other agricultural activities. Examples of materials that can be used to cultivate mushrooms are corncob, sawdust, paddy straw, wood, hay, and manures from farm animals. In Malaysia, some of the easy substrates are rubber sawdust, sago *hampas*, paddy rice straw, and residues from the oil palm, such as from its, empty fruit bunches and fronds [6, 7]. During the cultivation process, the mycelia of the mushroom will colonize the substrate, then various degrading enzymes including

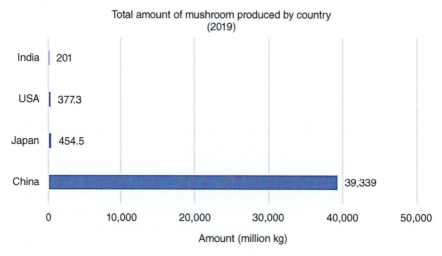

Figure 5.1 Total mushroom produced (million kg) by some countries in 2019. Source: Adapted and redrawn from Singh et al. [2].

the lignin-modifying enzymes (LMEs) are secreted extracellularly to breakdown the lignocellulosic substrate [7, 8]. These enzymes include heme peroxidases, laccases, lignin peroxidases, and cellobiose dehydrogenase [4].

SMS or spent mushroom compost (SMC) is the term that refers to the left substrate and mycelia after a few mushroom flushes. It should be noted that the contents of SMS depend on what kind of mushroom was cultivated. Due to the increasing demand for mushrooms, the industry now faces a problem of how to deal with the piling SMS. According to Lau et al. [9], for every 1 kg of mushroom, it may produce up to 5 kg of SMS, which results in approximately 100 million tons of SMS generated annually [1]. Common methods have been employed to deal with SMS including burning, composting, and disposal to landfills [6, 10]. Some of these strategies present a new threat of pollution to the soil and air, worsened by the unsupervised practice of SMS management. Hence, the effort of recycling and upcycling, or creative reusing of the SMS needs to be looked into. Research shows that SMS can be utilized for other beneficial activities, such as synthesizing enzyme cocktails from SMS for bioremediation, turning SMS into animal feed, and even isolation of pharmaceutical lead compounds for anticancer uses [11].

In the conversion process of the substrate into SMS, a depolymerization reaction by the degrading enzymes takes place. Through this process, long-chain polysaccharide polymers, such as celluloses are broken down into shorter-chain chemical molecules, such as glucose. The process of SMS formation is illustrated in Figure 5.2. Degradation of lignin by mushroom enzymes is especially advantageous since unlike cellulose and hemicellulose, lignin is typically unused and thrown away as a waste product due to its relatively inert nature. For example, mushrooms can degrade lignin into heterogeneous aromatic compounds, which are the precursors

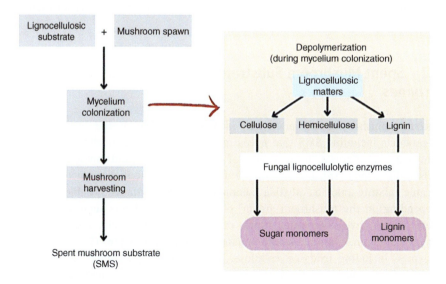

Figure 5.2 Illustration of the summary of the spent mushroom substrate (SMS) formation process. Source: Sabaratnam et al. (author).

of value-added products such as vanillic acid and syringic acid [12]. In addition, the LMEs produced from the lignin degradation can be extracted and used for various purposes. Examples of LMEs usage are for bioleaching, dye degradation, dye-contaminated water remediation, and in various industries [4]. Other than that, Grimm and Wösten [13], noted that the depolymerization of cellulose and hemicellulose yields sugar monomers, mostly glucose and xylose, which then can be used for a considerable amount of ethanol production. These examples showcase that upcycling SMS is essential to add more value to the usually discarded leftovers. These mushroom-derived enzymes are proposed as substitutes for certain materials of chemical products, which will be further discussed thoroughly in Sections 5.2.1, 5.2.2, and 5.3.2.

A circular economy is a term referring to a practice of consumption and production in which materials are recycled and reused for as long as possible. This concept has become more strongly encouraged in facing the ever-expanding population, specifically to tackle pollution problems and increase sustainability. The practice of mushroom cultivation by itself is a good method of promoting a circular economy due to its ability to process agricultural waste into something that can be commercialized, i.e. "from waste to wealth." Meyer et al. [14] noted that the utilization of mushrooms is also highly studied for its future prospects in the production of biodegradable fungi-derived plastics, biomaterials from fungal hyphae for furniture and building blocks, textiles, fuel, and even pharmaceuticals. The circular economy of SMS prospects continues to fully exploit the waste of the mushroom industry. The complete utilization of SMS will contribute even more income by turning the once-believed unusable by-product into wealth while improving the sustainability of the mushroom industry and environmental safety at the same time. The effort of reusing SMS the term that will be used in this chapter (or SMC) is also under some of the United Nations Sustainable Development Goals (SDGs) 8, 9, 12, and 15.

5.2 Spent Mushroom Substrate (SMS) as Source of Bulk Enzymes

Recently, mushroom cultivation has increased, leading to the production of by-products including SMS that needs to be managed. The mushroom industry generates 5 kg of the SMS from the yield of 1 kg of mushroom and annual production of 60 million tons [15]. Prior to valorizing SMS, it is essential to fractionate the organic mixture into its valuable components. SMS contains variable amounts (depending on the mushroom grown) of valuable ingredients such as carbohydrates, lignin, diverse enzymes, fungal metabolites, organic acid, and lactic acid, which may improve animal health and antioxidant capacity [16, 17]. Commercial enzymes, including oxidases, cellulose, lignin-degrading enzymes, lipases, and proteolytic enzymes can be extracted from SMS. These enzymes are widely used in textile, paper industries, food applications, and bioremediation processes, including textile dyes, herbicides, and waste degradation [17].

5.2.1 Enzymes Extracted from SMS for Bioremediation

In recent years, SMS-derived enzymes have been used in the bioremediation process. Applying SMS-derived enzymes in the bioremediation process is a green and sustainable technology. Ligninolytic mushroom species produce many laccase isoforms and over 100 laccase isoforms have been isolated from mushrooms [18]. They are dominant among the lignin-degrading enzymes in the soil environment [19]. Mushroom laccase is responsible for the oxidation of several aromatic compounds in the presence of oxygen. Laccase possesses high redox potential, involved in lignin degradation, detoxification, and bioremediation. Laccase is the key enzyme in SMS of the *Agaricus bisporus, Pleurotus sajor-caju, P. ostreatus, Flammulina velutipes, Hericium erinaceus*, and *L. edodes* [20–23]. Further, it has been reported that *Trametes versicolor* and *P. ostreatus* are suitable strains to investigate the laccases [24]. The crude enzyme extracted from SMS from *Agaricomycetes* cultivation contains polycyclic aromatic hydrocarbons (PAH)-degrading potential and is cost-effective compared to commercial enzymes [18]. Laccase enzymes from mushrooms and SMS degrade harmful hydrocarbons, mutagenic PAHs, industrial dyes, and various commercial pesticides [25, 26]. The inclusion of 10% of SMS of *P. ostreatus* can degrade petroleum hydrocarbon soil in a short period [18]. Thus, SMS can be a promising additive for PAH biodegradation in ex situ remediation. The multiple types of SMS enzymes, such as exocellulase, endocellulase, total cellulase, β-glucosidase, dextranase, amylase, and laccase are involved in the dye decolorizing process [26]. It has been reported that *Lenzites elegans* laccase can catalyze the decolorization of industrial dyes [27]. However, laccase from *Lentinus polychrous* SMS decolorized synthetic dyes, and the partially purified fractions showed high decolorizing ability [28].

Enzymes extracted from the spent compost of *P. sajor-caju* have potential to the decolorization of chemically different dyes [29]. This study was done on both reactive and dispersed dyes, as well as including the representatives of azo and anthraquinone dye groups. In addition, the effect of certain physical parameters on the rate of decolorization of each dye was also shown. As such, reactive dyes, such as reactive black 5 and reactive orange 16 were decolored in a higher percentage compared to the dispersed type of dyes [29]. Moreover, Singh et al. [29], also noted that there was an indication that the percentage of degradation of the dyes is related to the structure of the dye molecule, in which dyes with a favorable structure will be more decolored. For instance, reactive black 5 and reactive orange 16 both have a hydroxyl group at the ortho position as well as a sulfonate group at the meta position with respect to the azo bond were shown to be degraded in a percentage higher than the other tested dyes [29]. The study also provided insight into how the knowledge of these parameters can help in the optimization of processes employing enzymes for bioremediation on an industrial scale. Furthermore, the in vitro treatment of environmental pollutants, such as dyes with crude ligninolytic enzymes of white rot fungi represents a simpler and effective method as compared to the direct application of the fungi for removal or degradation by absorption effects or mineralization of the dyes pollutants by the mycelia. The utilization of

SMS for the extraction of enzymes offers an economical advantage of obtaining industrially important enzymes without long incubation periods and additional cost of specialized fermentations. The postulated mechanism of decolorization action by SMS enzymes is oxidation by the peroxidases [29].

Laccase is also a key enzyme involved in oxidation and decontaminating phenolic compounds [6]. Laccase detoxifies the plasticizer substances like phthalates and pentachlorophenol, they are widely distributed in the land and water ecosystem [30]. Enzymes extracted from edible mushrooms *Pleurotus eryngii*, *P. djamor*, *P. ostreatus*, and *Auricularia polytricha* SMS removed the phthalate [30]. Further, SMS enzymes have the potential to be applied in wastewater treatment [31]. *Pleurotus djamor* SMS enzymes potentially removed the emerging pollutants like organic UV filters 2-ethylhexyl salicylate (EHS) and homosalate (HMS) from wastewater [31]. *Pleurotus djamor* SMS enzymes potentially removed the emerging pollutants like organic UV filters EHS and HMS from wastewater [31]. Again, the chemical structure of the contaminants determines how the enzyme from SMS affects them. For example, Yang et al. [31], showed that both EHS and HMS, which have a similar structure, were degraded into benzoic acid, unlike ethylhexyl methoxycinnamate (EHMC), which are isomerized by the SMS extract [31]. Lastly, this study concluded by deeming SMS is better to be employed to degrade organic UV filters as compared to using the enzyme by itself. This is because by using the SMS directly as a whole, the enzymes can be continuously produced by the fungal mycelia, hence the whole process would be simpler without having to first extract the enzymes [31].

In terms of the enzyme extraction process, numerous methods can be utilized. However, the studies that are featured in this part of the chapter both used a mostly similar method with some minor setting and reagent differences. Singh et al. [29], extracted the crude enzyme from SMS by mixing and homogenizing it with pH 4 tap water, followed by a centrifugation process in which the resulting supernatant was collected for the crude enzyme extract. Similarly, Yang et al. [31] used a process similar to Singh et al.'s, however instead of tap water, they used sodium acetate buffer as the solvent for the extraction purpose.

5.2.2 Enzymes Extracted from SMS for Green Fuel Feedstock Production: Case Study

Large volumes of organic renewable liquid waste stream palm oil mill effluent (POME) are generated by the palm oil industry in Malaysia. Raw POME is produced from a combination of sterilizer condensate, separator sludge, and hydrocyclone wastewater in a proportion of 9 : 15 : 1, respectively [32]. Further, the high concentrations of lipids, minerals, carbohydrates, protein, and nitrogenous compounds in POME make it a suitable substrate for biofuel including biohydrogen production [33]. POME is recognized as one of the lignocellulosic biomass substrates in the biofuel industry for conversion into fermentable sugars [34]. The sugars can be utilized for a higher-value feedstock to produce biogas and biofuel. Further, enzymatic hydrolysis is deemed environmentally friendly. High-cost commercial enzymes have been used for POME hydrolysis, which converts the organic substances in POME into monomeric sugars [35–37]. Commercial enzymes are expensive. However, studies on the potential use of enzymes from SMS for the hydrolysis of POME have not been explored.

Numerous studies have reported that SMS may be an inexpensive source of LMEs and cellulolytic enzymes. The enzymes of SMS of *Pleurotus pulmonarius* have been studied for hydrolysis of POME to produce fermentable feedstock for biohydrogen production [38]. In this study by Yunan et al. [38], the enzymes from SMS were extracted using acidic tap water and centrifugation, in which the supernatant was collected and concentrated further by freeze drying, followed by the measurement of enzyme activity. The POME hydrolysis was done with a concentration of 100 ml of aqueous enzyme/100 ml of POME. Next, for the biohydrogen batch fermentation process, 80% of hydrolyzed POME was added to 20% of POME sludge, which acted as the inoculum for the fermentation [38]. The fermentation was performed in an anaerobic and mesophilic condition (37 °C), and the produced biogas was analyzed using gas chromatography for quantification.

Productivities of the enzymes for xylanase, laccase, endoglucanase, lignin peroxidase, and β-glucosidase were 2.3, 4.1, 14.6, 214.1, and 915.4 U/g, respectively, A maximum of 3.75 g/l of reducing sugar was obtained under optimized conditions (pH 4.8; 10% (v/v) enzyme loading; 15 hours incubation), which was consistent with the predicted reducing sugar concentration of 3.76 g/l [38]. In batch fermentation, the biohydrogen cumulative volume of 302.78 ml H_2/l POME and 83.52% biohydrogen gas were recorded. In this study, it was demonstrated that the aqueous bulk crude enzymes extracted from SMS of *P. pulmonarius* can be utilized to hydrolyze POME [38]. The biohydrogen production in the study by Yunan et al. [38] was 302.78 ml H_2/l POME. This was lower than the biohydrogen production of 1439 ml H_2/l of POME reported [37] (Table 5.1).

The high amounts of biohydrogen produced in the study [37] could be attributed to the use of purified and concentrated commercial enzymes such as *Celluclast* 1.5L and *Novozyme* 188. The hydrolysis achieved resulted in high feedstock sugars and

Table 5.1 Comparison in biohydrogen production after fermentation employing POME as a substrate.

Temp (°C), pH	Treatment of POME	Type of microorganisms	Biohydrogen production (ml H_2/l POME)	Biohydrogen productivity (ml H_2/l POME/h)	References
37, 6.0	Enzymatic hydrolysis	Fungal enzyme	1439.00	87.19	[37]
35, 6.5	Enzymatic hydrolysis	Plant enzyme	134.74	8.42	[39]
37, 5.5	NA	Bacteria	3195.00	1034	[40]
37, 5.5	Enzymatic hydrolysis	Fungal enzyme	302.78	12.62	[38]

NA, not available.
Source: Yunan et al. [38]/The Korean Society for Microbiology and Biotechnology.

higher biohydrogen production. However, the study [38] used a cheap source of enzymes. The need to concentrate the enzymes as well as optimize the parameters to produce biohydrogen.

Besides, the biohydrogen production in the study [37] was twofold higher than the one recorded by Garritano et al. [39]. In that study, plant enzyme preparation (PEP) was extracted from dormant castor bean seeds. Further, the high initial pH of 6.5 of the fermentation process may have contributed to higher levels of biohydrogen. Similar results were reported in a study by Chong et al. [40] where the highest hydrogen production was 3195 ml H_2/l POME at a pH of 5.5. Other than the production of biohydrogen, SMS is also proven in multiple research being able to be employed for the production of other bio-based fuels and compounds, which will be discussed further in Section 5.3.2 of the chapter.

5.3 Various Challenges and Future Prospects in the Use of SMS

5.3.1 Challenges in the Use of Enzymes in SMS

SMS is no longer regarded as a waste, but it is considered a renewable resource from the mushroom industry [6]. Not only can the SMS be employed in several green technology endeavors, but the enzymes recovered from it are also potentially useful for the bioremediation of pollutants and for other industrial biotechnology purposes. Having said that, the consistency in the performance of the SMS has to be evaluated and assured to realize the application of SMS and SMS-related products in the near future. Further, the many uses reviewed here are in lab-scale studies. The challenges faced for large-scale applications will be the availability of a consistent supply of SMS enzyme cocktails.

The SMS generated during the cultivation of mushrooms may vary based on the mushroom cultivated, the growing substrates utilized, as well as the duration of the cultivation process [41]. The oyster mushrooms (*Pleurotus* spp.) are popular commercial mushrooms and are one of the top cultivated mushrooms in the world. *Pleurotus* spp. (oyster mushrooms) grow readily in a hardwood that needs minimal pre-composting and produces a wide range of lignocellulolytic enzymes laccases and lignin peroxidases [22]. What needs to be developed is large-scale extraction techniques and processing plants to produce the bulk concentrated enzymes. Other than that, as mentioned earlier, different types of mushrooms cultivated determine the enzyme concentrations in the SMS. For instance, Yang et al. [31] noted that the enzyme extracted from the SMS of *A. polytricha* did not degrade the organic UV filters well due to the little amount of laccase present. Hence, a deeper understanding of the composition of each mushroom's varying SMS composition is crucial in the effort of utilizing it effectively.

Even though, generally, the usage of SMS is considered more cost-effective than conventional methods, some things should still be addressed. Firstly, the cost of the purification of enzymes from SMS is quite considerable compared to the direct implementation of SMS [41]. In a way, people might think using SMS directly might be better since it is simpler and cheaper, however, the usage of purified enzymes will ensure a more controlled and standardized yield since there are no other determining factors present unlike in the direct SMS application. Thus, such a dilemma must be considered, whether it is best to use the SMS directly or employ the purified enzyme. As discussed earlier, both sides have their pros and cons. The answer may be that how SMS is best used is different on a case-by-case basis, hence further research needs to be conducted for the realization of SMS practical application.

5.3.2 Future Prospects for Use of SMS for Production of Green Chemicals

The by-product of the mushroom industry, the SMS, is a renewable resource and contributes to a circular economy that converts agricultural residues [13]. Numerous strategies have been reported/explored for the sustainable utilization of SMS [42]. Conversion of SMS into biofuels, biofertilizers, and sugar recovery from fermentation processes are typical techniques of management and utilization of SMS [10]. According to studies, pretreating SMS with diluted acid and alkali reagents can enhance the conversion of cellulose to sugars [38]. The study [38] suggested that concentrated enzymes isolated from SMS maybe applied to hydrolyze organic components in wastewater (POMEs) from palm oil operations. Hence, enzymes extracted from SMS can be used to produce numerous biofuels. Dairy manure and SMS of *F. velutipes* and *P. eryngii* were used to produce methane, which can be then processed for green fuels. The combination produced more methane than SMS or dairy manure when used separately as a feedstock. This demonstrated the beneficial effects of SMS co-digestion [43]. It was also shown in another study, that 1 ton of SMS can yield up to 150 kg of ethanol. The digestion of *P. ostreatus* SMS containing spent sorghum husk generated 187 g of ethanol per kilogram of dry matter [44]. These studies indicate that the enzymes in SMS can contribute to generating feedstock, such as fermentable sugars.

In another study, the major ingredient in the SMS, the spent corncobs, were hydrothermally treated to grow enoki mushrooms, which were subsequently turned into xylooligosaccharides [45], a functional food that has gained rising attention for numerous uses including pharmaceuticals and animal feed [45]. Seekram et al. [46] showed that the SMS of *P. ostreatus* is suitable for the material of xylooligosaccharide production by treating it with xylanase enzymes from *Aspergillus flavus*, in which the xylan in the SMS is hydrolyzed by the enzymes. This also highlights the sustainability of mushroom cultivation in a way that the byproduct of one kind of mushroom cultivation can be turned into a value-added product by using the enzyme extracted from another mushroom.

Simple water extraction techniques may be employed to extract the enzymes from SMS, and these extracts can be further utilized alone or in combination with enzymes from other sources. The crude enzymes can then be used to convert lignocellulosic waste streams into sugars for use in second-generation biofuel [47]. Based on studies, it was also shown that SMS was a good source of ligninolytic enzymes, particularly laccase, a "green catalyst." The SMS crude extract may remove up to 90% of aflatoxin (AFB1), which could contaminate numerous grain products and is simple to acquire, nontoxic, affordable, and environmentally benign. Branà et al. [41] demonstrated that ligninolytic enzymes, namely laccase, and Mn-peroxidase, from *P. eryngii* have the capability to mitigate AFB1 toxin. They expanded their study further and noted that the optimal pH and temperature for the detoxification process are 8 and 25–37 °C, respectively [41]. To establish the lack of by-products with any residual toxicity, more investigation is required. Additional research is required to pinpoint and perfect SMS enzyme's useful uses in lowering AFB1 levels in animal diets. To determine the best soaking periods and conditions and to evaluate various SMS application techniques, including spraying and washing, "in vivo" testing on the main commodities impacted by AFB1 is required [41].

Utilizing waste mushroom substrate in solid-state fermentation, poly (L-diaminopropionic acid) and poly(L-lysine) were both produced simultaneously [47]. This was demonstrated by the study by Wang and Rong [48], in which they used the SMS from *P. ostreatus* cultivation in the solid-state fermentation of Streptomyces albulus. The production of poly(L-lysine) and poly(L-diaminopropionic acid) was increased by adding glycerol and corn steep liquor as well as optimizing the fermentation conditions, reaching 51.4 mg/g substrate and 25.4 mg/g substrate, respectively. Solid-state fermentation in several batches suggests that these chemicals might be generated continuously and effectively. Additionally, poly(L-diaminopropionic acid) and poly(L-lysine) applied together showed

Table 5.2 Various research showcasing the future prospects of SMS utilization.

Aim	SMS-derived product used	Result	References
Production of green fuels	*Pleurotus pulmonarius* SMS enzyme extract	Successfully hydrolyzed POME and increased biohydrogen production	[38]
	SMS of *Pleurotus eryngii* and *Flammulina velutipes* combined with dairy manure	Higher yield of methane with co-digestion compared to mono-digestion	[43]
	Sorghum chaff-derived SMS	Suitable for ethanol production	[44]
Production of functional foods	SMS of *Pleurotus ostreatus*	Suitable to be the raw material for xylooligosaccharide production	[46]
Toxin removal	Ligninolytic enzymes from *P. eryngii* SMS	Successfully removed 90% of aflatoxin (AFB1) from various commodities	[41]
Production of chemical compounds	SMS of *P. ostreatus*	The solid-state fermentation produced poly(L-diaminopropionic acid) and poly(L-lysine)	[48]

promising promise as a biological preservative. As a result, this fermentation technique not only increases output but also makes it possible to recycle waste biomass and diversify the products [48].

After mushroom harvest, the spend mushroom compost (SMC) must be disposed of and stored, which presents an economic and environmental challenge. Instead of disposing of the SMC waste, it can be used as a substrate for the solid-state fermentation (SSF) of fungi to produce cellulase, xylanase, amylase, and b-glucosidase, which are industrially important enzymes. *Trichoderma* and *Aspergillus* can be grown on SMC successfully without the need for additional (nutritive) elements. A mixture of cellulolytic enzymes hydrolyzed SMC, resulting in a 30% reduction in SMC [47]. Table 5.2 summarizes the discussed prospects of SMS utilization in this part of the chapter.

5.4 Conclusions

The study of enzymes from SMS for the treatment of POME is certainly limited. However, from the case study, hydrolyzing POME using the enzymes of SMS is possible. High productivities of concentrated enzymes including LiP (214.1 U/g) and β-glucosidase (915.4 U/g) indicated that the SMS possibly is a good source of ligninolytic and hydrolytic enzymes. The hydrolysis of POME using a 10% concentrated SMS enzyme cocktail gave the highest reducing sugar concentration (3.76 g/l) after 12 hours of incubation. The POME hydrolysis was significantly influenced by incubation time and enzyme loading ($p < 0.05$). The rate of biohydrogen production was 1.01 ± 0.05 ml H_2/h from the hydrolyzed POME was 25 times higher than raw POME, which was 0.04 ± 0.01 ml H_2/h after 24 hours of fermentation. Further, the fermentation of hydrolyzed POME resulted in 302.78 ml H_2/l POME, which was 23-fold higher than the raw POME (13.3878 ml H_2/l POME).

SMS has several useful components that can be utilized as a waste-to-wealth initiative, in this review it has been shown that SMS can be applied in bioremediation processes as well as green energy production. Both the SMS and the enzymes extracted from it demonstrate high potential to be upcycled for the production of value-added products, especially from agro residues. The enzymes extracted from the SMS were shown to be able to degrade various harmful contaminants both in the soil and in the water. This will prove to be a more cost-efficient and eco-friendly way of tackling environmental pollution. However, several factors need to be taken into consideration to ensure consistency in the production of the SMS and that the enzymes needed are yielded in sufficient quantities. The lack of a standardized large-scale method of SMS enzyme extraction, as well as the varying variables in the production process should be addressed to realize the practical application of SMS. Standard protocols must be in place to ensure the efficacy of bioremediation and hydrolysis of polymers to monomeric forms suitable for fermentation. Lastly, a deeper understanding of the effect of different kinds of mushrooms cultivated on the composition of the SMS extract must be covered to ensure an effective SMS upcycling. As the full potential of SMS has not been unfolded completely, further studies on the ins and outs of the

SMS usage are highly encouraged. Overall, the complete utilization of SMS will contribute to more sustainable chemical production and the further promotion of the concept of the circular economy.

References

1 Okuda, Y. (2022). Sustainability perspectives for future continuity of mushroom production: the bright and dark sides. *Front. Sustainable Food Syst.* https://doi.org/10.3389/fsufs.2022.1026508.
2 Singh, M., Kamal, S., and Sharma, V.P. (2022). Species and region-wise mushroom production in leading mushroom producing countries - China, Japan, USA, Canada and India. *Mushroom Res.* 30 (2): 99. https://doi.org/10.36036/mr.30.2.2021.119394.
3 Haimid, M.T., Rahim, H., and Dardak, R. (2013). Understanding the mushroom industry and its marketing strategies for fresh produce in Malaysia (Memahami industri cendawan dan strategi pemasaran untuk produk segar di Malaysia). *Econ. Technol. Manag. Rev.* 8: 27–37.
4 Iram, A., Berenjian, A., and Demirci, A. (2021). A review on the utilization of lignin as a fermentation substrate to produce lignin-modifying enzymes and other value-added products. *Molecules* 26 (10): 2960. https://doi.org/10.3390/molecules26102960.
5 Cagide, C. and Castro-Sowinski, S. (2020). Technological and biochemical features of lignin-degrading enzymes: a brief review. *Environ. Sustainability* 3 (4): 371–389. https://doi.org/10.1007/s42398-020-00140-y.
6 Phan, C.W. and Sabaratnam, V. (2012). Potential uses of spent mushroom substrate and its associated lignocellulosic enzymes. *Appl. Microbiol. Biotechnol.* 96 (4): 863–873. https://doi.org/10.1007/s00253-012-4446-9.
7 Aziera, N., Rasib, A., Zakaria, Z. et al. (2015). Characterization of biochemical composition for different types of spent mushroom substrate in Malaysia. *Malaysian J. Anal. Sci.* 19: 41–45.
8 Carrasco, J., Zied, D.C., Pardo, J.E. et al. (2018). Supplementation in mushroom crops and its impact on yield and quality. *AMB Exp.* 8 (1): https://doi.org/10.1186/s13568-018-0678-0.
9 Lau, K., Tsang, Y., and Chiu, S. (2003). Use of spent mushroom compost to bioremediate PAH-contaminated samples. *Chemosphere* 52 (9): 1539–1546. https://doi.org/10.1016/s0045-6535(03)00493-4.
10 Mohd Hanafi, F.H., Rezania, S., Mat Taib, S. et al. (2018). Environmentally sustainable applications of agro-based spent mushroom substrate (SMS): an overview. *J. Mater. Cycles Waste Manage.* 20 (3): 1383–1396. https://doi.org/10.1007/s10163-018-0739-0.
11 Zou, G., Li, B., Wang, Y. et al. (2021). Efficient conversion of spent mushroom substrate into a high value-added anticancer drug pentostatin with engineered *Cordyceps militaris*. *Green Chem.* 23 (24): 10030–10038. https://doi.org/10.1039/d1gc03594k.

12 Kamimura, N., Sakamoto, S., Mitsuda, N. et al. (2019). Advances in microbial lignin degradation and its applications. *Curr. Opin. Biotechnol.* 56: 179–186. https://doi.org/10.1016/j.copbio.2018.11.011.

13 Grimm, D. and Wösten, H.A.B. (2018). Mushroom cultivation in the circular economy. *Appl. Microbiol. Biotechnol.* 102 (18): 7795–7803. https://doi.org/10.1007/s00253-018-9226-8.

14 Meyer, V., Basenko, E.Y., Benz, J.P. et al. (2020). Growing a circular economy with fungal biotechnology: a white paper. *Fungal Biol. Biotechnol.* 7 (1): https://doi.org/10.1186/s40694-020-00095-z.

15 Atallah, E., Zeaiter, J., Ahmad, M.N. et al. (2021). Hydrothermal carbonization of spent mushroom compost waste compared against torrefaction and pyrolysis. *Fuel Process. Technol.* 216: 106795.

16 Chuang, W.Y., Liu, C.L., Tsai, C.F. et al. (2020). Evaluation of waste mushroom compost as a feed supplement and its effects on the fat metabolism and antioxidant capacity of broilers. *Animals* 10: 445. https://doi.org/10.3390/ani10030445.

17 Raman, J., Kim, J.S., Choi, K.R. et al. (2022). Application of lactic acid bacteria (LAB) in sustainable agriculture: advantages and limitations. *Int. J. Mol. Sci.* 23: 7784. https://doi.org/10.3390/ijms23147784.

18 Mohammadi-Sichani, M.M., Assadi, M.M., Farazmand, A. et al. (2017). Bioremediation of soil contaminated crude oil by *Agaricomycetes*. *J. Environ. Health Sci. Eng.* 15: 8. https://doi.org/10.1186/s40201-016-0263-x.

19 Baldrian, P. (2006). Fungal laccases – occurrence and properties. *FEMS Microbiol. Rev.* 30: 215. https://doi.org/10.1111/j.1574-4976.2005.00010.x.

20 Mayolo-Deloisa, K., Trejo-Hernández, M.R., and Rito-Palomares, M. (2009). Recovery of laccase from the residual compost of *Agaricus bisporus* in aqueous two-phase systems. *Process Biochem.* 44 (4): 435–439. https://doi.org/10.1016/j.procbio.2008.12.010.

21 Singh, A.D., Abdullah, N., and Vikineswary, S. (2003). Optimization of extraction of bulk enzymes from spent mushroom compost. *J. Chem. Technol. Biotechnol.* 78 (7): 743–752. https://doi.org/10.1002/jctb.852.

22 Ko, H.G., Park, S.H., Kim, S.H. et al. (2005). Detection and recovery of hydrolytic enzymes from spent compost of four mushroom species. *Folia Microbiol.* 50 (2): 103–106. https://doi.org/10.1007/bf02931456.

23 Yang, S.O., Sodaneath, H., Lee, J.I. et al. (2017). Decolorization of acid, disperse and reactive dyes by *Trametes versicolor* CBR43. *J. Environ. Sci. Health – Toxic/Hazard. Subst. Environ. Eng.* 52 (9): 862–872. https://doi.org/10.1080/10934529.2017.1316164.

24 Abadulla, E., Tzanov, T., Costa, S. et al. (2000). Decolorization and detoxification of textile dyes with a laccase from *Trametes hirsuta*. *Appl. Environ. Microbiol.* 66 (8): 3357–3362. https://doi.org/10.1128/AEM.66.8.3357-3362.2000.

25 Jin, X., Yu, X., Zhu, G. et al. (2016). Conditions optimizing and application of laccase-mediator system (LMS) for the laccase-catalyzed pesticide degradation. *Sci. Rep.* 6: 35787. https://doi.org/10.1038/srep35787.

26 Nakajima, V.M., Soares, F.E.F., and Queiroz, J.H. (2018). Screening and decolorizing potential of enzymes from spent mushroom composts of six different mushrooms. *Biocatal. Agric. Biotechnol.* 13: 58–61. https://doi.org/10.1016/j.bcab.2017.11.011.

27 Pandey, R.K., Tewari, S., and Tewari, L. (2018). Lignolytic mushroom *Lenzites elegans* WDP2: laccase production, characterization, and bioremediation of synthetic dyes. *Ecotoxicol. Environ. Saf.* 158: 50–58. https://doi.org/10.1016/j.ecoenv.2018.04.003.

28 Khammuang, S. and Sarnthima, R. (2013). Decolorization of synthetic melanins by crude laccases of *Lentinus polychrous* Lév. *Folia Microbiol.* 58: 1–7. https://doi.org/10.1007/s12223-012-0151-4.

29 Singh, A.D., Vikineswary, S., Abdullah, N. et al. (2011). Enzymes of spent mushroom substrates of *Pleurotus sajor-caju* for the decolorization and detoxification of textile dyes. *World J. Microbiol. Biotechnol.* 27: 535–545. https://doi.org/10.1007/s11274-010-0487-3.

30 Chang, B.V., Yang, C.P., and Yang, C.W. (2021). Application of fungus enzymes in spent mushroom composts from edible mushroom cultivation for phthalate removal. *Microorganisms* 9 (9): 1989. https://doi.org/10.3390/microorganisms9091989.

31 Yang, C.W., Tu, P.H., Tso, W.Y. et al. (2021). Removal of organic UV filters using enzymes in spent mushroom composts from fungicultures. *Appl. Sci.* 11: 3932. https://doi.org/10.3390/app11093932.

32 Wu, T.Y., Mohammad, A.W., Jahim, J.M. et al. (2010). Pollution control technologies for the treatment of palm oil mill effluent (POME) through end-of-pipe processes. *J. Environ. Manag.* 91 (7): 1467–1490. https://doi.org/10.1016/j.jenvman.2010.02.008.

33 Sompong, O., Prasertsan, P., Karakashev, D. et al. (2008). Thermophilic fermentative hydrogen production by the newly isolated *Thermoanaerobacterium thermosaccharolyticum* PSU-2. *Int. J. Hydrogen Energy* 33 (4): 1204–1214. https://doi.org/10.1016/j.ijhydene.2007.12.015.

34 Silvamany, H., Harun, S., Mumtaz, T. et al. (2015). Recovery of fermentable sugars from palm oil mill effluent via enzymatic hydrolysis. *Jurnal Teknologi* 77 (33): 115–121. https://doi.org/10.11113/jt.v77.7016.

35 Seong, K.T., Hassan, M.A., and Ariff, A.B. (2008). Enzymatic saccharification of pretreated solid palm oil mill effluent and oil palm fruit fiber. *Pertanika J. Sci. Technol.* 16 (2): 157–169.

36 Khaw, T.S. and Ariff, A.B. (2009). Optimization of enzymatic saccharification of palm oil mill effluent solid and oil palm fruit fibre to fermentable sugars. *J. Trop. Agric. Food Sci.* 37 (1): 85–94.

37 Khaleb, N.A., Jahim, J.M., and Kamal, S.A. (2012). Biohydrogen production using hydrolysates of palm oil mill effluent (POME). *J. Asian Sci. Res.* 2 (11): 705–710.

38 Yunan, N.A.M., Shin, T.Y., and Sabaratnam, V. (2021). Upcycling the spent mushroom substrate of the grey oyster mushroom *Pleurotus pulmonarius* as a source of lignocellulolytic enzymes for palm oil mill effluent hydrolysis. *J. Microbiol. Biotechnol.* 31 (6): 823–832. https://doi.org/10.4014/jmb.2103.03020.

39 Garritano, A.D., de Sa, L.R.V., Aguieiras, E.C.G. et al. (2017, 2017). Efficient biohydrogen production via dark fermentation from hydrolized palm oil mill

effluent by non-commercial enzyme preparation. *Int. J. Hydrogen Energy* 42 (49): 29166–29174. https://doi.org/10.1016/j.ijhydene.2017.10.025.

40 Chong, M.L., Rahim, R.A., Shirai, Y. et al. (2009). Biohydrogen production by *Clostridium butyricum* EB6 from palm oil mill effluent. *Int. J. Hydrogen Energy* 34 (2): 764–771. https://doi.org/10.1016/j.ijhydene.2008.10.095.

41 Branà, M.T., Sergio, L., Haidukowski, M. et al. (2020). Degradation of aflatoxin B_1 by a sustainable enzymatic extract from spent mushroom substrate of *Pleurotus eryngii. Toxins* 12 (1): 49. https://doi.org/10.3390/toxins12010049.

42 Rinker, D.L. (2017). Spent mushroom substrate uses. In: *Edible and Medicinal Mushrooms: Technology and Applications* (ed. C.Z. Diego and A. Pardo-Giménez), 427–454. Hoboken: Wiley https://doi.org/10.1002/9781119149446.ch20.

43 Luo, X., Yuan, X., Wang, S. et al. (2018). Methane production and characteristics of the microbial community in the co-digestion of spent mushroom substrate with dairy manure. *Bioresour. Technol.* 250: 611–620. https://doi.org/10.1016/j.biortech.2017.11.088.

44 Ryden, P., Efthymiou, M.N., Tindyebwa, T.A.M. et al. (2017). Bioethanol production from spent mushroom compost derived from chaff of millet and sorghum. *Biotechnol. Biofuels* 10: 195. https://doi.org/10.1186/s13068-017-0880-3.

45 Sato, N., Shinji, K., Mizuno, M. et al. (2010b). Improvement in the productivity of xylooligosaccharides from waste medium after mushroom cultivation by hydrothermal treatment with suitable pretreatment. *Bioresour. Technol.* 101 (15): 6006–6011. https://doi.org/10.1016/j.biortech.2010.03.032.

46 Seekram, P., Thammasittirong, A., and Thammasittirong, S.N. (2021). Evaluation of spent mushroom substrate after cultivation of *Pleurotus ostreatus* as a new raw material for xylooligosaccharides production using crude xylanases from *Aspergillus flavus* KUB_2. *3 Biotech* 11 (4): 176. https://doi.org/10.1007/s13205-021-02725-8.

47 Marica, G., Biljana, D., Ivana, P. et al. (2015). Spent mushroom compost as substrate for the production of industrially important hydrolytic enzymes by fungi *Trichoderma* spp. and *Aspergillus niger* in solid state fermentation. *Int. Biodeterior. Biodegradation* 104: 290–298. https://doi.org/10.1016/j.ibiod.2015.04.029.

48 Wang, M. and Rong, C. (2022). Poly(ε-L-lysine) and poly(L-diaminopropionic acid) co-produced from spent mushroom substrate fermentation: potential use as food preservatives. *Bioengineered* 13 (3): 5892–5902. https://doi.org/10.1080/21655979.2022.2040876.

6

Essential Oil from Pineapple Wastes

Mohamad F. Ibrahim, Nurshazana Mohamad, Mariam J. M. Fairus, Mohd A. Jenol, and Suraini Abd-Aziz

Universiti Putra Malaysia, Department of Bioprocess Technology, Faculty of Biotechnology & Biomolecular Sciences, Serdang, 43400, Selangor, Malaysia

6.1 Introduction

The pineapple industry generates a notable amount of waste, throughout the processing of pineapples, including peels, cores, stems, crowns, and pulp removed during planting, harvesting, and manufacturing. This can result in the disposal of 30–50% of the total fruit weight during canning [1]. The increasing production of pineapples means that the amount of waste generated is likely to increase, highlighting the need for effective management of pineapple waste. Inappropriate disposal of this waste can cause environmental issues and microbial spoilage due to its high sugar and moisture content [1]. The recycling of pineapple waste usually involves shredding or plowing it back into the soil, but in peat soil, this is not possible, leading to the practice of open burning [2]. Peels make up the largest portion of the waste generated, accounting for approximately 30–42% (w/w) of the total waste. Pineapple peels contain essential oils, esters, and organic acids, which give them an aromatic flavor. Esters make up the majority of pineapple leftovers, including peels and leaves, followed by ketones (26%), alcohols (18%), aldehydes (9%), acids (3%), and other chemicals (9%) [3]. These substances imply that it maybe possible to extract the fragrant essence of pineapple peels in the form of essential oils.

Plants produce concentrated volatile aromatic compounds known as essential oils, which give them their delightful fragrances. These chemicals are often present in specific cells, glands, or ducts in various plant parts, including flowers, stems, leaves, seeds, roots, bark, fruit rinds, or resin [4]. Due to their flavor, therapeutic, or olfactory qualities, essential oils are frequently utilized in a range of products, including food, cosmetics, and medicine. Due to the increasing demand of perfumes and flavors in food and drinks, personal care products, and aromatherapy, it is anticipated that the global market for essential oils would exceed US$ 8.8 billion in 2022 and continue to expand [5]. Lemon, geranium, ylang-ylang, bergamot, chamomile, patchouli, cedarwood, clove, eucalyptus, tea tree, clove, orange, lime, lavender, and peppermint are

Chemical Substitutes from Agricultural and Industrial By-Products: Bioconversion, Bioprocessing, and Biorefining, First Edition. Edited by Suraini Abd-Aziz, Misri Gozan, Mohamad Faizal Ibrahim, and Lai-Yee Phang.
© 2024 WILEY-VCH GmbH. Published 2024 by WILEY-VCH GmbH.

just a few of the many available essential oils in the market. Even though there are pineapple essential oils on the market, the majority of them are fragrance oils. A mixture of synthetic and natural components is used to create pineapple fragrance oil, which has a light pineapple scent. It is frequently used to make scented candles, bath salts, body lotions, and other personal care items [6].

To produce pure pineapple essential oils, it can be extracted from pineapple plant material using a variety of techniques. The most often employed extraction processes are conventional ones, such as steam distillation, hydro-distillation (HD), and cold-pressing [7, 8]. These traditional methods are preferred due to their simplicity and low-cost setup. Additionally, there are innovative, environmentally friendly green extraction methods that have recently been developed, including microwave-assisted extraction [9], CO_2 supercritical fluid extraction (SFE) [10], and enzyme-assisted extraction [11]. This trend of innovation is driven by economic demands, industry competition, sustainability, and the need for high-quality industrial production.

6.2 Pineapple Wastes

Significant amounts of waste are generated by the pineapple industry during processing, with as much as 75–80% of the waste being solid [12]. In the canning process, the wastes consist of peels, cores, stems, crowns, and pulp that is typically discarded, representing 30–50% of the total fruit weight. Generally, the solid waste is crushed to be used as cost-effective animal feed, and the liquid waste is utilized in the production of alcohol. The high moisture and sugar content in pineapples results in various concerning consequences, including microbial spoilage and environmental issues, and disposal of pineapple trash. The dumping of waste in landfills is not advised, and open burning of waste can cause air pollution. As an alternative, the residue that is rich in vitamins, carbohydrates, fiber, and other ingredients can be recycled for use in food.

Pineapple residues, including the peel and pulp generated during mechanical peeling, are typically discarded as waste. However, the peel, which constitutes the largest portion of the waste at around 30–42% (w/w), can be used to produce various value-added products, just like other agricultural wastes. For instance, ferulic acid derived from pineapple peel has been used in the bioconversion of vanillic acid and vanillin [1], as well as bromelain extract [13]. The cellulose, hemicellulose, and carbohydrates abundant in pineapple peel can be utilized in the production of paper, banknotes, and cloth. Although producing animal feed is the most typical use for pineapple waste, various other valuable products can be produced from these residues.

The pineapple's flavor came from the essential oils, esters, and organic acids that are distributed throughout the pulp and peel in a particular amount. The type and quantity of sugars found in pineapple depend primarily on the fruit's maturity and degree of acidity [14]. Table 6.1 summarizes the chemical composition of solid pineapple waste.

Table 6.1 Chemical composition of pineapple wastes.

Composition (%)	Pineapple cannery waste [13]	Pineapple processing waste [15]	Pineapple peel [10]	Fresh [16]
Moisture	87.5	92.8	—	71.1
Total solid	12.5	7.8	—	29.0
Ash	4.1	10.6	—	3.9
Organic carbon	—	51.9	—	—
Total carbohydrates	—	35.0	—	—
Reducing sugar	20.9	—	—	27.8
Glucose	8.2	—	3.1	—
Fructose	12.2	—	3.4	—
Sucrose	0.0	—	5.2	—
Cellulose	—	19.8	19.0	11.2
Crude fiber	10.56	—	—	—
Hemicellulose	—	11.7	22.0	7.0
Total soluble	—	30.0	—	—
Total nitrogen	0.8	1.0	—	—
Crude protein	5.9	—	—	—
Fat	0.2	—	—	—

Note:
1. Regarding the type of solid waste used in respective references, they are not specifically mentioned type of waste used, however, from the explanation, it seems like the mixture waste of peel in bulk of pineapple.
2. No specific characteristics of the desired compounds were performed during extraction process. Essential oils are composed of a complex mixture of natural compounds, including terpenes, phenols, alcohols, ketones, aldehydes, ethers, and esters. The specific compounds present in an essential oil will depend on the type of plant material from which it is extracted.

6.3 Pineapple Essential Oil

Pineapple essential oil is a much sought-after product due to its unique scent and potential therapeutic properties. It is derived from the pineapple fruit, which is abundant in tropical regions all over the world. The oil is rich in active compounds, such as bromelain, alpha-pinene, and beta-pinene that have been shown to have anti-inflammatory, antimicrobial, and antioxidant properties in laboratory studies. These qualities make pineapple essential oil a versatile and potentially valuable natural product. Aromatherapy is one of its primary uses as its distinct aroma is believed to have a calming effect on both the mind and body, making it a popular ingredient in diffusers, candles, and other aroma products. Pineapple essential oil is also widely used in personal care products such as soaps, lotions, and perfumes due to its refreshing and tropical scent. Furthermore, the food and beverage industry uses pineapple essential oil as a flavoring agent, particularly in candies, baked goods, and beverages, providing a unique and tropical taste to these.

Recent studies have investigated the medicinal potential of pineapple essential oil. Animal studies have shown that bromelain, a proteolytic enzyme found in pineapple essential oil, can accelerate wound healing and reduce inflammation [17]. Pineapple essential oil may also have analgesic properties due to the presence of alpha-pinene and beta-pinene, making it a potential natural remedy for pain relief [18]. Moreover, the antioxidant properties of pineapple essential oil may provide benefits for overall health and well-being.

Numerous commercial suppliers offer pineapple essential oils that they claim are pure and natural. This type of essential oil usually appears as a pale-yellow color with a citrusy aroma. Manufacturers commonly advertise the anti-inflammatory, antiseptic, antiviral, antibacterial, antidepressant, purifying, and warming properties of pure pineapple essential oil, which is utilized in the production of various personal care products such as body lotion, bath salts, and scented candles [6]. However, it is crucial for users to verify the purity of the pineapple essential oil they purchase. The reason for this is that some of the oils marketed as pineapple essential oil are fragrance oils, which are synthetic and differ significantly from essential oils. Essential oils possess healing properties and are frequently diffused or utilized for specific purposes, while fragrance oils are primarily used for superficial purposes.

6.4 Extraction of Essential Oils

Essential oils can be extracted from plant materials in a variety of ways, including HD, solvent extraction, and enzyme-assisted extraction. The distillation method employed has a significant impact on the quality of commercial essential oils. Numerous studies have recently concentrated on creating environmentally friendly, economically viable, sustainable, and high-quality essential oil extraction methods. The production of essential oils with comparable or better quality and yield while consuming less energy, emitting less CO_2, and, in some cases, not producing any harmful co-extracts is the aim of researchers working to achieve these goals. To this end, they have been rethinking and enhancing traditional methods. Microwave-assisted [9], CO_2 supercritical fluid [10], and enzyme-assisted extractions [11] are a few of the often-employed green extraction methods for essential oil extraction.

6.4.1 Distillation

Distillation is the technique most frequently utilized to extract essential oils from plant materials. The raw plant material is boiled in water during this hydroextraction process to create steam. For materials that are sensitive to temperature, such as natural aromatic compounds, which are easily decomposed at prolonged high temperatures, this approach is preferable. The boiling point of these compounds is lowered by the addition of water, allowing for their evaporation at lower temperatures ideally, below those at which the material starts to degrade. Following distillation, the vapors are condensed, resulting in a two-phase system comprising easily separable water and organic molecules [4]. There are three different types of distillation processes: steam distillation, HD, or water distillation, as well as water and steam distillation. Steam distillation includes passing steam through a bed of the extraction material, whereas water distillation involves submerging the plant material in water and boiling it at a particular temperature.

6.4.1.1 Hydro-distillation

The process of obtaining essential oils from plant materials using water distillation, commonly referred to as HD, is simple and inexpensive. The plant material is submerged in water during this process and then cooked. The distillation system is used to separate the oil from the water once the steam and oil vapors have been condensed. Water distillation stills are simple and appropriate for small-scale farming. However, if the open fire is not closely supervised, overheating and burning of the plant material may occur. Oils produced using conventional stills can be heated with steam produced in a separate boiler, although this needs additional capital equipment and degrades the oil's quality. For the continued extraction of essential oils from barks like cinnamon and sandalwood. Since HD has a long history of use in the field and is a tried-and-true technology, it is a dependable and reputable way to extract essential oils. It is an effective approach for enterprises that need significant volumes of essential oils since it can be used to extract those oils from enormous amounts of plant material. In comparison to other extraction methods, HD equipment is comparably straightforward and inexpensive, making it affordable for small-scale businesses and private individuals who desire to extract essential oils from plants.

One of the major disadvantages of HD is the high energy consumption required for the generation of steam [19]. In addition, the extraction time for HD can be lengthy, ranging from several hours to days, depending on the plant material being used. This limitation can be a significant challenge for industries that require rapid extraction of essential oils. Another limitation of HD is its potential to degrade heat-sensitive compounds, such as terpenes and phenols. This degradation can lead to the loss of the aroma and flavor of the essential oils. Furthermore, HD can also result in the formation of unwanted by-products such as PAHs [20], which can be harmful to human health. Alternative techniques, such as SFE, ultrasound-assisted extraction, and microwave-assisted extraction, have been developed to solve the drawbacks of HD. The essential oil's heat-sensitive components are maintained using these techniques, which also allow quicker extraction times and less energy use.

6.4.1.2 Soxhlet Extraction

Soxhlet extraction is a method of extracting compounds, such as essential oils, from solid materials using a solvent. The method was first developed by Franz von Soxhlet in 1879 has since gained significant prominence and widespread application across various industries, including the chemical, food, and pharmaceutical sectors.

The method involves placing the solid material, such as a plant material, in a thimble made of filter paper or a porous material. The thimble is then placed in a special apparatus known as a Soxhlet extractor. The extractor is made up of a round-bottom flask, a condenser, and a siphon. A solvent, such as hexane, is placed in the round-bottom flask, and the flask is heated. As the solvent boils, it vaporizes and rises into the condenser, where it is cooled and condensed back into a liquid [21]. The liquid then drips down into the thimble containing the solid material, dissolving the desired compounds. Once the solvent has dissolved the compounds from the solid material, it flows back into the round-bottom flask due to the siphoning effect. The solvent is then heated again, vaporizing it and starting the cycle once again. This cycle is repeated multiple times, with the solvent cycling through the solid material several times to increase the extraction efficiency. The extracted compounds, including the essential oils, are then collected in the round-bottom flask, while the solid material remains in the thimble.

The pineapple waste is ground to a fine powder and packed into the thimble. Hexane is used as the solvent, as it has a low boiling point and can effectively extract essential oils from pineapple waste. In comparison with HD, the main difference between the two methods is that HD requires the use of water as a solvent, while Soxhlet extraction can use a variety of solvents, including hexane, ethanol, and methanol. HD is generally considered to be a gentler method that is more suitable for fragile or heat-sensitive plant material, while Soxhlet extraction can extract more compounds due to its repeated cycling of solvent through the plant material. HD is a one-time procedure, but Soxhlet extraction necessitates several cycles to extract the essential oil. Soxhlet extraction can be finished more rapidly, but HD frequently necessitates a lengthier processing period. HD often yields less than Soxhlet extraction in terms of yield. However, because it can create oil of a higher grade with a more complex scent profile, HD is frequently favored. The comparison of the setup between Soxhlet and HD is shown in Figure 6.1.

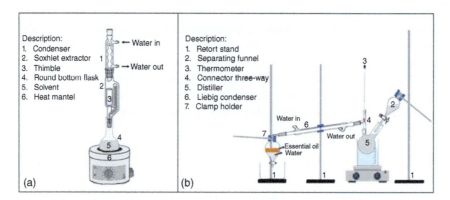

Figure 6.1 Apparatus setup of (a) Soxhlet extraction versus [21] (b) hydro-distillation [19]. Source: Drawn by Ibrahim et al.

6.4.2 Enzyme-assisted Extraction

One recent method for overcoming the constraint in the extraction of essential oils from plant sources, which is related to the low-yield extraction is enzyme-assisted extraction. This technique is used as a combination with other methods in order to enhance the yield of essential oil released. To weaken the botanical material's structural integrity and improve the extraction of the required components, especially essential oils, a variety of enzymes are utilized. Enzymes have the power to break down or damage membranes and cell walls resulting in improved essential oil release and more effective extraction. Using pretreatment of the substrate with cellulase enzyme prior to distillation as an example, the enzyme will work on the substrate's cell wall, hydrolyze it, and so enhance permeability, allowing more essential oil to be released. Many earlier research studies have demonstrated the efficacy of enzymatic pretreatment prior to distillation to increase the extraction yield of essential oils [22]. Enzyme-assisted extraction has been combined with SFE to extract compounds from various plant materials.

Prior to distillation, enzymatic pretreatment improved the yield and quality of the essential oil extracted from cardamom seeds. In this investigation, a variety of enzymes including Celluclast, Pectinex, Vicsozyme, and Protease were employed. The essential oil production increased after the enzyme pretreatment from 6.73% of the control sample to 7.23–8.3%, with pretreatment with Viscozyme yielding the maximum yield. Contrarily, Boulila et al. [23] carried out an enzyme-assisted extraction to examine the possibility of using several enzymes, including xylanase, cellulase, hemicellulase, and the ternary mixture of them to enhance the efficiency of the extraction of essential oil from bay leaves. This research discovered that the treated sample's essential oil output increased by 0.54–1.25% compared to the untreated sample. Table 6.2 summarizes several other studies utilizing the enzyme pretreatment approaches in the extraction of the essential oil.

6.4.3 Supercritical Fluid Extraction

Due to the anticipated benefits of the supercritical extraction method, the use of supercritical fluids, particularly carbon dioxide (CO_2), in the extraction of volatile plant components has expanded in recent years. A quick, accurate, and practical method for sample preparation before component analysis in the volatile plant product matrices is SFE. Additionally, SFE is a virtually solvent-free, quick, easy, and simple sample pretreatment method [10]. When a specific temperature and pressure are met, certain gases have the ability to behave as nonpolar solvents, and this is how the SFE principle works. By adjusting the temperature and/or pressure, it is possible to control the solvating power of supercritical fluid, allowing extractions to be specifically designed to extract only the required components while leaving undesirable molecules behind. Figure 6.2 displays the SFE's schematic flow diagram.

Due to its nontoxic and noncombustible characteristics, CO_2 is most frequently employed in extraction since it is environmentally beneficial. Compared to most other gases, supercritical CO_2 has lower critical values and a larger density (and,

Table 6.2 Previous studies on enzyme-assisted extraction of essential oil from various substrates.

Substrate	Methods	Enzyme used	Untreated yield (%)	Pretreated yield (%)	References
Black Pepper	Hydro-distillation with enzyme-assisted	Lumicellulase (a mixture of cellulase, β-glucanase, pectinase, and xylanase)	0.9	1.8 (increased by 50%)	[22]
Cardamom	Hydro-distillation with enzyme-assisted	Lumicellulase (a mixture of cellulase, β-glucanase, pectinase, and xylanase)	1.9	2.5 (increased by 24%)	[22]
Celery	Hydro-distillation with enzyme-assisted	Cellulase, pectinase, protease, and viscozyme	1.8	2.2–2.3 (increased by 20%)	[23]
Cumin Seeds	Hydro-distillation with enzyme-assisted	Cellulase, pectinase, protease, and viscozyme	2.7	3.2–3.3 (increased by 17%)	[23]
Garlic	Hydro-distillation with enzyme-assisted	Cellulase, pectinase, protease, and viscozyme	0.3	0.5–0.6 (increased by 45%)	[24]
Pineapple peels	Supercritical fluid extraction with enzyme-assisted	Cellulase	—	0.17	[10]
Cardamom	Supercritical fluid extraction with enzyme-assisted	α-Amylase	—	Increased by 29.55% of yield	[25]

consequently, solubility). The energy cost of supercritical CO_2 extraction is less than those of other fluids, with a critical temperature of 31 °C and a critical pressure of 73 atm, as shown in Figure 6.3 [27]. High-purity CO_2 is widely accessible and reasonably priced to acquire. The most widely used and least expensive solvent in business today is supercritical CO_2.

CO_2 is an environmentally friendly solvent because it is nonflammable, nontoxic, and nonpolluting. Compared to other solvents, such as ethanol or hexane, CO_2 is a safer and more sustainable option. Supercritical CO_2 has a higher density and solubility than other gases, making it an effective solvent for extraction. The critical temperature of CO_2 is 31 °C and the critical pressure is 73 atm. This means that CO_2 can exist in both liquid and gas states, depending on its pressure and

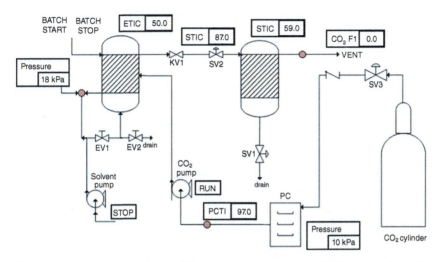

Figure 6.2 Schematic diagram of SC-CO_2 extraction apparatus (DEVEN Supercritical, Pvt. Ltd., India). Source: Drawn by Ibrahim et al.

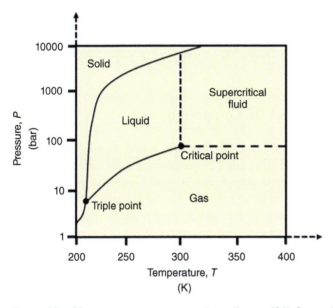

Figure 6.3 CO_2 temperature–pressure phase diagram [26]. Source: Drawn by Ibrahim et al.

temperature. Supercritical CO_2 extraction is also energy-efficient compared to other extraction methods. It requires lower temperatures and pressures, which reduces energy consumption and lowers costs. Additionally, supercritical CO_2 extraction produces high-quality extracts with minimal damage to the original material.

Temperature and pressure are just a couple of the variables that will determine if this operation is successful. In SFE, temperature is crucial because it determines

Table 6.3 Extraction of essential oil using supercritical fluid extraction.

Substrate	Temperature (°C)	Pressure (bar)	Essential oil yield (%)	References
Marchantia convoluta (Liverwort)	35–65	50–200	0.87–4.69	[28]
Pineapple peels	50	180	0.17	[10]
Salvia officinalis L (Garden sage)	50	65–160	0.7–1.40	[29]

how soluble the target chemicals are in the supercritical fluid. The target molecule will typically become more soluble in the supercritical fluid as the temperature rises, improving the extraction process's effectiveness. However, there is a maximum temperature at which the supercritical fluid maybe heated before becoming too thick and viscous and losing its capacity to efficiently permeate the matrix and extract the desired component. Another crucial SFE component is pressure. The density and solubility of the supercritical fluid can be manipulated by changing the pressure. The density of the supercritical fluid likewise rises as the pressure does, increasing its solvent power. However, there is a pressure threshold at which the supercritical fluid starts to behave more like a liquid than a gas, decreasing the extraction process's efficiency. As shown in Table 6.3, several investigations on the application of SFE for essential oil extraction employing a variety of extraction temperatures and pressures were carried out.

6.5 Extracted Essential Oil Compounds

The essential oil emitted from pineapple peel was found to have the greatest composition of propanoic acid ethyl ester. Propanoic acid ethyl ester, also known as ethyl propionate, is a volatile ester found in many essential oils, including pineapple essential oil. This ester contributes to the characteristic fruity and sweet aroma of pineapple essential oil, making it a valuable ingredient in the fragrance industry. This ester has also been discovered as one of the most prevalent molecules in a number of other essential oils, including Lippia chevalieri oil and Lippia multiflora oil. Along with ethyl acetate, ethyl butyrate, methyl butyrate, and ethyl butanoate, They are key components of citrus fruit flavor. In addition to its fragrance properties, propanoic acid ethyl ester has been reported to have various biological activities. A study reported volatile compounds in pineapple essential oil, showed antimicrobial properties against several strains of bacteria, including *Escherichia coli*, *Salmonella enterica*, and *Staphylococcus aureus* [30]. Meanwhile, ethyl propionate exhibited anti-inflammatory properties by minimizing the pro-inflammatory cytokines production in immune cells. Propanoic acid ethyl ester has been used as a food flavoring agent due to its fruity and sweet aroma. It is approved by the United States Food and Drug Administration (FDA) for use in food products as a flavoring agent and is classified as "generally recognized as safe" (GRAS).

Besides, this main component of essential oil is also a good solvent for many organic compounds, which means it can help dissolve and carry other components of the essential oil. This can improve the oil's overall effectiveness and absorption. Propanoic acid ethyl ester is a natural compound that can help stabilize and preserve essential oils. It works as an antimicrobial agent, inhibiting the growth of bacteria and other microorganisms that can cause spoilage and degradation of the oil. By reducing the presence of these harmful organisms, propanoic acid ethyl ester can help prolong the shelf life of the essential oil. Additionally, propanoic acid ethyl ester has a low volatility, which means that it evaporates slowly and can help maintain the concentration and composition of the other volatile components in the essential oil. This can also contribute to the preservation of the oil's fragrance and therapeutic properties. To take advantage of propanoic acid ethyl ester's preserving properties, essential oil manufacturers may add it as a natural preservative during the production process. It can help to ensure that the oil remains fresh and effective for longer periods of time, reducing the need for additional preservatives or additives [31].

Lactic acid ethyl ester, also known as ethyl lactate, a monobasic ester produced by the interaction of lactic acid and ethanol, was another abundant component found in this pineapple peel essential oil. It can also be found in various other fruits and flowers essential oil and distillate, such as melon [32]. The comparison of the volatile compounds obtained in the pineapple peel essential oil with the other aromatic compounds often discovered in pineapple is presented in Table 6.4. Several similar compounds that also being detected in various pineapple parts and products such as pineapple residues distillate, pineapple pulp and pineapple juice, and water. It can be said that the essential oil from pineapple peel contained the common compounds that are available generally in pineapple.

Table 6.4 Similar compounds in pineapple peel essential oil compared to various pineapple parts.

Compound	Products composition (%)			
	Pineapple peel essential oil [10]	Pineapple resides distillate [3]	Pineapple juice and water [33]	Pineapple pulp [31]
Propanoic acid ethyl ester	40.25	1.80	—	—
Benzaldehyde	1.25	0.04	1×10^{-9}	—
1-pentanol	1.92	0.36	1×10^{-7}	—
1-octanol	0.15	0.14	—	—
Diethyl succinate	0.12	0.05	2×10^{-7}	—
Furfural	0.14	0.04	1.6×10^{-8}	—
2(3H)-furanone	0.06	—	—	7.647×10^6
Butanoic acid ethyl ester	1.58	—	—	1.848×10^6

6.5.1 Essential Oils and Hydrosols

Essential oils, often known as the easily evaporable essences that give plants their lovely aromas, are concentrated volatile aromatic chemicals generated by plants. Common locations for these chemicals in plants include flowers, stems, leaves, bark, roots, seeds, resin, and fruit rinds. They are also frequently found in specialized cells, glands, or ducts in other sections of the plant. Essential oils are secondary metabolites made by a wide range of plant species that are linked to a number of different plant functions, including animal repellent, wound healing, defense against dangerous insects and microbial attacks, and pollinator attraction. The majority of the more than 200 components that makeup essential oils are terpenes and phenylpropanic derivatives, with just minor chemical and structural variations. Essential oils, often known as the easily evaporable essences that give plants their lovely aromas, are concentrated volatile aromatic chemicals generated by plants. Common locations for these chemicals in plants include flowers, leaves, stems, roots, seeds, bark, resin, and fruit rinds. They are also frequently found in specialized cells, glands, or ducts in other sections of the plant. Essential oils are secondary metabolites made by a wide range of plant species that are linked to a number of different plant functions, including animal repellent, wound healing, defense against dangerous insects and microbial attacks, and pollinator attraction. The majority of the more than 200 components that make up essential oils are terpenes and phenylpropanic derivatives, with just minor chemical and structural variations. The volatile fraction and nonvolatile residue can be divided into two groups. About 90–95% of the weight of the oil is made up of the volatile fraction, which also includes monoterpene and sesquiterpene oxygenated derivatives, hydrocarbons, aliphatic aldehydes, alcohol, and esters. On the other hand, the nonvolatile residue, which includes 1–10% of the oil, contains carotenoids, hydrocarbons, sterols, waxes, fatty acids, and flavonoids [34].

Prior to the distillation process of essential oil, a small portion of oxygenated, polar, odor-giving, water-soluble oil components can escape into the distillation water. Hydrosol is the liquid that is produced during the distillation process of essential oil and contains dissolved oil components [35]. Hydrosols are solutions that are highly diluted, with the volume of water much greater than the dissolved essential oil constituents. They have a mild or pleasant scent and are typically acidic, with a pH range of 3.5–6.5. Hydrosols are complex mixtures containing varying amounts of essential oil and other volatile compounds, water-soluble substances, and other secondary metabolites [10].

6.5.2 Applications of Essential Oils and Hydrosols

There are few documented scientific studies on the application of essential oil specifically from pineapple, but the applications of essential oil in general have been widely reported. Pineapple essential oil has been introduced as antimicrobial agent against foodborne pathogens in chocolate [36]. Generally, essential oils are used in aromatherapy, which involves inhaling the aroma of essential oils to

Table 6.5 Applications of essential oils.

Essential oil	Common uses	References
Citrus	Industrial chemicals, cleaning agent fragrances, and flavoring	[38]
Spearmint	Toothpaste, mouthwash, and flavoring for candies	[39]
Peppermint	Food flavoring, chewing gum, toothpaste, mouthwash, cosmetics, and tobacco	[39]
Lavender	Cosmetics and toiletries	[40]
Eucalyptus	Cold and flu medications, solvents, cleaning products, and flavoring	[41]
Tea tree	Cosmetics, hygiene products, and insect repellents	[42]
Blackcurrant bud	Antimicrobial agent and chemotaxonomy marker	[43]
Pineapple	Antimicrobial agent for foodborne pathogens	[36]

stimulate the olfactory system and promote physical and emotional well-being. The molecules in essential oils can interact with the limbic system, which is responsible for regulating emotions, memory, and behavior, and influence the autonomic nervous system, which controls many bodily functions, such as heart rate, blood pressure, and digestion. Scientific research has shown that essential oils may have anti-inflammatory, analgesic, antimicrobial, anxiolytic, and sedative effects, among others, and maybe helpful in treating various health conditions, such as anxiety, depression, pain, and insomnia [37], although more studies are needed to confirm their efficacy and safety. These are essentially beneficial in several industries, including pharmaceuticals, cosmetics, food, fragrances, soap, and confectionery. Table 6.5 lists some significant applications of essential oils.

The food industry has also made extensive use of nanoemulsion due to its capacity to resolve issues including limited essential oils stability and strong odors that negatively affect food's organoleptic properties. But more crucially, due to their nonmetered size and enhanced diffusion, nanoemulsions can be used as a delivery mechanism to enhance essential oil stability, retain the activity of bioactive components, and reduce their sensory effects on food. Active packaging containing essential oils, which can be related directly to packaged meat to minimize or avoid protein/lipid oxidation and to moderate the usage of chemicals that have been associated with many health disorders [44].

In medicinal applications, the anti-inflammatory and antispasmodic effects of essential oils are quite potent, and they can also be employed as local anesthetics. The major purposes of essential oil encapsulation are to control the rate at which drugs are released into the body, shield the active ingredients from interactions with the environment, lessen their volatility, and increase their biological activity [45]. Other applications of essential oils are listed in Figure 6.4.

Although hydrosols have lesser interest than essential oil, hydrosols have gained popularity in Western countries for aromatherapy and other purposes. They

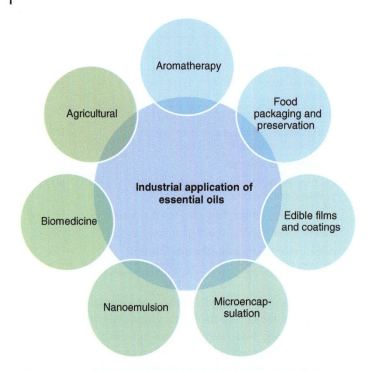

Figure 6.4 Applications of essential oils. Source: Drawn by Ibrahim et al.

have become attractive to global consumers, aroma therapists, and researchers as affordable flavorings and perfumery ingredients. Hydrosols are extensively used as floral and herbal water in Europe, Asia, and Africa, they are used for medicinal purposes and food flavorings. In India, farmers use hydrosols to repel insect pests and disease-causing organisms on crops. Additionally, orange and rose flower hydrosols are used to flavor household confectionery [46].

6.5.3 Market Analysis of Essential Oils and Hydrosols

The output of essential oils in the agriculture sector is significantly rising. The global market for essential oils was valued at US$ 21.8 billion in 2022 and is expected to growing at a compound annual growth rate (CAGR) of 7.9% from 2023 to 2030 [47]. This expansion is anticipated to be fueled by the rising demand for flavors and fragrances in the food and beverage sector as well as the expanding use of essential oils in medical, spa, and relaxation services and cleaning and home appliances. In 2021, the US essential oils market size by applications has reached US$ 72.4 billion, which is significantly higher than the market size in 2020, which was at US$ 68.1 billion as shown in Figure 6.5.

Numerous businesses have entered the essential oil sector as it expands globally. The Essential Oils of New Zealand, Farotti Essenze, Moksha Lifestyle Products, doTerra, Sydella Laboratoire, Young Living Essential Oils, West India Spices Inc., Falcon, and Ungerer Limited are among the companies that produce essential oils

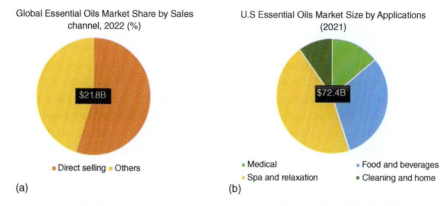

Figure 6.5 (a) Global essential oils market share by sales channel 2022 (%) (b) U.S essential oils market size by application (2021) [48]. Source: Drawn by Ibrahim et al.

and are considered the leading companies in the industry [47]. However, due to their special qualities, producing essential oils can be expensive and necessitates a sizable capital investment in cutting-edge machinery, which frequently leads to the dominance of multinational firms in the sector.

Orange oil is the most significant essential oil segment, generating US$ 1356.2 million in 2015 due to its numerous applications and therapeutic properties such as antidepressants, diuretics, anti-inflammatory, aphrodisiac, and antiseptic. Corn mint oil is also expected to have significant growth, potentially due to its high menthol content and associated health benefits. While Tea tree oil, eucalyptus, jasmine, rose, geranium, lemon, rosemary, orange, frankincense, and sandalwood are more frequently used domestically, citronella, orange, corn eucalyptus, mint, peppermint, and lemon are among the most frequently produced essential oils for industry purposes. Brazil, China, the United States, Indonesia, India, and Mexico are the top producers of essential oils, and the United States, the European Union (particularly the United Kingdom, Germany, and France), and Japan are the top consumers [47].

Hydrosols are typically discarded after the distillation process, but they have significant economic potential that is often unrecognized. In India, the hydrophilic essential oil fractions that escape into the hydrosols were estimated to be worth US$ 50–100 million [49]. Despite this, hydrosols are gaining popularity in the aromatherapy industry, and various companies are advertising and trading hydrosols of different plant species. The top hydrosol brands are Tianyuan Wusha, Fangcj, Bogazy, Bay House, Oshadhi, Florial, Herbal Aroma Hydrosol, and Oshadhi.

6.6 Conclusions

In conclusion, pineapple wastes are a potential source of essential oils with a variety of biological activities and prospective uses in the food, cosmetic, and pharmaceutical industries. The terpenes, ketones, and esters found in the essential oils derived from pineapple wastes help to give them their distinctive scent and flavor as well as their biological qualities. These properties make them suitable for use as natural preservatives, flavoring agents, and therapeutic agents. Several extraction methods have been used to extract essential oils from pineapple wastes, including distillation, enzyme-assisted extraction, and SFE. Among these methods, distillation, specifically HD is the most commonly used method due to its simplicity and low cost. However, enzyme-assisted extraction and SFE have been reported to produce higher yields of essential oil. To maximize the extraction processes and explore the full potential of pineapple essential oils in various applications, further study is required. In the beginning, essential oils are majorly used in aromatherapy, however, with recent technology, their applications have been widely broadened to medical applications, food packaging and preservation, edible films and coatings, nanoemulsion, biomedicine as well as agriculture. The by-product of essential oils, known as hydrosol has also gained popularity in aromatherapy, medicinal purposes, and flavoring as it is more affordable and cheaper in comparison to essential oils. The essential oils market is expected to gradually boost over the year, at a CAGR of 7.9% from 2023 to 2030, fueled by the increasing demand of flavors and fragrances in the food and beverages industry. Besides, the utilization of hydrosol, which was commonly discarded, as an alternative fragrance and flavor enhancer in the aromatherapy industry is highly beneficial due to the cost reduction. Hence, a circular economy and trash reduction can both benefit from the development of efficient, sustainable systems for extracting essential oils from pineapple wastes. Overall, using pineapple waste as a source of essential oils offers a bright prospect for environmentally friendly and sustainable production in the long run.

References

1 Lun, O.K., Wai, T.B., and Ling, L.S. (2014). Pineapple cannery waste as a potential substrate for microbial biotransformation to produce vanillic acid and vanillin. *Int. Food Res. J.* 21 (3): 953–958.
2 Ahmed, O.H., Husni, M.H.A., Anuar, A.R., and Hanafi, M.M. (2002). Effect of residue management practices on yield and economic viability of Malaysian pineapple production. *J. Sustain. Agric.* 20 (4): 83–93.
3 Barretto, L.C.D.O., De Jesus, J., Antônio, J. et al. (2013). Characterization and extraction of volatile compounds from pineapple (*Ananas comosus* L. Merril) processing residues. *Food Sci. Technol.* 33 (4): 638–645.
4 Kumar, R. and Tripathi, Y.C. (2011). *Getting Fragrance from Plants Training Manual*, 77–102. Dehra Dun: Indian: Chemistry Division, Forest Research Institute.
5 ReportLinker (2022). Essential oils market by product type, application, source, method of extraction and region – global forecast to 2027. https://www.reportlinker.com/p04988913/Essential-Oils-Market-by-Product-Type-Method-of-Extraction-Application-and-Region-Global-Forecast-to.html (accessed on 19 March 2023).

6 Fuller, J. (2016). Pineapple essential oil benefits. A massage therapy and wellness blog. https://foryourmassageneeds.com/pineapple-essential-oil-benefits (accessed on 19 March 2023).

7 Abdul-Majeed, B.A., Hassan, A.A., and Kurji, B.M. (2013). Extraction of oil from *Eucalyptus camadulensis* using water distillation method. *Iraqi J. Chem. Pet. Eng.* 14 (2): 7–12.

8 Galadima, M.S., Ahmed, A.S., Olawale, A.S., and Bugaje, I.M. (2012). Optimization of steam distillation of essential oil of *Eucalyptus tereticornis* by response surface methodology. *Niger. J. Basic Appl. Sci.* 20 (4): 368–372.

9 Cansu, T.B., Yocel, M., Sinex, K. et al. (2011). Microwave assisted essential oil analysis and antimicrobial activity of *M. alpestris* subsp. alpestris. *Asian J. Chem.* 23 (3): 1029–1031.

10 Mohamad, N., Ramli, N., Abd-Aziz, S., and Ibrahim, M.F. (2019). Comparison of hydro-distillation, hydro-distillation with enzyme-assisted and supercritical fluid for the extraction of essential oil from pineapple peels. *3 Biotech* 9 (6): 234.

11 Ghandahari Yazdi, A.P., Barzegar, M., Sahari, M.A., and Ahmadi Gavlighi, H. (2019). Optimization of the enzyme-assisted aqueous extraction of phenolic compounds from pistachio green hull. *Food Sci. Nutr.* 7 (1): 356–366.

12 Abdullah, A. and Mat, H. (2008). Characterisation of solid and liquid pineapple waste. *Reaktor.* 12 (1): 48.

13 Gil, L.S. and Maupoey, P.F. (2018). An integrated approach for pineapple waste valorisation. Bioethanol production and bromelain extraction from pineapple residues. *J. Clean. Prod.* 172: 1224–1231.

14 Raji, Y.O., Jibril, M., Misau, I.M., and Danjuma, B.Y. (2012). Production of vinegar from pineapple peel. *Int. J. Adv. Sci. Res. Technol.* 3 (2): 656–666.

15 Bardiya, N., Somayaji, D., and Khanna, S. (1996). Biomethanation of banana peel and pineapple waste. *Bioresour. Technol.* 58 (1): 73–76.

16 Upadhyay, A., Chompoo, J., Araki, N., and Tawata, S. (2012). Antioxidant, antimicrobial, 15-lox, and ages inhibitions by pineapple stem waste. *J. Food Sci.* 77 (1): 9–15.

17 Fathi, A.N., Sakhaie, M.H., Babaei, S. et al. (2020). Use of bromelain in cutaneous wound healing in streptozocin-induced diabetic rats: an experimental model. *J. Wound Care.* 29 (9): 488–495.

18 Salehi, B., Upadhyay, S., Orhan, I.E. et al. (2019). Therapeutic potential of α-and β-pinene: a miracle gift of nature. *Biomolecules.* 11: 738.

19 Gavahian, M., Farhoosh, R., Farahnaky, A. et al. (2015). Comparison of extraction parameters and extracted essential oils from *Mentha piperita* L. Using hydrodistillation and steamdistillation. *Int. Food Res. J.* 22 (1): 283–288.

20 Khajeh, M., Yamini, Y., Bahramifar, N. et al. (2005). Comparison of essential oils compositions of *Ferula assa-foetida* obtained by supercritical carbon dioxide extraction and hydrodistillation methods. *Food Chem.* 91 (4): 639–644.

21 Salve, R.R. (2020). Comprehensive study of different extraction methods of extracting bioactive compounds from pineapple waste – a review. *Pharma Innov. J.* 9: 327–340.

22 Chandran, J., Amma, K.P.P., Menon, N. et al. (2012). Effect of enzyme assisted extraction on quality and yield of volatile oil from black pepper and cardamom. *Food Sci. Biotechnol.* 21 (6): 1611–1617.

23 Sowbhagya, H.B., Srinivas, P., Purnima, K.T., and Krishnamurthy, N. (2011). Enzyme-assisted extraction of volatiles from cumin (*Cuminum cyminum L.*) seeds. *Food Chem.* 127 (4): 1856–1861.

24 Sowbhagya, H.B., Purnima, K.T., Florence, S.P. et al. (2009). Evaluation of enzyme-assisted extraction on quality of garlic volatile oil. *Food Chem.* 113 (4): 1234–1238.

25 Dutta, S. and Bhattacharjee, P. (2017). Microencapsulation of enzyme-assisted supercritical carbon dioxide extract of small cardamom by spray drying. *J. Food Meas. Charact.* 11: 310–319.

26 Knez, Ž., Pantić, M., Cör, D. et al. (2019). Are supercritical fluids solvents for the future? *Chem. Eng. Process. - Process Intensif.* 141: 107532.

27 Sapkale, G.N., Patil, S.M., Surwase, U.S., and Bhatbhage, P.K. (2010). A review: supercritical fluid extraction. *Int. J. Chem. Sci.* 8 (2): 729–743.

28 Jian, B.X., Jing, W.C., and Xu, M. (2007). Supercritical fluid CO_2 extraction of essential oil from *Marchantia convoluta*: global yields and extract chemical composition. *Electron. J. Biotechnol.* 10 (1): 141–148.

29 Fellah, S., Diouf, P.N., Petrissans, M. et al. (2005). Supercritical CO_2, hydrodistillation extractions of *Salvia officinalis* L. influence of extraction process on antioxidant properties. *Proc. 10th Eur. Meet. Supercrit. Fluids, React. Mater. Nat. Prod.*, (Natural Products processing N17) 1–8.

30 Nahid Hasan, A., Saha, T., and Ahmed, T. (2021). Antibacterial activity of the extracts of pineapple and pomelo against five different pathogenic bacterial isolates. *Stamford J. Microbiol.* 11 (1): 1–6.

31 Zheng, L.Y., Sun, G.M., Liu, Y.G. et al. (2012). Aroma volatile compounds from two fresh pineapple varieties in China. *Int. J. Mol. Sci.* 13 (6): 7383–7392.

32 Hernandez-Gomez, L.F., Ubeda-iranzo, J., Garcia-Romero, E., and Brinoes-Perez, A. (2005). Food chemistry comparative production of different melon distillates: chemical and sensory analyses. *Food Chem.* 90: 115–125.

33 Elss, S., Preston, C., Hertzig, C. et al. (2005). Aroma profiles of pineapple fruit (*Ananas comosus* [L.] Merr) and pineapple products. *LWT - Food Sci. Technol.* 38 (3): 263–274.

34 Lingan, K. (2018). A review on major constituents of various essential oils and its application. *Transl. Med.* 8 (1): 1000201.

35 Aswandi, A. and Kholibrina, C.R. (2021). Healthy hydrosol production from styrax resin harvesting waste using steam distillation. In IOP Conference Series:. *Earth Env. Sci.* 782 (4): 1–6.

36 Kotzekidou, P., Giannakidis, P., and Boulamatsis, A. (2008). Antimicrobial activity of some plant extracts and essential oils against foodborne pathogens in vitro and on the fate of inoculated pathogens in chocolate. *LWT Food Sci. Tech.* 41 (1): 119–127.

37 Hikal, W.M., Said-Al Ahl, H.A.H., Tkachenko, K.G. et al. (2022). Sustainable and environmentally friendly essential oils extracted from pineapple waste. *Biointerface Res. Appl. Chem.* 12 (5): 6833–6844.

38 Jing, L., Lei, Z., Li, L. et al. (2014). Antifungal activity of citrus essential oils. *J. Agric. Food Chem.* 62 (14): 3011–3033.

39 Elansary, H.O. and Ashmawy, N.A. (2013). Essential oils of mint between benefits and hazards. *J. Essent. Oil-Bearing Plants* 16 (4): 429–438.

40 Huang, S.-H., Fang, L., and Fang, S.-H. (2014). The effectiveness of aromatherapy with lavender essential oil in relieving post arthroscopy pain. *JMED Res.* 2014: 1–9.

41 Kumar Tyagi, A., Bukvicki, D., Gottardi, D. et al. (2014). Eucalyptus essential oil as a natural food preservative: in vivo and in vitro antiyeast potential. *Biomed Res. Int.* 2014: 969143.

42 Yadav, E., Kumar, S., Mahant, S. et al. (2017). Tea tree oil: a promising essential oil. *J. Essent. Oil Res.* 29 (3): 201–213.

43 Dordevic, B.S., Pljevljakušič, D.S., Šavikin, K.P. et al. (2014). Essential oil from blackcurrant buds as chemotaxonomy marker and antimicrobial agent. *Chem. Biodivers.* 11 (8): 1280–1240.

44 Smaoui, S., Ben Hlima, H., Tavares, L. et al. (2022). Application of essential oils in meat packaging: a systemic review of recent literature. *Food Control.* 132: 108566.

45 Ni, Z.J., Wang, X., Shen, Y. et al. (2021). Recent updates on the chemistry, bioactivities, mode of action, and industrial applications of plant essential oils. *Trends Food Sci. Technol.* 110: 78–89.

46 Rao, B.R.R. (2013) Hydrosols and water-soluble essential oils: medicinal and biological properties, in Recent Progress in Medicinal Plants Essential Oils 1 (eds. J.N. Govil, and Sanjib, B.), Studium Press LLC,U.S.A, India, 120–140.

47 IBISWorld (2022). Essential Oil Manufacturing Industry in the US – Market Research Report. https://www.ibisworld.com/united-states/market-research-reports/essential-oil-manufacturing-industry (accessed on 19 March 2023).

48 Grand View Research (2023). Essential Oils Market Size, Share & Trends Analysis Report By Product (Orange, Cornmint, Eucalyptus), by Application (Medical, Food & Beverages, Spa & Relaxation), by Sales Channel, by Region and Segment Forecasts, 2023–2030. https://www.grandviewresearch.com/industry-analysis/essential-oils-market#:~:text=Report%20Overview,7.9%25%20from%202023%20to%202030 (accessed on 19 March 2023).

49 Pangarkar, V.G. (2008) Microdistillation, thermomicrodistillation and molecular distillation techniques, in *Extraction Technologies for Medicinal and Aromatic Plants.*, edited by Handa, S. S., Khanuja, S. P. S., Longo, G. and Rakesh, D. D., ICS UNIDO, Trieste, Italy, 129–143.

7

Chicken Feather as a Bioresource to Produce Value-added Bioproducts

Kai L. Sim[1], Radin S. R. Yahaya[2], Suriana Sabri[2], and Lai-Yee Phang[1]

[1]*Universiti Putra Malaysia, Department of Bioprocess Technology, Faculty of Biotechnology & Biomolecular Sciences, Serdang, 43400, Selangor, Malaysia*
[2]*Universiti Putra Malaysia, Department of Microbiology, Faculty of Biotechnology & Biomolecular Sciences, Serdang, 43400, Selangor, Malaysia*

7.1 Introduction

Chicken is one of the top items on people's shopping lists nowadays. Due to the high demand, chicken meat consumption is increasing faster than any other meat on the globe. For health-conscious individuals, chicken is high in protein; low in calories; rich in essential vitamins like iron, zinc, and vitamin B; and low in fat and cholesterol. Subject to accessibility, availability, and affordability, chicken meat can be considered the cheapest protein source, which can be easily found everywhere around the market. According to the Food and Agriculture Organization of the United Nations database, world poultry consumption per capita was 14.8 kg in 2019 compared to 9.8 kg in 2000. The world's consumption of chicken meat is increasing and is expected to continue to increase in near future. Figure 7.1 shows the projected poultry meat consumption worldwide from 2021 to 2031. According to the report, the projected global consumption of poultry meat will amount to 153.85 metric kilotons by 2031.

Due to the high amount of chicken consumption, chicken feathers, which are made up of 5–7% of the total weight of the chicken have resulted in significant feather waste from broiler slaughterhouses or poultry processing plants. The yearly estimate of chicken feathers generated worldwide is around 15 billion tons [2]. In the United States, annually two to three billion tons of feathers are produced according to the Department of Animal and Food Science from the College of Agriculture, Food and Environment. To date, the Republic of South Africa currently contributes two hundred and fifty-eight million tons to the global generation of chicken feather waste from poultry slaughterhouses [2]. India, which is the best livestock wealth country, produced around seven hundred million tons of feathers waste annually [3]. While in Malaysia, according to annual chicken production of two hundred and ninety-five million heads [1], an estimated number of twenty million tons of feathers waste are produced annually.

Chemical Substitutes from Agricultural and Industrial By-Products: Bioconversion, Bioprocessing, and Biorefining, First Edition. Edited by Suraini Abd-Aziz, Misri Gozan, Mohamad Faizal Ibrahim, and Lai-Yee Phang.
© 2024 WILEY-VCH GmbH. Published 2024 by WILEY-VCH GmbH.

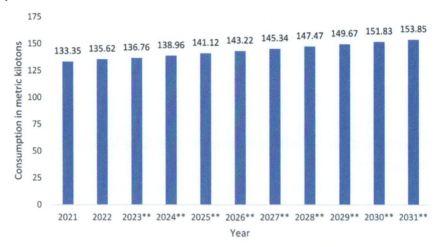

Figure 7.1 Projected poultry meat consumption worldwide from 2021 to 2031 *(in metric kilotons)*. Source: Adapted from Shanbandeh [1].

Till now, to handle the massive feather produced from poultry slaughterhouses, the most seen disposal methods for feather waste are incineration and land fillings. These disposing methods not only pollute the environment and at the same time damage the ecosystem, causing air and water pollution as well as global warming and also pathogen production [3, 4]. Feathers composed of 91% recalcitrant structural protein named keratin and its presence as the main component has made feathers to be recalcitrant in nature [5]. Although it is hard to degrade the feathers, researchers have been always finding ways to make use of feather wastes.

Research activities continue to explore the value of keratin from feathers via innovative strategies and approaches applied to the valorization process (bioprocess/transformation process) and upstream process (genetic modification). The aim of this chapter is to provide the approaches and treatments including physical, chemical, and biological methods to degrade or hydrolyze the feathers. These treatment methods can be applied as a single treatment or combined treatment primarily to enhance the feathers processing performance. The bioprocessing of chicken feathers into biofertilizer, feather meal, bioplastics, and keratinase production using feathers as substrate in a bioprocess will be described. This chapter also discusses the upstream process involved in the valorization of feathers, focusing on the molecular approaches to increase keratinase production or its characteristics, either through engineering the native keratinolytic strain itself, cloning and expressing recombinant keratinases in various heterologous hosts, or subjecting rational design on the keratinase structure.

7.2 Valorization of Chicken Feathers

By the year 2030, all countries are committed to achieving sustainable development in its three dimensions – economic, social, and environment. All countries must

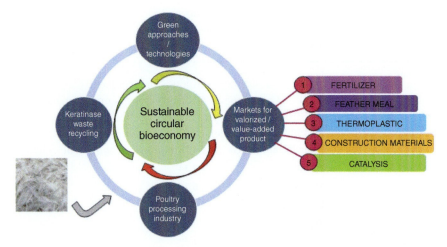

Figure 7.2 Recycling of poultry waste (feathers) toward a sustainable circular economy. Source: Lai-Yee et al., (author).

contribute to changing unsustainable consumption and production patterns, especially to strengthen developing countries' scientific, technological, and innovative capacities to move toward more sustainable patterns of consumption and production. Therefore, it is high time that the poultry processing industry identifies and implements an appropriate strategy to recycle feather keratin toward a sustainable circular bioeconomy as shown in Figure 7.2. Any green approach or biological route that produces eco-friendly products that are harmless to our mother nature is prioritized in order to achieve the Sustainable Development Goals (SDGs). Through a circular bioeconomy, the feathers are managed more efficiently through recycling and reprocessing of this waste material, to form various value-added products that can fulfill the market demand for chemical substituents.

7.2.1 Feather Composition

Chicken feathers are composed of over 90% protein, keratin being the main component is a fibrous and insoluble protein highly cross-linked with disulfide covalent bonds. This allows the feathers to be water and organic solvent impermeable as well as providing a rigid structure for the feathers. Feather keratin shows an elevated content of amino acids, such as serine, glutamic acid, leucine, proline, valine, and cysteine. Other amino acids are also present but in a lower amount especially some essential amino acids namely methionine, histidine, and tryptophan as shown in Table 7.1.

7.2.2 Types of Feathers Treatment

The treatments applied to valorize the feathers can be categorized into four types, chemical, physical (mechanical or thermal), biological, and combined treatment.

Table 7.1 Amino acid composition of chicken feathers.

	Amino acid	%
Essential	Leucine	8.75
	Valine	7.55
	Isoleucine	4.97
	Threonine	3.45
	Phenylalanine	4.75
	Lysine	2.34
	Methionine	0.71
	Histidine	0.75
	Tryptophan	0.38
Nonessential	Serine	11.97
	Glutamic acid	9.45
	Glycine	7.13
	Arginine	7.03
	Aspartic acid	5.97
	Alanine	5.69
	Cysteine	7.53
	Tyrosine	3.09
	Proline	8.50

Source: Adapted from Adler et al. [6].

The purpose of applying various types of treatment on the feathers is to (i) reduce the size of feathers, (ii) disrupt the feather keratin structure (iii) hydrolyze the feathers and form protein hydrolysate, or (iv) produce enzymes or other chemical compounds. A combination of two treatment methods or more methods is applicable for further improvement of the yield and quality of the valorized products. Mechanical treatments such as grinding and milling are usually used to reduce the size of the feathers via machines. An increase in the surface area of the feathers facilitates the degradation of chicken feathers in the subsequent process compared to untreated feathers. These physical methods require high energy consumption and are costly. On the other hand, physical methods such as the milling and grinding process are used to reduce the size of feathers only, and hence, subsequent process/processes is/are necessary to transform the feather keratin into a valuable product. Thermal treatment is related to heat and is good when comes to solubility as it can make the feather to be more water soluble and increase the feather hydrolysis. However, some amino acids might be degraded at high temperatures, which leads to products containing low amino acid content. Upon hydrolysis or degradation of keratin by thermal, chemical, or biological treatment, a product known as feather hydrolysate containing amino acids, is formed. The quality or value of this hydrolysate is highly dependent on the yield of amino acids from the keratin feather.

Chemical treatment for hydrolysis of chicken feathers is commonly conducted using either acid or alkali to break the disulfide bond allowing the feather keratin structure to be more water-soluble. Hydrochloric acid (HCl), sulfuric acid (H_2SO_4), phosphoric acid (H_3PO_4), sodium hydroxide (NaOH), potassium hydroxide (KOH), and lime are the chemicals commonly used. This method is known to have a simple procedure and fast reaction rate, whereby the feathers could be treated in less than 24 hours while giving a higher hydrolysis yield than biological treatment [6, 7]. Prolonging chemical treatment might even lead to a complete degradation and solubilization of feathers into protein hydrolysate [6]. The use of strong acids and/or alkali breaks the disulfide bonds of feathers into smaller peptide bonds, however, some essential amino acids denatured during the hydrolysis leading to protein hydrolysate with low amino acid content. Biological method utilizes microorganisms or enzymes to degrade the feathers under certain conditions.

The biological treatment operates in a milder condition with lesser by-products compared to other routes. This milder process can minimize the loss of essential amino acids as well. Till now, there are many types of keratinolytic microbes being discovered and isolated where most of them being bacteria and actinomyces. The percentage of feather degradation could reach up to 90% by *Pseudomonas* sp. P5 with an incubation time of 5 days in which approximately 301 mg/l free amino acid and 6.2 g/l soluble protein were recovered from the fermentation process [8]. Nonetheless, this method is time-consuming as it involves living microorganisms in the process and contamination might occur if the process is not well handled and prepared.

Combined treatment can be defined as the combination of at least two treatments from chemical, biological, or physical treatments [7]. Combined treatment has the same objective as those single treatments where they can be used to degrade the feathers and recover the soluble protein and amino acids. This treatment can be divided into two categories namely single-stage combined treatment and two-stage combined treatment. For single-stage treatment, examples are microwave-chemical treatment and hydrothermal treatment where these chemical and physical treatments are carried out simultaneously. Feather hydrolysis via hydrothermal treatments using 0.1 g of $Ca(OH)_2$ heated to 150°C for 25 minutes produced hydrolysate with 80% of soluble protein [9]. This method involved high temperature and pressure incorporated with chemical hydrolysis using diluted acid or alkaline. However, this method might degrade some proteins and essential amino acids.

The two-stage combined treatment hydrolyses feathers via two steps mechanism where the first step can be named pretreatment. This pretreatment plays an important role in breaking down the bonds between the keratin before undergoing the second treatment step. Pretreatment step can enhance the degradation rate of the feathers and product yield as the feathers are mostly partially hydrolyzed via chemical, physical, or even combination treatment. After the pretreatment step, the feathers will be subjected to the following stage, which is normally a biological treatment where the enzyme is involved to valorize the partially hydrolyzed

Table 7.2 Advantages and disadvantages of various feather hydrolysis treatments.

Treatments	Advantages	Disadvantages
Chemical	High product yield	Have pollution risk
	Fast reaction time	Loss of essential amino acids
Physical	No pollution risks	High energy consumption
	Can reduce particle sizes	Low product yield
		Costly
Biological	Eco-friendly production	Low reaction rate
	No destruction of amino acids	Time-consuming process
Combined treatment		
Single-stage	Fast process	Destruction of amino acids
	High product yield	
Two-stage	Fast process	Costly
	High product yield	More units of treatment
	No destruction of amino acids	

Source: Adapted from Cheong et al. [7].

feathers. For example, Akhter et al. [10] showed that by pretreating the feathers using autoclaving method with conditions at 120 °C for 20 minutes before treating them with crude keratinase, the feather degradation rate was improved as compared to raw feathers without pretreatment. The two-stage combined treatment can be a solution to solve the problem of loss of essential amino acids and excessive protein degradation. However, the cost for this method is relatively higher as the number of unit operations increases. Table 7.2 summarizes the advantages and disadvantages of the single and combined treatments applied in feather hydrolysis.

7.3 Bioprocessing of Chicken Feathers into Chemical Substitutes

Microorganisms that secret keratinase and degrade feathers are known as keratinolytic microbes. Feather can be used as a substrate by keratinolytic microbes to produce enzymes and amino acids as products. The use of single culture in the bioprocessing of the feathers waste is preferable as compared to a co-culture or mixed culture system due to ease of handling. Co-culture system requires two or more inoculum to be prepared before starting the fermentation process. Peng et al. [11] reported that by using a co-culture system, the efficiency of the chicken feathers degradation was improved compared to using a single culture.

Bacillus licheniformis BBE11-1 and *Stenotrophomonas maltophilia* BBE11-1 were used to hydrolyze the feathers and could produce a significant amount of amino acids and also achieved a high conversion rate of up to 70%. Nasipuru et al. [12] reported that the co-culture system aids in enhancing keratin degradation as compared to the single species degrader of the individual co-cultures, *Xanthomonas retroflexus*. The degradation was enhanced by 30% compared to monoculture.

Enzymatic hydrolysis is another option for the biological treatment of feathers. This method uses semi-purified or purified keratinase to hydrolyze the feathers. The performance of enzymatic hydrolysis of feathers is dependent on several factors, such as pH, temperature, and reaction time. Optimization is a crucial process in fermentation to determine the optimum conditions that are favorable to the microbes to grow well and produce the target products in the shortest time. Temperature, pH, incubation time, inoculum size, fermentation medium, and also substrate concentration are the common parameters tested in the optimization process. Dhiva et al. [13] showed that the keratinase production and the feathers degradation by *Pseudomonas aeruginosa* SU-1 could be maximized under optimum fermentation conditions. Keratinase operated optimally in alkaline conditions with a pH range of 5.0–9.0 [14]. Nonetheless, the optimum conditions vary based on the characteristics of the keratinase enzymes produced by the microorganisms. For example, the feather degradation carried out by *Pseudomonas* species was optimum at alkali pH (pH 8) and the temperature of 37 °C, whereas the optimum feathers degradation conditions by *Bacillus* species were at pH 7 and a temperature of 30.

In short, optimization of the fermentation process is able to improve the hydrolysis of the chicken feathers and increase the production of soluble proteins, digestible proteins, and amino acids in the hydrolysates. Moreover, this process is important as it can help to increase the quality and functionality of the processed feathers to be further used in industries such as feed industry, biopolymer industry, and biofertilizer industry.

7.3.1 Feather Meal

Although chicken feathers can be used as animal feed, unfortunately, chicken feathers could not be directly utilized by animals due to their low digestibility with the high percentage of keratin [5]. Thus, chicken feathers must be treated before feeding the animals. After the treatment, the protein hydrolysate (mixture of peptides and amino acids) is further processed to make the feather meal as animal feed. Animal feed produced from feather meal is relatively cheap and also high in protein content. Feather meal has been explored as a fish meal and soybean meal substitute in animal feed formulation, and fortunately, it has been showing promising results in animal growth performance [15]. Keratinases might be added along with feed products to improve food assimilation by animals and their digestibility [16]. Besides, the use of keratinase-complemented diets with high content of amino acids can help to reduce feed requirements [16].

Recently, researchers are focusing on feather meals with antioxidants and antimicrobial properties as these protein resources have many advantages, especially in the feed and food industries as additives. On the other hand, feather meals with antimicrobial properties can be applied to plants as plant bio-stimulants to influence plant growth [17]. The protein hydrolysate produced is readily applied to leaves or to the root system of the plants to stimulate plant growth. Compared to chemically synthesized fertilizer, bio-stimulant is more environmentally friendly because it is biodegradable and does not release hazardous compounds into the soil [17].

7.3.2 Bioplastic

Plastics are made from a wide range of synthetic or semisynthetic polymers such as polyethylene, nylon, polyester, polystyrene, polyvinyl chloride, and polypropylene that can be molded into shape while soft, and then set into a rigid or slightly elastic form. However, plastics are nonbiodegradable in nature, they are polluting the environment and resulting in hazardous damage to the Earth. Poultry waste can be used as raw materials to produce thermoplastic resins, a bioplastic, through an extrusion process. This step is started by cleaning the feathers followed by pulverizing them into powder form. Then, the feathers are dissolved in sodium sulfide (0.5 M in 1 l) with pH 7.5 at 50 °C for 24 hours where the supernatants are separated after this. The keratin protein sediments were collected and mixed with glycerol before the aliquot was spread over in an aluminum weighing boat and dried in an oven in order to produce the biofilm [18]. This biopolymer or bioplastic can be used to make a variety of products including plastic utensils, containers, and also furniture.

Keratin-based plastics exhibit better strength and tear resistance compared to other bioplastics that are derived from modified starch and plant protein. Ramakrishnan et al. [18] reported that synthesizing bioplastics using keratin from chicken feathers is workable by mixing keratin solution and different concentrations of glycerol with continuous stirring of five hours at 60 °C to produce plastic films. It was reported that the bioplastic produced from feathers with 2% of glycerol exhibited promising mechanical and thermal properties. The bioplastic produced is also proven to be biodegradable through the biodegradability test. This is a significant output as it can become a possible application to replace fossil-based materials in near future, therefore reducing the destruction of nature.

7.3.3 Biofertilizer

Keratin contains around 15% nitrogen, allowing it to become a suitable candidate for fertilizer or soil amendment. Chicken feathers that are rich in keratin have a robust structure that can resist degradation by soil microorganisms. With this characteristic, feather keratin can be transformed into a biofertilizer with the slow-releasing nitrogen characteristic. Thus, this environmentally friendly product could release its nitrogen into the soil slowly and it is very suitable to be used in the nursery and greenhouse industries. Rai and Mukherjee [19] reported that adding iron-oxide magnetic nanoparticles (MNP) bound β-keratinase, which is also known as alkarnase produced from alkaline β-keratinase bacterium into the soil could show significant growth of Bengal gram and seed germination, as well a rise in the soil microbial population. The release of low volatile compounds post degradation recommended that it could be an ecological method of green nature fertilization. Studies proved that plants treated with valorized feathers showed better growth results compared to plants treated with mulberry, urea, and cow dung. Feather hydrolysate resulted from valorization of feathers by *Amycolatopsis* sp. MBRL 40 promoted the rice plants to grow at all levels and the growth enhancement was comparable to the rice plants treated with urea fertilizers [16]. Thus, it offers significant biotechnological application of chicken-feather hydrolysate, which could serve as a cheap source of liquid organic fertilizer. Moreover, feather hydrolysates not only enhance plant growth but improve soil microbial activity.

7.3.4 Keratinase

Enzymes are secreted by microorganisms to facilitate the hydrolysis of feathers in the microbial degradation process. The native keratin is initially attacked by the disulfide reductase in a process called sulfitolysis. Subsequently, keratinase plays its role in proteolysis in which the reduced keratin molecules are converted into soluble peptides and free amino acids. Keratinase is an extracellular enzyme, which comes under the category of protease that is also considered one of the products formed during the valorization of chicken feathers. Keratinase can be used as a bioadditive in industries such as detergent, pharmaceutical, cosmetics, leather, and especially in the animal feed industry [20]. Table 7.3 shows the fermentation conditions used to produce keratinase that is applied in different industries.

Keratinase could degrade recalcitrant keratinous materials such as chicken feathers, which could alleviate current environmental issues and confer industrial importance at the same time by converting the hydrolysates into better quality bio-products [23, 24]. However, most studies reported insufficient keratinase

Table 7.3 Different fermentation conditions used to produce keratinase that is applied to various Industries.

Industries	Conditions for keratinase production via fermentation	References
Animal feed	60 min, 60 °C, 1000 g/kg moisture, and 10 g/kg protease	[15]
Feather meal	Substrate 1% (w/w); inoculum size 4% (v/v); pH 10; 200 rpm at 37 °C (with feathers as sole carbon and nitrogen source)	[21]
Detergent	Substrate 1% (w/w); inoculum size 4% (v/v); pH 10; 200 rpm at 37 °C) (221.44 µg/ml tryptophan, 15.0 µg/ml isoleucine, 10.81 µg/ml lysine and 7.24 µg/ml methionine with feathers as sole carbon, and nitrogen source)	[21]
Leather	3% inoculum, 39 °C, pH 9.50, 250 rpm for 96 hours	[10]
Bioplastics	25 g of chopped feathers in sodium sulfide (0.5 M in 1 l) with pH 7.5 at 50 °C for 24 hours	[18]
Cosmetic	Feather meal in 20 mM potassium phosphate buffer at pH 7.2 at 0.0125 g/ml, incubated overnight at 4 °C	[22]

production or keratinase activity from native keratinolytic hosts [20, 24, 25]. Furthermore, prominent native keratinolytic microorganisms were reported to be pathogenic, thus raising concerns for further downstream applications [26]. This impedes further biotechnological applications of using keratinases from the native hosts. Recently, research studies have been focusing on the molecular approaches to increase keratinase production of the native host and recombinant keratinase production, which will be discussed in Sections 7.4 and 7.5, respectively.

7.4 Molecular Approaches to Improve Keratinolytic Propensity of Native Host

Progress in molecular approaches to improve keratinase production from the native keratinolytic host has been gaining ground recently [27, 28]. The main issue to address is to mitigate the insufficient keratinase production from the keratinolytic host because large quantities are required to meet industrial demands. Several approaches have been conducted to improve keratinolytic propensity as illustrated in Figure 7.3.

7.4.1 Overexpression of Keratinase from Native Host

It is imperative to overexpress recombinant keratinases in a prominent keratinolytic host to further increase its keratinolytic propensity in order to cater for better downstream industrial applications [24, 26]. Generally, keratinase production occurs under certain conditions affected by nutritional stresses where readily metabolizable substrates are not present, hence, the keratinolytic host must metabolize alternative substrates, such as keratins to support their growth. Additionally, another factor is substrate concentration, where it was reported that high substrate concentration that exceeded the optimal threshold would impede keratinase production [25]. Current research aims to overexpress keratinase in native keratinolytic hosts to overcome these constraints. Overexpression could be achieved by cloning the host's keratinase in an expression vector and retransform it into the keratinolytic host (Figure 7.4). As the recombinant keratinase is now inducible via promoter of the vector, its overexpression could be facilitated. The recombinant keratinase could be overexpressed functionally, as the keratinolytic host already possess efficient expression, translation, and secretion systems required to regulate such task. Various vectors are available that could tailor to each species such as pHT, pWHK, pSUGV4, and pUB810 that are used for *Bacillus* sp. and pPICZα, or pPIC9, which are suited for *Pichia pastoris*. Yang et al. [30] reported six-fold overexpression of keratinase (pUB110-*kerK*) when the keratinase was cloned and retransformed back into *Bacillus amyloliquefaciens* K1. The strain showed promising potential in preparing better feather meals due to the feather hydrolysates containing a high amount of essential amino acids and could also be applied in keratin waste disposal due to its prominent feather-degrading capacity.

7.4 Molecular Approaches to Improve Keratinolytic Propensity of Native Host

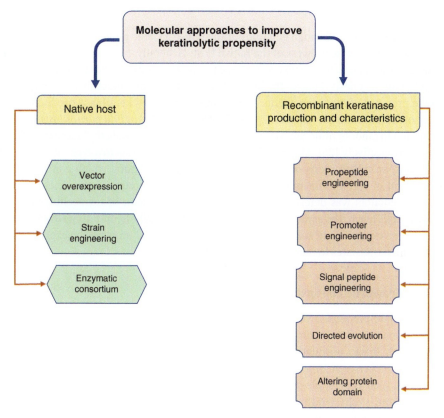

Figure 7.3 Molecular approaches to improve keratinase production. Source: Lai-Yee et al., (author).

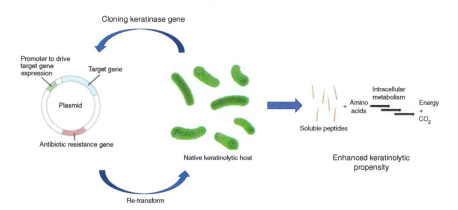

Figure 7.4 General illustration of plasmid overexpression in native keratinolytic host. Source: Redrawn and modified from Lange et al. [29].

7.4.2 Strain Engineering

Engineering the host through genome editing is also a viable molecular approach to improve keratinolytic prowess of the native host. Generally, random or site-specific mutagenesis is done to increase keratinase production from the keratinolytic host. Random mutagenesis involves the host's chromosome being randomly mutated due to the presence of mutating agents. For example, increased keratinase productions were observed when *Brevibacillus* sp. strain ASS10II and *Candida parapsilosis* were subjected to random mutagenesis via UV radiation exposure and ethyl methanylsulfonate, respectively [26]. Meanwhile, site-specific mutagenesis involves insertions, deletions, or substitutions of gene sequences. Gene regulators, which control the expression of certain enzymes are usually targeted. For example, the deletion of sRNA regulators was determined to improve the expression of extracellular proteases [31]. Hence, the same approach could be applied where mutating sRNA regulating the expression of keratinase in the native host could improve the expression rate, which translates to increased production.

7.4.3 Enzymatic Consortium

Keratinolytic propensity could be escalated by expressing enzymatic consortium or keratinase cocktail using the native host as an expression host as shown in Figure 7.5. It was postulated that the degradation of keratin requires synergistic cooperation between different types of keratinases as well as other supporting enzymes, such as disulfide reductase or lipo-monooxygenase [29]. Currently, keratinase cocktail is achieved through microbial co-culture in which several keratinolytic microorganisms are grown together to carry out the degradation of chicken feathers. Each respective keratinolytic microorganism secretes different types of keratinases or provides necessary by-products to boost the keratinolytic propensity of other microorganisms [11, 12]. For instance, the co-culture of *B. licheniformis* BBE11-1 and *Stenotrophomonas maltophilia* BBE11-1 conferred better feather degradation and higher conversion rate to oligopeptide and soluble peptides compared to individual strains, which proved to be beneficial in animal feed industry [11]. Another example is enhanced keratinolytic propensity and keratin biofilm productions from co-culture community of *Stenotrophomonas rhizophila*, *Microbacterium oxydans*,

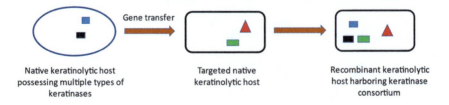

*Different colors indicate different types of keratinases

Figure 7.5 Overview of creating a keratinolytic host capable of producing keratinase cocktail/consortium. Source: Lai-Yee et al., (author).

Xanthomonas retroflexus, and *Paenibacillus amylolyticus* [12]. High keratinolytic performance was achieved through co-culture of keratinolytic microorganisms, which produce various types of keratinases. It is compelling to theorize that creating a recombinant keratinolytic host possessing a keratinase cocktail could confer a decisive advantage in terms of efficiency compared to co-culture approach. For the native keratinolytic host to be a suitable expression host to produce keratinase cocktail, several augmentations need to be carried out such as deleting intracellular and extracellular proteases that degrade foreign proteins, codon optimization of foreign keratinase genes to suit the respective keratinolytic host and genome editing to silence virulence genes to reduce its pathogenic risk [27, 28, 32]. One study reported the presence of 10 potential extracellular keratinases comprises S8 subtilisin and metalloproteases as well as having intracellular keratinases [33]. This strain could be utilized to create a keratinase cocktail to have better efficiency in degrading chicken feathers. Creating a keratinase cocktail-producing host will be cost-effective as only a single culture needs to be maintained and different types of keratinases could be selected to be cloned and expressed in accordance with desired bioproducts.

7.5 Molecular Approaches to Improve Recombinant Keratinase Production and Characteristics

Apart from trying to improve keratinase production from a native keratinolytic host, research emphasizing the improvement of recombinant keratinase in heterologous hosts is also gaining a significant trend [28]. Due to native keratinolytic host usually having limited keratinase production, thus interest in cloning and expressing keratinase in heterologous hosts are preferred since further protein rational design could be implemented to improve the keratinase attributes, such as catalytic efficiency, thermostability, pH stability, and halotolerance [34–36]. As shown in Figures 7.3 and 7.6, various molecular approaches could be conducted to improve the production or characteristics of recombinant keratinase in heterologous hosts, such as *Escherichia coli*, *Bacillus* sp., and *Pichia pastoris* [28]. As it is feasible to engineer recombinant keratinase in heterologous hosts, thus its applications in biotechnological sectors hold concrete potential.

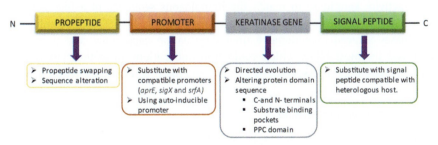

Figure 7.6 Strategies to improve recombinant keratinase characteristics in heterologous hosts. Source: Lai-Yee et al., (author).

Most molecular research focuses on cloning and expressing keratinases from native keratinolytic hosts in heterologous hosts, such as *E. coli*, *Bacillus* sp., and *P. pastoris*. These heterologous hosts are commonly used as they are model microorganisms, and each has its own advantages in terms of the research purpose. *E. coli* has been extensively used in cloning and expression of recombinant keratinases due to the availability of numerous strains and vectors that could be tailored to suit certain purposes, such as having better keratinase activity, purification yield, and stable expression [30, 34, 37]. Meanwhile, *Bacillus* sp. was also frequently utilized to express recombinant keratinase [20, 38]. Being genetically stable, having a less pathogenic tendency, possessing a higher protein quality control system, and being flexible to genetic engineering are the reasons *Bacillus* sp. is used as an expression host [39]. Some examples of *Bacillus* sp. used are *B. subtilis* WB600, *B. subtilis* DB104, and *B. megaterium* MS941 [28]. Then, *P. pastoris* such as *P. pastoris* GS115 or *P. pastoris* X33 was also utilized as heterologous hosts to express recombinant keratinases. The main advantages of using *P. pastoris* as a heterologous host are allowing stable chromosomal integration, a higher chance of stable secretion of functional proteins, and the usage of alternative expression inducers, such as alcohol rather than antibiotics [32].

7.5.1 Propeptide Engineering

One of the molecular approaches to improve keratinase characteristics is through propeptide engineering, in which the gene sequence encoding propeptide of the respective keratinase is altered, replaced, or swapped. Propeptide acts as a precursor to start the folding process of an enzyme. Hence, engineering the propeptide would cause different folding pathways to occur, thus giving various structural impacts on the enzyme with altered characteristics. For example, propeptide swapping between keratinases *KerBP* and *KerBL* resulted in recombinant *KerBL* attaining better thermostability and higher keratinase activity [40]. Furthermore, altering the gene sequence of the propeptide is also a viable approach, demonstrated by Su et al. [41] where saturation mutagenesis of the propeptide cleavage site of keratinase *kerBp* conferred activity increase by three-fold. Peng et al. [11] obtained similar findings employing the same approach, which increase the activity of *KerZ1* by 86% compared to the wild type. Then, changing the propeptide sequence of a low-activity keratinase to match the homologous sequence of a high-activity keratinase was reported to improve the former's activity [42]. These recombinant keratinases showed exceptional feather degradation capabilities and possessed improved thermostability, which could find applications in leather tanning as industrial tanning processes require alkaline and high-temperature environment.

7.5.2 Promoter Engineering

Keratinase activity could also be improved through promoter engineering. The induction of keratinase is mostly regulated by a promoter; however, native promoter might not induce substantial expression of keratinase as induction only occurs under nutritional stress [26, 28]. Thus, several studies have replaced the native promoter of the keratinase with other various promoters to improve the expression of recombinant keratinase. For instance, keratinase *kerBv* exhibited a significant increase of activity by 16-, 13-, and 10-folds, when the native promoter was substituted with *aprE, sigX,* and *srfA* promoters, respectively [39]. The resulting recombinant strain also showed a high feather degradation percentage and could be applied in animal feed production as the feather hydrolysates contained considerable amount of essential amino acids, such as phenylalanine and tyrosine. Another alternative is replacing the inductive native promoter with an autoinducible promoter. Since keratinase expression is induced under nutritional stress, substituting the native promoter with an autoinducible promoter will allow automatic production of recombinant keratinase [43]. For example, promoter P_{srfA} is an autoinducible promoter, which is activated through quorum sensing when the recombinant host reaches late log phase. Hence, utilizing this promoter to induce the expression of recombinant keratinase automatically will prove to be advantageous to suit industrial demands.

7.5.3 Signal Peptide Engineering

Engineering the signal peptide could improve recombinant keratinase production in heterologous hosts [44]. Native signal peptide is usually not compatible with the heterologous host's secretion system, thus substituting the native signal peptide with the ones compatible with the heterologous host could strengthen recombinant keratinase production. For example, the expression of keratinase isolated from *Bacillus* sp. LCB12 in *B. subtilis* SCK6 was improved by two-folds when the native signal peptide was replaced with SP_{LipA} [38]. The mentioned keratinase was reported to be prominent in dehairing goatskins, hence, improving its production in heterologous host, surely raises its potential applications in leather and textile sector. Similar trend was reported by Dong et al. [37] where the production of keratinase *KerP* was elevated when the native signal was replaced with SP_{aprE}. This evidence showed that optimizing the native signal peptide in accordance with the heterologous hosts could improve the production of recombinant keratinase in heterologous host.

7.5.4 Directed Evolution

Directed evolution of recombinant keratinase could also improve its characteristics, such as having better catalytic efficiency, thermostability, and pH stability. Most of the directed evolution approaches are done through DNA shuffling or rearrangement, error-prone PCR, cross-extension, and recombination of nonhomologous sequences. For instance, the thermostability half-life of keratinase isolated from *B. subtilis* was exponentially improved by 200-folds through directed evolution approach. Furthermore, catalytic efficiency, thermostability, and optimal pH of *kerBp* were significantly improved through error-prone PCR to suit industrial conditions. Moreover, the feather hydrolysates contained substantial number of oligopeptides, amino acids, and short peptides, which confers valuable applications in animal feed industry [45].

7.5.5 Alteration of Protein Domains

Altering the gene sequence of important protein domains of keratinase such as the C- and N-terminal, and the substrate binding pockets, could strengthen keratinase characteristics to suit industrial processing conditions and to create better bioproducts [11, 46]. It is imperative to analyze domain sequences that would impact keratinase characteristics, such as metal-binding affinity, thermostability, and activity. For instance, amino acid substitution N122Y at the backbone region of keratinase from *B. licheniformis* BBE11 managed to improve its activity by 5.6-folds [47]. The same study also reported that N122Y/N217S/A193P/N160C amino acid substitutions resulted in enhanced thermostability of the recombinant keratinase. Another example is alteration of amino acids sequence at C- and N-terminals of the keratinase, where replacing the C- and N-terminal sequences with homogenous keratinase sequence could elevate keratinase activity of recombinant KerSMD by two-folds [42]. Mutagenesis of substrate binding pockets (termed as S1, S2, S3, and S4) of respective keratinase could also amplify catalytic efficiency of recombinant keratinase. For instance, amino acid substitutions of A218S and A218G of S1 binding pocket of keratinase FDD improved its thermostability and activity by 30% as compared to the wild type [46]. Then, mutagenesis of PPC domain also played crucial role in tailoring substrate specificity of recombinant keratinase to suit certain industrial processes. For example, the detrimental collagen-degrading KerSMD limited its applications in leather industry, hence, truncation of C-terminal of PPC domain vastly reduces its collagen-degrading ability while maintaining dehairing activity [42]. This broadened its potential applications in leather and textile industry where it could dehair goatskins without destroying the collagens. The same investigation also reported C-terminal improved its optimal temperature but decreases its stability. This could be useful to be applied in food treatment processing where enzyme inactivation at high temperature is required. The mentioned approaches could be applied to tailor the recombinant keratinase characteristics to suit the industry of interest.

7.6 Challenges and Future Perspectives

The main concerns of a conversion process of waste materials are process efficiency and feasibility that will influence the overall processing/production cost. Low degradation rate or conversion rate, low protein yield, high by-products, and loss of essential amino acids are the challenges present in valorization of feathers. The degradation process is accompanied by bacterial metabolism (maintaining the growth), which prevents the amino acids in the feather hydrolysate from reaching the desired concentration. Low protein yield in the hydrolysate with the presence of by-products (resulting from the treatment applied) has hindered the scale-up production because a downstream process involving more unit operations such as filtration, fractionation, and purification, is required to further improve the amino acids composition. The number of unit operations used in downstream process will have significant impact on the production cost that eventually affects the feasibility and economics of the bioprocess of feathers. The keratin degradation performance observed from single culture is not at par with that of other known degradation or the increased industrial demand. Higher degradation rates of feathers are expected to be achieved using the mixed species community [11, 12]. However, research studies on the screening of mixed species communities that can degrade feathers efficiently are limited. On the other hand, keratinase enzyme is produced using inexpensive keratinous substrates, such as feathers, and subsequently applied to various industries. Low enzyme stability and conversion rate are the limiting factors for the bioprocessing of feathers or valorization of feathers via enzymatic reaction. Moreover, the mechanisms of keratin degradation by diverse microorganisms in nature are not well documented in which only three mechanisms have been proposed to be common for the keratinolytic microorganisms [16]. The involvement of other enzymes, other than keratinase in the hydrolysis of keratins has been reported and the role of these enzymes is yet to be investigated.

Several challenges exist in the keratinase field, which mostly are inadequate keratinase production to meet industrial demands and pathogenicity concerns from the native keratinolytic host, unsuitable keratinase characteristics to meet industrial processing conditions, insufficient diversity of keratinase data, and limited information of combination of different types of keratinases to achieve substantial keratinolytic performances [26, 28, 47]. The major discovered keratinases have been purified from mesophilic bacteria; however, several industries need highly thermostable enzymes [16]. Generally, native keratinolytic hosts only produced scant amount of keratinase under nutritional stress, but also requires the need for optimization to be carried out, which starts from one factor at a time optimization followed by statistical optimization, such as central-composite design or response surface technology [24]. Even after optimization has been carried out, the amount of keratinase produced is still insufficient to compensate industrial demands. Apart from that, since most prominent keratinolytic hosts are categorized as *Bacillus* spp., thus pathogenicity of the strain has become a concerning issue when considering its applicability in industrial processing [28]. Then, the diversity of keratinase data is still lacking, this limits the extent of phylogenetic analyses that can be done, hence, giving difficulty in identifying important conserved residues and discovering novel keratinases with significant characteristics [48]. Furthermore, literature regarding combinations of different types of keratinases along with other keratinolytic-assisted enzymes is still not comprehensive enough to develop a framework of which combinations best suit to create specific bioproducts. Currently, only combinations of S8 subtilisin, M4 and M28 metalloproteases, lipo-monooxygenase, as well as disulfide reductase play important roles in ensuring significant keratinolytic performance [29]. More combinations need to be tested in order to broaden the practicability of applying keratinase to create better bioproducts on an industrial scale.

The product yield and the overall feather hydrolysis performance carried out by microorganisms or enzymes are influenced by physical and biochemical factors. Hence, optimization of the bioprocess is conducted to maximize the yield. With the help of simulation and programming using software, the bioprocess conditions can be varied and further studied to monitor the outcomes of the products. Besides, the process changes can be evaluated/estimated to enable a better prediction and better understanding of the processes that allow overall cost analysis and projection at a large scale. The valorization of feathers toward a sustainable circular economy can be realized with the cooperations of industries, stakeholders, research institutions, local authorities, and government agencies that play vital roles in various aspects. Novel approaches for probing keratinase-producing strains should be developed, for example, biosensors or rapid test strips based on functionalization of selected keratinous materials to uncover the microorganisms that possess novel keratinases with new characteristics. Future studies should focus on identifying more prominent keratinolytic hosts and novel keratinases to develop a comprehensive library to allow extensive analyses to be done. When sufficient data is present, accurate analyses could be done to identify which molecular approaches work best to improve keratinase production from the respective host, determining important regions for protein engineering to carry out and choosing optimal keratinase combinations to achieve significant keratinolytic performance. Scale-up production of those novel and potential keratinases is imperative and would be focused on molecular cloning and heterologous expression of keratinase in suitable industrial hosts.

7.7 Conclusions

Feather keratins are conventionally processed into feather meal, biofertilizer, and bioplastic. Moreover, feather keratin is also the substrate for keratinase enzyme production through a fermentation process. And, this enzyme can be applied in the detergent, cosmetic, and leather industries. The application of biological treatment involving microorganisms or enzymes is always an attractive option for processing feathers into chemical substitutes. Keratinases have potential applications in various sectors, not only because of their catalytic efficiency, but also due to sustainable production on a cost-effective renewable resource, i.e. chicken feathers. These described molecular approaches have been conducted to increase keratinase production from native keratinolytic hosts as well as improve the characteristics of the keratinase. The improvement of keratinase production in heterologous hosts is gaining the interest of researchers due to the advantages served by heterologous hosts. More studies on developing robust industrial hosts via molecular approaches are required. The challenges remain as bottlenecks for current research to tackle, in which solving these issues will realize the implementation of keratinases in various biotechnological sectors.

References

1 Shahbandeh, M. (2023). Projected poultry meat consumption worldwide from 2021 to 2031 [Infographic]. Statista Search Department. https://www.statista.com/statistics/739951/poultry-meat-consumption-worldwide/ (accessed on 31 July 2022).

2 Fagbemi, O.D. and Sithole, B. (2021). Evaluation of waste chicken protein hydrolysate as a bio-based binder for particleboard production. *Curr. Res. Green Sustainable Chem.* 1 (4): 100168.

3 Zul SM, Iwamoto K, Rahim MA, Abdullah N, Mohamad SE, Shimizu K, Hara H. Production of liquid fertilizer from chicken feather waste by using subcritical water treatment for plant and algal growth. In IOP Conference Series: Earth and Environmental Science 2020 (Vol. 479, No. 1, p. 012033). IOP Publishing.

4 Ali, M.F., Hossain, M.S., Moin, T.S. et al. (2021). Utilization of waste chicken feather for the preparation of eco-friendly and sustainable composite. *Cleaner Eng. Technol.* 4: 100190.

5 Li, Q. (2019). Progress in microbial degradation of feather waste. *Front. Microbiol.* 10: 2717.

6 Adler, S.A., Slizyte, R., Honkapää, K., and Løes, A.K. (2018). In vitro pepsin digestibility and amino acid composition in soluble and residual fractions of hydrolyzed chicken feathers. *Poult. Sci.* 97 (9): 3343–3357.

7 Cheong, C.W., Lee, Y.S., Ahmad, S.A. et al. (2018). Chicken feather valorization by thermal alkaline pretreatment followed by enzymatic hydrolysis for protein-rich hydrolysate production. *Waste Manage. (Oxford)* 79: 658–666.

8 Stiborova, H., Branska, B., Vesela, T. et al. (2016). Transformation of raw feather waste into digestible peptides and amino acids. *J. Chem. Technol. Biotechnol.* 91 (6): 1629–1637.

9 Coward-Kelly, G., Chang, V.S., Agbogbo, F.K., and Holtzapple, M.T. (2006). Lime treatment of keratinous materials for the generation of highly digestible animal feed: 1. Chicken feathers. *Bioresour. Technol.* 97 (11): 1337–1343.

10 Akhter, M., Wal Marzan, L., Akter, Y., and Shimizu, K. (2020). Microbial bioremediation of feather waste for keratinase production: an outstanding solution for leather dehairing in tanneries. *Microbiol. Insights* 13: 1178636120913280.

11 Peng, Z., Mao, X., Zhang, J. et al. (2019 Dec). Effective biodegradation of chicken feather waste by co-cultivation of keratinase producing strains. *Microb. Cell Fact.* 18 (1): 1–1.

12 Nasipuri, P., Herschend, J., Brejnrod, A.D. et al. (2020). Community-intrinsic properties enhance keratin degradation from bacterial consortia. *PLoS One* 15 (1): e0228108.

13 Dhiva, S., Ranjith, K.R., Prajisya, P. et al. (2020). Optimization of keratinase production using *Pseudomonas aeruginosa* Su-1 having feather as substrate. *Biointerface Res. Appl. Chem.* 10: 6540–6549.

14 Nnolim, N.E. and Nwodo, U.U. (2020). *Bacillus* sp. CSK2 produced thermostable alkaline keratinase using agro-wastes: keratinolytic enzyme characterization. *BMC Biotech.* 20 (1): 1–4.

15 Poolsawat, L., Yang, H., Sun, Y.F. et al. (2021). Effect of replacing fish meal with enzymatic feather meal on growth and feed utilization of tilapia (*Oreochromis niloticus*× *O. aureus*). *Anim. Feed Sci. Technol.* 274: 114895.

16 Hassan, M.A., Abol-Fotouh, D., Omer, A.M. et al. (2020). Comprehensive insights into microbial keratinases and their implication in various biotechnological and industrial sectors: a review. *Int. J. Biol. Macromol.* (154): 567–583.

17 Gupta, S., Kulkarni, M.G., White, J.F. et al. (2021). Categories of various plant biostimulants–mode of application and shelf-life. In: *Biostimulants for Crops from Seed Germination to Plant Development* (ed. S. Gupta and J. Van Staden), 1-60. Academic Press.

18 Ramakrishnan, N., Sharma, S., Gupta, A., and Alashwal, B.Y. (2018). Keratin based bioplastic film from chicken feathers and its characterization. *Int. J. Biol. Macromol.* 111: 352–358.

19 Rai, S.K. and Mukherjee, A.K. (2015). Optimization for production of liquid nitrogen fertilizer from the degradation of chicken feather by iron-oxide (Fe_3O_4) magnetic nanoparticles coupled β-keratinase. *Biocatal. Agri. Biotechnol.* 4 (4): 632–644.

20 Vidmar, B. and Vodovnik, M. (2018). Microbial keratinases: enzymes with promising biotechnological applications. *Food Technol. Biotechnol.* 56 (3): 312–328.

21 Reddy, M.R., Reddy, K.S., Chouhan, Y.R. et al. (2017). Effective feather degradation and keratinase production by *Bacillus pumilus* GRK for its application as bio-detergent additive. *Bioresour. Technol.* (243): 254–263.

22 Pongkai, P., Saisavoey, T., Sangtanoo, P. et al. (2017). Effects of protein hydrolysate from chicken feather meal on tyrosinase activity and melanin formation in B16F10 murine melanoma cells. *Food Sci. Biotechnol.* 26 (5): 1199–1208.

23 Nnolim, N.E., Udenigwe, C.C., Okoh, A.I., and Nwodo, U.U. (2020). Microbial keratinase: next generation green catalyst and prospective applications. *Front. Microbiol.* (11): 580164.

24 Mohamad, N., Phang, L.Y., and Abd-Aziz, S. (2017). Optimization of metallo-keratinase production by *Pseudomonas* sp. LM19 as a potential enzyme for feather waste conversion. *Biocatal. Biotransform.* 35 (1): 41–50.

25 Yusuf, I., Ahmad, S.A., Phang, L.Y. et al. (2016). Keratinase production and biodegradation of polluted secondary chicken feather wastes by a newly isolated multi heavy metal tolerant bacterium-*Alcaligenes* sp. AQ05-001. *J. Environ. Manage.* 183: 182–195.

26 Li, Q. (2021). Structure, application, and biochemistry of microbial keratinases. *Front. Microbiol.* 1510.

27 Li, Q. (2022). Perspectives on converting keratin-containing wastes into biofertilizers for sustainable agriculture. *Front. Microbiol.* 13.

28 Yahaya, R.S., Normi, Y.M., Phang, L.Y. et al. (2021). Molecular strategies to increase keratinase production in heterologous expression systems for industrial applications. *Appl. Microbiol. Biotechnol.* 105 (10): 3955–3969.

29 Lange, L., Huang, Y., and Busk, P.K. (2016). Microbial decomposition of keratin in nature—a new hypothesis of industrial relevance. *Appl. Microbiol. Biotechnol.* 100 (5): 2083–2096.

30 Yang, L., Wang, H., Lv, Y. et al. (2016). Construction of a rapid feather-degrading bacterium by overexpression of a highly efficient alkaline keratinase in its parent strain *Bacillus amyloliquefaciens* K11. *J. Agric. Food. Chem.* 64 (1): 78–84.

31 Hertel, R., Meyerjürgens, S., Voigt, B. et al. (2017). Small RNA mediated repression of subtilisin production in *Bacillus licheniformis*. *Sci. Rep.* 7 (1): 1–1.

32 Aggarwal, S. and Mishra, S. (2020). Differential role of segments of α-mating factor secretion signal in *Pichia pastoris* towards granulocyte colony-stimulating factor emerging from a wild type or codon optimized copy of the gene. *Microb. Cell Fact.* 19 (1): 1–6.

33 Yahaya, R.S., Phang, L.Y., Normi, Y.M. et al. (2022). Feather-degrading *Bacillus cereus* HD1: genomic analysis and its optimization for keratinase production and feather degradation. *Curr. Microbiol.* 79 (6): 1–5.

34 Nnolim, N.E., Mpaka, L., Okoh, A.I., and Nwodo, U.U. (2020). Biochemical and molecular characterization of a thermostable alkaline metallo-keratinase from *Bacillus* sp. Nnolim-K1. *Microorganisms* 8 (9): 1304.

35 Fang, Z., Zhang, J., Liu, B. et al. (2014). Cloning, heterologous expression and characterization of two keratinases from *Stenotrophomonas maltophilia* BBE11-1. *Process Biochem.* 49 (4): 647–654.

36 Lin, H.H., Yin, L.J., and Jiang, S.T. (2009). Functional expression and characterization of keratinase from *Pseudomonas aeruginosa* in *Pichia pastoris*. *J. Agric. Food. Chem.* 57 (12): 5321–5325.

37 Dong, Y.Z., Chang, W.S., and Chen, P.T. (2017). Characterization and overexpression of a novel keratinase from *Bacillus polyfermenticus* B4 in recombinant *Bacillus subtilis*. *Bioresour. Bioprocess.* 4 (1): 1–9.

38 Tian, J., Long, X., Tian, Y., and Shi, B. (2019). Enhanced extracellular recombinant keratinase activity in *Bacillus subtilis* SCK6 through signal peptide optimization and site-directed mutagenesis. *RSC Adv.* 9 (57): 33337–33344.

39 Gong, J.S., Ye, J.P., Tao, L.Y. et al. (2020). Efficient keratinase expression via promoter engineering strategies for degradation of feather wastes. *Enzyme Microb. Technol.* 137: 109550.

40 Rajput, R., Tiwary, E., Sharma, R., and Gupta, R. (2012). Swapping of pro-sequences between keratinases of *Bacillus licheniformis* and *Bacillus pumilus*: altered substrate specificity and thermostability. *Enzyme Microb. Technol.* 51 (3): 131–138.

41 Su, C., Gong, J.S., Sun, Y.X. et al. (2019). Combining pro-peptide engineering and multisite saturation mutagenesis to improve the catalytic potential of keratinase. *ACS Synth. Biol.* 8 (2): 425–433.

42 Fang, Z., Zhang, J., Liu, B. et al. (2016). Enhancement of the catalytic efficiency and thermostability of *Stenotrophomonas* sp. keratinase KerSMD by domain exchange with KerSMF. *Microb. Biotechnol.* 9 (1): 35–46.

43 Tran, D.T., Phan, T.T., Doan, T.T. et al. (2020). Integrative expression vectors with Pgrac promoters for inducer-free overproduction of recombinant proteins in *Bacillus subtilis*. *Biotechnol. Rep.* 28: e00540.

44 Freudl, R. (2018). Signal peptides for recombinant protein secretion in bacterial expression systems. *Microb. Cell Fact.* 17 (1): 1.

45 Zhang, J., Su, C., Kong, X.L. et al. (2022). Directed evolution driving the generation of an efficient keratinase variant to facilitate the feather degradation. *Bioresour. Bioprocess.* 9 (1): 1–3.

46 Fang, Z., Zhang, J., Du, G., and Chen, J. (2017). Rational protein engineering approaches to further improve the keratinolytic activity and thermostability of engineered keratinase KerSMD. *Biochem. Eng. J.* 15 (127): 147–153.

47 Liu, B., Zhang, J., Fang, Z. et al. (2013). Enhanced thermostability of keratinase by computational design and empirical mutation. *J. Ind. Microbiol. Biotechnol.* 40 (7): 697–704.

48 De Oliveira Martinez, J.P., Cai, G., Nachtschatt, M. et al. (2020). Challenges and opportunities in identifying and characterising keratinases for value-added peptide production. *Catalysts* 10 (2): 184.

8

Bio-bleaching Agents Used for Paper and Pulp Produced from the Valorization of Corncob, Wheat Straw, and Bagasse

Kanya C. H. Alifia[1], Tjandra Setiadi[2], Ramaraj Boopathy[3], Hendro Risdianto[4], Muhammad Irfan[5], and Ibnu M. Hidayatullah[1,6]

[1] *Universitas Indonesia, Department of Chemical Engineering, Faculty of Engineering, Depok, 16424, West Java, Indonesia*
[2] *Institut Teknologi Bandung, Department of Chemical Engineering, Faculty of Industrial Technology, Bandung, 43710, West Java, Indonesia*
[3] *Nicholls State University, Department of Biological Sciences, Thibodaux, LA 70310, USA*
[4] *Center for Industrial Standardization and Services of Cellulose, Agency for Industrial Standardization and Services Policy, Ministry of Industry of the Republic of Indonesia, Bandung, 40258, West Java, Indonesia*
[5] *University of Sargodha, Department of Biotechnology, Sargodha, 40100, Punjab, Pakistan*
[6] *Research Center for Biomass Valorization Universitas Indonesia (RCBV-UI), Universitas Indonesia, Depok, 16424, West Java, Indonesia*

8.1 Introduction

The growth of food and agriculture industry, valued at US$ 75 billion in 2017, leads to a growing problem of agricultural waste [1]. With 5 million tons of agriculture biomass waste and nearly 7 million tons of food waste produced annually, these will be a significant burden on the environment if not treated adequately [1, 2]. Traditionally, these are treated by incineration or dumped into a landfill but both methods will cause air and land pollution [2], but the creation of high-value products from wastes is possible by utilizing lignocellulosic biomass waste as cost-effective raw materials with "waste valorization" concept [1].

According to Ravindran et al., different countries have different types of main agro-industry wastes depending on their main agricultural products [1]. Some of these include corncob, wheat straw, and sugarcane bagasse, which have been researched for their potential to be valorized by identifying their chemical compositions. These agro-industry wastes are rich in carbohydrates, fiber, protein, and lignin, which are highly nutritious for microbes to consume and convert into various products such as enzymes, biofuel, and other biochemical products. If we specifically consider corncobs, wheat straws, and sugarcane bagasse, these agricultural wastes comprise 75–98% of the whole plant because only a fraction of the plant is used in the food industry [1, 2].

Implementing waste valorization can fulfill the demand for industrial enzymes in a cost-effective way because using agro-industry wastes will reduce the enzyme's production cost significantly [3]. Sharma et al. predicted the growth of the global

Chemical Substitutes from Agricultural and Industrial By-Products: Bioconversion, Bioprocessing, and Biorefining, First Edition. Edited by Suraini Abd-Aziz, Misri Gozan, Mohamad Faizal Ibrahim, and Lai-Yee Phang.
© 2024 WILEY-VCH GmbH. Published 2024 by WILEY-VCH GmbH.

market for industrial enzymes would reach US$ 6.2 billion in 2021 and is projected to keep increasing, as industries favor them over chemical catalysts due to the high substrate specificity [3]. One of the industries with a growing demand for industrial enzymes is the pulp and paper industry. The worldwide pulp and paper industries market was predicted to reach nearly US$ 373 billion by 2029, enrolling a CAGR of 0.72% from 2022 [4], due to the increasing demand for paper-based hygiene products, product packaging, and paper-based medicinal materials [5].

Pulp and paper industries include companies that utilize wood as the raw material to produce pulp, paper, paperboard, and cellulose derivatives products. Pulping process starts with wood chips that come from woodyard after debarking process. The main component of wood is cellulose, hemicellulose, lignin, and extractives. Lignin acts as the cementing material between the plant cells, thus it needs to be removed in the pulping process through a bleaching process to obtain the wood fiber [6]. There are several important emerging technologies for pulp bleaching: digester delignification, oxygen delignification, bleaching with chlorine dioxide, bleaching with oxygen, ozone, and hydrogen dioxide. However, these technologies release Adsorbable Organic Halide (AOX), which is bioaccumulative, carcinogenic, and persistent [7].

Bio-bleaching is one of the proposed methods in Totally Chlorine Free (TCF) bleaching and elementally chlorine-free (ECF) bleaching to reduce the AOX in the effluent. Bio-bleaching refers to the pretreatment of pulp with microbes or enzymes to delignify the pulp and to facilitate subsequent bleaching with chemicals. Xylanase, cellulase, hemicellulase, and lignin-degrading enzymes (laccases, etc.) are various enzymes that can be utilized in the bio-bleaching processes [8]. To fulfill the demands for enzymes, there are numerous studies about the production of industrial enzymes, which mainly employ solid substrate fermentation method of lignocellulosic biomass because it is the cheapest, most effective, and most environmentally friendly method [8].

This chapter identifies the characteristics of agricultural biomass as substrate for bio-bleaching enzyme production and the usage of bio-bleaching enzymes such as xylanase, cellulase, and laccase in pulp and paper industry. The microbes (fungi, yeast, and bacteria) that are being used as the producer of bio-bleaching enzymes are summarized and compared. A process flowsheet is designed in SuperPro Designer to illustrate the upstream, midstream, and downstream processing of agricultural wastes for bio-bleaching enzyme production. The techno-economic evaluation is conducted with SuperPro to estimate the process feasibility. Future challenges and outlooks are described to suggest the research direction of bio-bleaching enzyme development.

8.2 Characteristics of Biomass Substrate for Bio-bleaching Enzyme Production

Corncob, wheat straw, and bagasse are solid wastes generated from agricultural industrial activities. Solid agricultural waste is typically stored in one location and dried before being turned into various forms such as green manure. In the twenty-first century, researchers are innovating to turn these three biomass sources into valuable products. Chemical products with high economic value cannot be separated from the characteristics of the raw materials, in this case, the content of lignocellulosic material is one of the determining factors in the acquisition of a product. In the process of producing bio-bleaching agents for the pulp and paper industry, it is not only organisms that produce bio-bleaching agents that are determined, but also the accumulation of lignocellulose from biomass sources (as a cultivation medium). Corncob, wheat straw, and sugarcane bagasse are three types of biomass that have the potential to be employed as growth media for bio-bleaching enzyme-producing organisms (Figure 8.1).

Corn is a plant of the Poaceae family, a staple food for many countries worldwide. The high demand for corn is proportional to the increase in the amount of agricultural waste produced. Corncob comprises 75–85% of the overall fresh corn content [10]. It was reported that global corn production reached 1144.63 MT in 2019, meaning that there is a potential abundance of corncob produced from corn farming as much as 858.47–972.93 MT [11].

Wheat is a food crop that has been cultivated by humans for thousands of years. In some parts of the continent (Europe, East Asia, and North Africa), wheat is the basic ingredient for staple foods that they process into various types of food. According to the Food and Agricultural Organization, wheat is the most needed and used food crop worldwide, followed by rice, corn, and potatoes. Studies conducted by Khan and Mubee [12], wheat plants consist of internodes (68.5%); leaf sheath (20.3%); leaf blades (5.5%); nodes and fines (4.2%); and grains (1.5%). From 1 ton of wheat harvest, at least it produces 0.985 tons of wheat straw. This amount is certainly very important to be used as a valuable product.

Sugarcane (genus *Saccharum*) is an edible plant whose juice is typically used to produce sugar and additions for food and beverage goods. This plant can be found in numerous nations. Brazil has the biggest sugarcane production capacity in the world, surpassing India by a wide margin, which produces over double the amount annually [13]. Bio-bleaching enzymes can be produced from microbial cultivation through corncob, wheat straw, and sugarcane bagasse media. At least, it uses holocellulose, which is abundant in the three biomass, as a carbon source from biological

Figure 8.1 Types of biomass that have the potential as a growth medium for organisms that produce bio-bleaching enzymes: (a) corncob (own photo), (b) wheat straw [9], and (c) sugarcane bagasse.

Table 8.1 Lignocellulosic content in biomass used for bio-bleaching enzyme production.

Biomass	Cellulose (%)	Hemicellulose (%)	Insoluble lignin (%)
Corncob	32–34	55–60	3–10
Wheat straw	34–45	20–32	11–24
Sugarcane bagasse	47	31–38	17–22

Source: Ravindran et al. [1], Saleem Khan and Mubeen, [12], and Hernández et al. [14].

agents that produce bio-bleaching enzymes. The lignocellulosic composition of the three biomass is shown in Table 8.1.

Lignin deconstruction is an essential step in the production of holocellulose, which is easily accessible to hydrolase enzymes. After decomposition, holocellulose can be utilized as a carbon source for organisms that generate bio-bleaching enzymes. Lignin can be deconstructed using dilute acid, steam explosion, liquid hot water, alkaline, ammonia fiber explosion (AFEX), and other ways [15]. A suitable lignin deconstruction approach will result in excellent sugar recovery, allowing bio-bleaching enzyme-producing organisms to proliferate.

8.3 Microbial Sources of Bio-bleaching Enzymes

Microbes are the ideal cell systems for generating industrial enzymes due to the simplicity of microbial cultivation in comparison to plant and animal cells [16]. Bio-bleaching enzymes can be derived from a variety of microorganisms, including fungi, yeast, and bacteria. Various microbial species can produce different enzymes, including cellulase, xylanase, and laccase. This enzyme is utilized by microbes to digest the substrate, and this process becomes a fundamental concept of enzyme manufacturing. Numerous studies have examined microbial strains capable of generating enzymes on medium containing specific carbon sources. The following section will be followed with a variety of microorganisms capable of producing enzymes such as xylanase, laccase, and cellulase.

8.3.1 Fungi

Fungi are multicellular microorganisms that secrete intracellular and intercellular enzymes for digestion. Fungi are abundant because they are the dominant decomposers in the ecosystem, which are able to obtain nutrients from organic matter in their surroundings, specifically lignocellulose. This ability is a result of fungi's ability to create lignocellulose-degrading enzymes. Fungi-produced degrading enzymes are classified into two categories: hydrolytic and oxidative. Hydrolytic enzymes breakdown polysaccharides, whereas oxidative enzymes breakdown lignin and opens the phenyl ring. Fungi-produced enzymes are applicable and effective in the pharmaceutical industry, industrial operations, bioremediation, and agricultural applications [17]. Optimization will boost the rate of fungus growth and enzyme activity.

Enzyme synthesis is greatly controlled by substrate composition and environmental conditions.

Laccase is one of the enzymes produced by the lignin-decomposing fungus. Laccase enzyme is generated by *Neurosporacrassa* and several other fungi, including *Marasmius* sp., *Junghuhnia nitida*, *Pycnoporus sanguineus*, and others. Laccase formation in fungi is primarily influenced by plant pathogenic fungi, pigmentation, detoxification, and lignin degradation [18]. It is possible to transfer laccase-producing genes from one fungal species into fungal cells of a different species in order to enable laccase production by fungi whose wild strains lacked laccase activity. Fungi also manufacture cellulase enzymes. *Aspergillus* and *Trichoderma* are xylanase and cellulase-producing fungi [19, 20].

8.3.2 Yeast

Yeast is a microbe that belongs to the fungus classification, but it is unicellular. Yeast has several benefits over multicellular fungi, including simpler genome structures that enable genetic alteration easier, point centromeres, relatively few complicated gene sequences, and the capacity to be cultivated as haploid organisms [21]. *Candida* sp. is one of the most common yeasts that produce the enzymes cellulase and xylanase [22]. Carbon sources, carboxymethylcellulose, and glucose are utilized in the cultivation of yeast that produces cellulase enzymes. Xylanase enzyme producers cultivate yeast with xylan carbon sources from beechwood, oat spelt, bamboo leaves, and so on.

Laccases have been found as being manufactured by different forms of yeast, and some types of yeast can produce laccase naturally. Efforts have been made through genetic modification to enable the development of laccase enzymes from other fungi in yeast. *Yarrowia lopolytica* is an instance of a laccase producer [18]. Laccase expression in yeast is exceedingly uncommon; laccase production often requires genetic modification. *Pichia pastoris* is the most prevalent yeast utilized as a host cell for laccase enzyme expression [23].

8.3.3 Bacteria

Bacteria are unicellular prokaryotic organisms. Bacteria are ubiquitous and are even capable of surviving in harsh environments. This ability is a result of the bacterial genome, which is more compact and includes less intergenic DNA than that of eukaryotes. This genetic trait provides an excess of short cell division times for its quick repeatability and minimizes the energy required for nucleotide synthesis [23].

Bacteria generate enzymes to degrade complex molecules that provide nutrients for their growth. There are multiple types of bacteria that produce laccase, xylanase, and cellulase naturally. Laccase-producing bacteria include *Escherichia* species, *Bacillus* species, and *Streptomyces* species [24]; *Bacillus, Cellulomonas, Micrococcus, Staphylococcus*, and other bacteria can produce xylanase [25]; while *Pseudomonas, Ruminococcus, Clostridium*, and *Acidothermus* are bacteria that are able to generate cellulase [26]. Commonly, genes are modified to enable the expression of enzymes in nonproducing strains.

The ideal ranges of activity for the resulting cellulase, xylanase, and laccase enzymes differ. At 80 °C and a pH of 6.5, xylanase from *Escherichia coli* expressing the genome of *Caldicoprobacter algeriensis* is the most active [27]. Compared to enzymes from fungi, the enzymes generated by these bacteria are significantly more stable at high temperatures. The research on thermophilic enzyme stability enables the use of enzymes in processes involving high temperatures.

8.4 Bio-bleaching Enzymes and Their Usage in Pulp and Paper Industry

8.4.1 Xylanase

Xylanases are classified as hemicellulolytic enzymes which can breakdown the β-1,4 backbone of polysaccharide xylan, a component constituting lignocellulosic plant cell walls. Xylan is a heteropolysaccharide containing O-acetyl arabinosyl and 4-O-methyl-D-glucuronic substituents. Xylan is springier than cellulose due to its structure; a twofold extended ribbon-like linear polymer chain with intrachain hydrogen bonding, in which the repeating xylopyranosyl groups are substituted at different carbon positions with multiple sugars and/or acidic compounds [28].

Xylanases are categorized according to their molecular weight and isoelectric points, crystal structure, and kinetic properties, as well as the substrate specificity and product profile [28]. Due to the variety of xylanases, these enzymes are classified in different glycoside hydrolase (GH) families GH5, GH8, GH10, GH11, GH30, GH43, and other families in the Carbohydrate-Active Enzymes (CAZy) database, but the best-characterized family is GH11 due to their high substrate selectivity and specificity, wide range of optimum pH from 2.0 to 9.0, and wide range of optimum temperature from 35 to 85 °C [29].

Fungal Xyl-11 is typically acidophilic while bacterial Xyl-11 ranges between mesophilic and alkalophilic. However, the optimum pH depends on several structural and electrostatic features in Xyl-11, which makes it challenging to determine a general correlation between the enzyme structure and the optimum pH. As for the enzyme's optimum temperature, this can be illustrated by the Lumry–Eyring model in Eq. (8.1) about enzyme thermostability [29]. In the first reversible reaction, the enzyme becomes unfolded and turned inactive as the temperatures increases past its optimum point. However, prolonged high temperature will lead to enzyme aggregation and irreversible denaturation [29].

$$E^{\text{native}}_{\text{ACTIVE}} \underset{k_2}{\overset{k_1}{\rightleftarrows}} E^{\text{unfolded}}_{\text{INACTIVE}} \xrightarrow{k_3} E^{\text{aggregated}}_{\text{INACTIVATED}} \tag{8.1}$$

Thermostable Xyl-11 typically contains one or several of these factors: (i) presence of disulfide bridges, (ii) ionic and aromatic pairs or interactions, (iii) deep water molecules, (iv) high packing of the enzyme, and (v) oligomerization of the chain [29]. Due to the hemicellulolytic properties of xylanases, this enzyme is significantly utilized in the pulp and paper industry to boost the bleaching process with bio-bleaching [30]. Xylanases degrade the hemicellulosic fraction in the lignocellulosic pulp and catalyze the fibrillation of the pulp [28].

The reaction between xylanase and the lignocellulosic paper pulp begins with removing chromophoric groups from the pulp, then the partial hydrolysis of the lignin–carbohydrate complexes, which split the linkage between the residual lignin and carbohydrate. If there is cellulose present in the pulp, the bonding strength between fibers is decreased without lowering the mechanical strength. However, if there is no cellulose, then xylanase will increase the hydrolysis of hemicellulose to enhance the lignin removal. The removal of xylan in the pulp helps to increase the pulp's porosity, allowing free diffusion of bleaching chemicals, which ended up boosting the subsequent bleaching stages [28]. The reaction of lignin removal is dynamic at higher temperatures and basic pH, thus illustrating the importance of engineering xylanase properties to be more thermostable and alkalophilic [28].

There are multiple benefits to incorporating xylanase in the bio-bleaching treatment. The hemicellulosic degradation of the pulp leads to the forming of fibers within the pulp, as known as the fibrillation process. This increased the pulp's porosity to improve the diffusion rate of bleaching chemicals into the pulp, which leads to enhanced pulp and paper brightness and strength while consuming 20% less bleaching chemicals, which results in lowering the toxicity of the effluent and pollution load. Currently, there are numerous enzyme manufacturers in the world that produce xylanase for pulp bleaching as shown in Table 8.2 [28].

8.4.2 Cellulase

Cellulases are synergetic enzymes with the capability to degrade cellulose into glucose and/or other oligosaccharide compounds. They are classified into endoglucanase, cellobiohydrolase/exoglucanase, and β-glucosidase [31]. Exoglucanases mainly react with the cellulose chain and produce β-cellobiose. Meanwhile,

Table 8.2 List of global xylanase producers for pulp bleaching.

Producer	Product name
Biocon, India Bangalore	Bleachzyme F
Sandoz, Charlotte N.C.	Cartzyme
Clarient, U.K	Cartzyme MP
Genercor, Finland	Irgazyme 10 A, Irgazyme 40-4X
Novo Nordisk, Denmark	Pulpzyme (HA, HB, HC)
Voest Alpine, Austria	VAI Xylanase
Thomas Swan, U.K.	Ecozyme
Rohm, Germany	Rholase 7118
Alko Rajamaki, Finland	Ecopulp

Source: Adapted from Walia et al. [28].

endoglucanases give sporadic attacks to the internal O-glycosidic bonds to produce glucan chains at varying lengths. Lastly, β-glucosidases act specifically on the β-cellobiose disaccharides to produce glucose [32]. From the endo–exo energy model, it is shown that endoglucanases attack multiple points of the lignocellulosic chains until new sites are exposed to be attacked by cellobiohydrolases, which produce cellobiose as the main product [31].

Considering the mechanism of action of these three subgroups of cellulases, a combination of different cellulases can efficiently degrade cellulosic biomass by converting cellulose into cellobiose, then converting β-cellobiose disaccharides to produce glucose [31, 33]. Cellulases can also be paired with hemicellulases to modify the fiber properties [33]. Different cellulases are typically combined during enzyme treatment on lignocellulosic biomass to express higher enzymatic activities. The cellulase activity is described as its capability to digest crystalline cellulose into its derivatives [31].

The conventional pulping processes are conducted in a mechanical way, such as grinding wood into wood pulp and refining the wood pulp. This process uses high energy for the wood chopper and grinder, while, on the other hand, biochemical pulping with cellulases will reduce the energy cost by 20–40% as well as improve the handsheet strength properties. When combined with xylanases, cellulases can be applied for the de-inking process of paper wastes. Utilizing enzymes for the de-inking process can reduce the alkali usage, improve fiber brightness, enhance the strength properties, increase the pulp freeness and cleanliness, and reduce the fine particles in the pulp. However, enzymes should only be used moderately to avoid a significant reduction of the fiber bond strength due to the excessive hydrolysis of fine wood pulp [1, 2, 33]. Overall, cellulases are a desirable group of enzymes to be used in the pulp and paper industry because of their ability to modify the lignocellulosic fiber in paper processing [30], modify biochemical pulping of coarse pulp, improve the paper's strength, remove inks and coatings from papers, and assist the manufacturing process of biodegradable paper products [cardboard, sanitary paper, and paper towels) [33].

8.4.3 Laccase

Laccases are monomeric glycoproteins belonging to the oxidoreductase group, its official name is EC 1.10.3.2, benzenediol: oxygen oxidoreductases [34]. These enzymes belong in the multicopper oxidase family as they contain four copper centers: one type 1 Cu (T1), one type 2 Cu (T2), and a coupled binuclear type 3 (T3) Cu center. These copper centers play a role in becoming the active binding site with the respective substrates. The T2 and T3 sites form a trinuclear Cu cluster for the reduction of oxygen, while the T1 Cu atom oxidizes the reducing substrate and transfers electrons to the T2 and T3 Cu atoms [35].

There are approximately 90 laccase structures discovered through X-ray crystallography, including the native or wild-type enzymes, mutated structures, and enzymatic complexes with substrates, inhibitors, and products. The typical substrates for laccase's oxidation are a variation of compounds coupled to the four-electron reduction of molecular oxygen to water [34]. The laccase mechanism of action involves two individual sites that bind the reducing substrate and oxygen with four catalytic copper atoms configured as trinuclear clusters of the T1Cu (where the substrate oxidation takes place), the T2Cu, and the two T3Cu. The oxygen is then reduced to two molecules of water as the substrate is converted [34].

The classification of laccases is divided according to their functionality, substrates, and redox potential. The biological function of laccase's varies according to their origin and the organism source's life stages [34]. Laccase is generally found in higher plants and fungi, such as basidiomycetes, white rot fungi, and ascomycetes. The white rot basidiomycetes are the most efficient degraders of lignin and enzymes, which are implicated in the lignin degradation are lignin peroxidase, manganese-dependent peroxidase, and laccase [36]. Fungal laccases are involved in morphogenesis, lignin degradation, stress defense, and fungal plant–pathogen/host interactions [34]. Fungal laccases have higher redox potential than bacterial or plant laccases (up to +800 mV). Thus, fungal laccases are involved in the degradation of lignin or in the removal of potentially toxic phenols arising during lignin degradation [36].

Meanwhile, plant laccases are developed for lignin polymerization and give appropriate wound responses [34]. In plants, laccases play an important role in lignification whereas, in fungi, laccases have been implicated in many cellular processes including delignification, sporulation, pigment production, fruiting body formation, and plant pathogenesis [36]. Lastly, bacterial laccases play several roles in pigmentation, morphogenesis, toxic oxidation, and protection against oxidizing agents and UV light [34].

Laccase substrates include aromatic compounds, metal ions, and organometallics. Moreover, the scope of laccase substrates can be widened to higher-redox potential compounds with the help of diffusible electron carriers (laccase redox mediators) that constitute the laccase-mediator system (LMS) [34]. In the context of their redox potential at the paramagnetic type 1 copper T1Cu (E00 T1), laccases are usually classified as low-, medium- or high-redox potential. The high-redox potential laccases (HRPLs) generate the most interest as they can oxidize a wider range of substrates than their low and medium counterparts [34].

The oxidation rates of laccases are also more than 10-fold of the oxidation rates of lignin peroxidase or manganese peroxidase. However, laccase itself needs the presence of radical mediators such as 1-hydroxybenzotriazole (HBT) to create LMS to effectively demethylate kraft pulps. When only laccase is involved, side-chain oxidation and oxidative coupling reactions occurred but with LMS there were formations of o- and p-quinones, demethylation, aromatic-ring cleavage, and oxidative coupling reactions. The LMS reaction pathway was driven toward side-chain oxidation and oxygen addition products, while suppressing the formation of condensed structures [35].

There are numerous benefits of using laccase in the pulp and paper industry, which include the increased throughput in mechanical pulping, enhanced paper strength, and reduced pitch problems [34]. Laccase can enhance pulp delignification process by employing the enzyme in the pretreatment of wood pulp [29, 34].

It can also be used as an enzymatic bio-bleaching agent of flax pulp, bio-based decolorization agent, and induce the grafting of phenols into flax fibers for paper production [34]. Laccases are generally thermostable and are low in substrate specificity, making it suitable to oxidize an array of aromatic substrates at various temperatures [35]. The combination of these enzyme properties makes laccases suitable to be applied in various sectors, such as bioremediation, biofuel cells and biosensors, textiles, pulp and paper, and food industries [35].

8.5 Bioprocessing of Agricultural Wastes for Bio-bleaching Enzyme Production

8.5.1 General Block Flow Diagram

In general, the production of bio-bleaching enzymes from agricultural waste is separated into upstream and downstream processing as shown in the block flow diagram (Figure 8.2). Upstream processes include the biomass pretreatment (washing, drying, size reduction, and sterilization) and the inoculum preparation (starter agar cultivation and inoculum preparation), which will enter the main fermenter together with additional air supply and nutrients. The fermenter is a bioreactor where the main fermentation happens, the microbes feed on the biomass and other nutrients as the substrate to produce enzymes [37, 38].

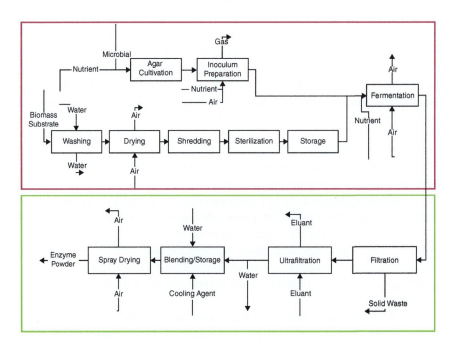

Figure 8.2 Block flow diagram of agricultural waste bioprocessing into bio-bleaching enzyme. Source: Redrawn by Alifia, adapted and modified from Ref. [37, 38].

The upstream processing line is separated into two lines prior to entering the fermenter. The first line is for the biomass pretreatment. In this chapter, our focused biomass is agricultural wastes from the food industry such as corncobs, wheat straws, and sugarcane bagasse. The specific case study being analyzed is wheat straw valorization for enzyme production, but the general processes are applicable for corncobs and sugarcane bagasse as well. The wastes need to undergo washing process to clean out the impurities from the farming process and the food processing industry. After washing, the biomass is dried out to reduce the water content before the shredding process. Some larger biomass such as sugarcane bagasse and corncobs may need multiple steps of shredding as compared to wheat straws. Once the biomass are shredded into smaller particles, they undergo sterilization in an autoclave. This is an important step to avoid wild microorganisms and impurities that can hinder the fermentation step [37, 38].

The second line is the inoculum preparation. Before preparing the inoculum, careful considerations should be taken when choosing the microorganism. Bacteria, yeast, and fungi all have different advantages and disadvantages as the agent for bio-bleaching enzyme production from agricultural waste valorization process. The main considerations for choosing the microorganism include what enzyme is being produced, what substrate is being used, and how big is the production scale. The microbe will then be cultivated in an agar plate along with all the necessary nutrients to increase its reproduction rate. This step will be repeated in a small-scale fermenter for the inoculum preparation step. After it reached a certain cell concentration, the cells and nutrients can be transferred into the fermenter to be mixed with the shredded biomass pellets under a certain operating condition. One batch of fermentation typically lasts for a week and the product will be a mixture of water, cells, nutrient, biomass substrate, and enzyme product [37].

Once the fermentation batch is over, the product will be transferred to the downstream processing section. The downstream processes include the plate and frame filtration, ion-exchange chromatography, blending and storage, and spray dryer. Filtration is the first step to separate the wheat chaff pieces from the soluble mixture. Since there are two fermentation techniques available for enzyme production, namely submerged and solid-state fermentation, each technique will have different requirements for the downstream filtration. Submerged fermentation uses higher amount of water as opposed to solid-state fermentation, thus the workload for the plate and frame filtration is higher than the solid-state fermentation. Multiple filtration steps maybe necessary to separate impurities at different particle sizes, therefore this process employs both plate–frame filtration and ultrafiltration [37].

The water-soluble mixture of cells, nutrients, biomass, and enzymes is passed through the ultrafiltration to separate the enzyme product from other impurities. The remaining cells, nutrients, and biomass maybe transferred to a waste treatment plant or recycled back to the main process. Ultrafiltration was chosen over ion-exchange chromatography, despite it being the most common method for enzymes, peptides, proteins, and other biopolymers separation. This was because industrial-rate enzymes have less purity specifications compared to medical-rate or food-rate enzymes, thus highly specific separation method, such as chromatography is not necessary [32].

Once the enzymes are purified, they will be blended with water and coating agent (maltodextrin) to form microcapsules. These microcapsules will be turned into powder in the spray dryer, removing nearly 100% of its water content to stabilize the enzyme activity in their powder form. For further understanding of each processing unit, a flowsheet according to the general block flow diagram was modeled in SuperPro Designer version 12.0 as shown in Figure 8.3. This software is a powerful tool for modeling and simulation of bio-based processes, it has an extensive option

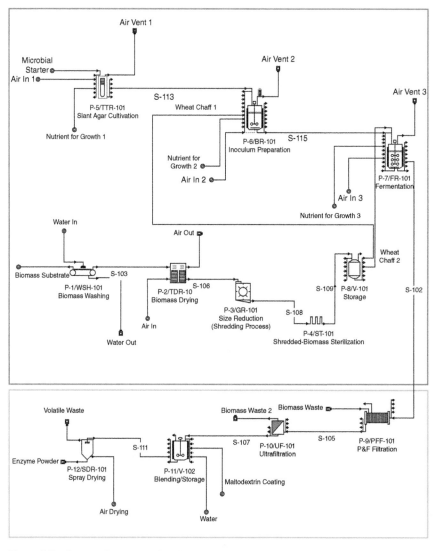

Figure 8.3 Process flowsheet of agricultural waste valorization for bio-bleaching enzyme. Source: Redrawn by Alifia, adapted and modified from Refs. [37, 38].

of processing units from upstream to downstream as well as a complete database of raw materials. It can also analyze the stream mass balances and their economic feasibility.

8.5.2 Upstream Processing

The detailed description of the raw materials and other streams as well as the operating conditions of the upstream processing units is benchmarked from Jovanović et al. [37] and Risdianto et al. [38]. In the study by Jovanović et al., a laboratory experiment of enzyme production from the fermentation of wheat chaff by the fungi *Trichoderma reesei* QM 9414 was conducted in two cultivation methods (submerged and solid-state fermentation). The results from the experiment are used to model the process on a larger scale using SuperPro Designer version 6.0 to compare the techno-economic feasibility of both fermentation methods. The enzymes being produced from this process are amylase, cellulase, and xylanase [37]. For our case study, the chosen method is submerged fermentation and it is assumed that amylase and xylanase have similar properties with cellulase so that the biochemical processes and reactions can be modeled in SuperPro Designer version 12.0.

The wheat chaff in the study is assumed to be a combination of dried wheat husk and chopped wheat straw chaff, which was mentioned in the section about potential agriculture waste. For a base case scenario, the wheat chaff being used is 375 kg in 12 500 kg liquid medium in the main fermenter (assumed to have the same density as water to give 12,500 l working volume) [37]. The biomass should be pretreated first by washing, drying, shredding, and sterilization. The drying process is used to reduce 98–99% of the water content from the wheat chaff as well as the remaining water from the washing process. The shredding process is mainly for the straw chaff because the wheat husk is already small enough at 1–1.5 cm in size. The straw chaff is shredded into 1 cm short pieces by the shredder [38]. The biomass is sterilized in an autoclave at 121 °C and 2.1 bar for 30 minutes [37].

In the inoculum preparation line, the fungi *T. reesei* QM 9414 is cultured in a PDA (Potato Dextrose Agar) medium along with additional nutrients (0.5% $(NH_4)_2SO_4$ and 1.36% K_2HPO_4) in 1000 ml of distilled water. This step is conducted at 28 °C for three to four days until the cell concentration reached 10^6 spores per gram in the culture [37]. The next step is inoculum preparation in a larger vessel (5–1000 l). The contents of the vessel include the transferred fungi from the agar culture, additional nutrients (0.5% $(NH_4)_2SO_4$ and 1.36% K_2HPO_4), shredded wheat chaff (30 g/l), 1% NaOH as buffer to maintain the pH at 4.5 ± 0.1, and water [37]. Typically, there are 2–3 fermenters for the inoculum preparation but in the case study only one 100 l inoculum preparation fermenter is used to simplify the process flowsheet. In actual implementation, two fermenters maybe used to gradually scale up the process from 1 l fermenter, for example, with 100 l fermenter then 1500 l fermenter before entering the 12,500 l main fermenter.

It is assumed that the fungi cells reproduce according to a stoichiometric growth reaction by converting the wheat chaff substrate into cell biomass, but the operating condition in the inoculum fermenter is maintained to limit the reaction rate for the fermentation process until it is negligible. By calculating the carbon, hydrogen, and oxygen amount in wheat chaff according to its constituent (54% cellulose, 18% water, 16% lignin, 7% ash, and 5% protein) [37], the molecular formula of wheat chaff can be estimated. Once the molecular formula is known, the coefficients for ammonia, oxygen, cell biomass, carbon dioxide, and water can be calculated. The stoichiometric growth reaction is shown below (Eq. (8.2)) [39].

$$C_{20}H_{27}O_{11} + NH_3 + 21O_2 \longrightarrow CH_2ON + 19CO_2 + 14H_2O \qquad (8.2)$$

In the 12,500 l main fermenter, the sterilized wheat chaff is mixed with the product stream from the inoculum preparation fermenter. The assumption that was made in the main fermenter is that fermentation (Eq. (8.3)) occurs in the vessel alongside cell growth (Eq. (8.2)). The simplified stoichiometric fermentation reaction is mentioned in research by Jourdier et al. [39] about a study of cellulase production using *T. reesei*. In this study, the chemical components of the fungal cell were estimated, while the mass content of the secreted enzymes was measured to give an estimate of the enzyme's chemical formula and stoichiometry coefficient [39]. With some adjustments to the coefficient shown in the reaction equation, the result of this analysis is portrayed below (Eq. (8.3)).

$$10CH_2ON + 5O_2 + NH_3 \longrightarrow 3C_2H_5O_2N + 4CO_2 + 6H_2O \qquad (8.3)$$

The fermentation process in the main fermenter is carried out at 30 °C, 200 rpm of agitation speed, and lasts for seven days per batch. The temperature is maintained at 30 °C because the production of enzymes such as cellulase and xylanase is in line with the fungi cell growth at room temperature (±30 °C) without the presence of light [38]. Maintaining the temperature is important to avoid enzymes and cell denaturation when it surpasses the optimum temperature, as well as enhancing the fermentation reaction rate to produce the enzymes [38].

8.5.3 Downstream Processing

The stream from the main fermenter contains a mixture of *T. reesei* cells, biomass substrate (fermented wheat chaff), enzymes, and nutrients dissolved in the water. Most of these constituents are water-soluble except the remains of fermented wheat chaff, therefore it should be separated first by filtration. Plate and frame microfiltration is the chosen method because the particle size of fermented wheat chaff is estimated to be bigger than 10 µm, industrial enzymes also need low requirement of purity [32]. The biomass substrate remains are collected as solid wastes, while the water mixture is transferred to the ultrafiltration step to separate the enzymes, cells, and nutrients from a large amount of water content. Approximately, 2–5% of the enzymes will denature after plate and frame filtration and ultrafiltration due to shear stress [37]. The filtrate from the ultrafiltration contains water, enzymes, and some soluble impurities, such as nutrients and cells.

Once the enzymes are purified, they will be blended with water and coating agent to form microcapsules in a liquid mixture. This liquid mixture will be transferred to the spray dryer. Spray drying is a cost-effective convective drying method for heat-sensitive substances, such as enzymes and bacteria, the low to medium temperature reduces the risk of severe heat damage that will decrease the enzyme's activity [40]. According to a study by Schutyser et al. the drying conditions for β-galactosidase (a member of cellulases family) include the usage of single droplet drying method with 8 μl droplets, the air temperature is 90 °C, air flow rate at 1 m/s, the solid carrier concentration at 20%, and 7% residual moisture content [40]. This data maybe used for spray drying of enzymes from the cellulases family. The powdered enzyme product can be packaged accordingly in the packaging line. In the industry, a quality control step is necessary after the product is finished. The quality of the enzymes will be assessed by undergoing an enzymatic assay to analyze the enzyme's activity, ensuring that the enzyme is still active after the spray drying process. Using the available tools in SuperPro Designer version 12.0, the mass balance analysis of all streams and the economic evaluation of the production process are explained further in the following subchapter.

8.6 Techno-economic Evaluation

8.6.1 Technical Analysis

The purpose of this technical analysis is to identify which unit contributes to the process bottleneck and propose a strategy to optimize the process. The mass balances of all streams and processing units can be automatically calculated by SuperPro Designer. This process operation is batch, the total occupation time per cycle is 394 hours (more than 16 days), with eight days of which happening in the biomass storage and seven days happening in the main fermenter. The annual number of batches is 61. Thus, the process bottlenecks are the biomass storage and the main fermenter since they limit the maximum possible batches per year [37].

The overall batch begins with the wheat chaff biomass preparation (washing in P-1, drying in P-2, grinding in P-3, and sterilization in P-4). At the same time, the agar cultivation of *T. reesei* begins in P-5 where it is cultivated for three days in PDA medium. After that, the inoculum is upscaled in the bioreactor P-6 for five days to increase the cell numbers. Meanwhile, the processed wheat chaff is held in the biomass storage for three days until 25% of the content is ready to be transferred to the bioreactor P-6, and for another five days before the remaining 75% transferred to the main fermenter (P-7). The fermentation process in the main fermenter takes seven days, then the product is separated from the liquid mixture in the downstream processing units, such as plate and frame filtration (P-9) and ultrafiltration (P-10). The enzyme is mixed with water and coating agent (maltodextrin) in a blending vessel (P-11) and then transferred to the spray dryer (P-12).

The Gantt chart for the occupation time of each equipment in a single batch is shown in Figure 8.4, while the detailed breakdown of operations in each unit is shown in Figure 8.5. To address the issue of process bottleneck, the two approaches that can be used include scheduling adjustment into staggered mode to minimize idle time of the processing units, as well as adding more unit(s) to operate simultaneously [37]. In this case study, two main fermenters and two biomass storage are used simultaneously. When the scheduling is adjusted into staggered mode, the annual number of batches increases as shown in Figure 8.6. Although it still takes

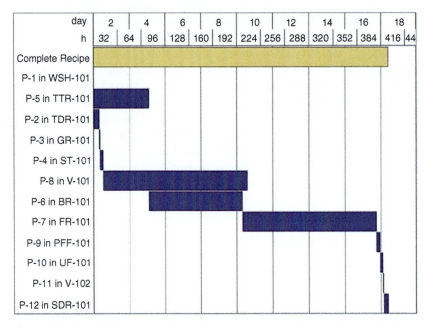

Figure 8.4 Overall Gantt chart of a single batch enzyme production from agricultural waste. Source: Alifia (author).

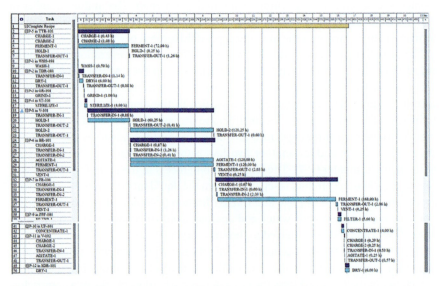

Figure 8.5 Gantt chart and operations breakdown of a single batch enzyme production. Source: Alifia (author).

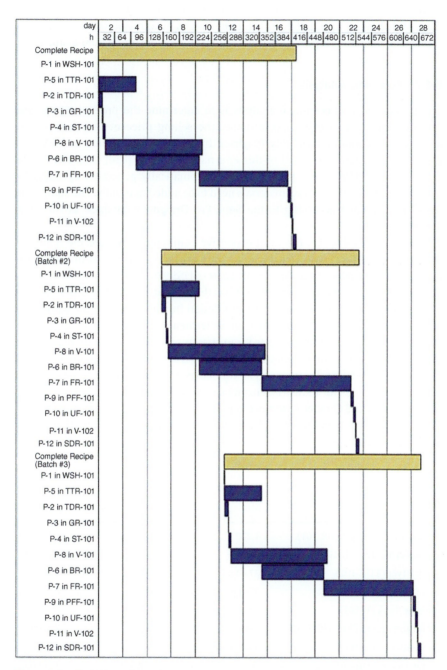

Figure 8.6 Gantt chart for multiple batches in a staggered mode for enzyme production. Source: Alifia (author).

more than 16 days to complete one cycle of batch recipe, operating in stagger mode allows three cycles to complete in less than 28 days. As a result, this will lower the capital and operating expenditure (OPEX) for this enzyme production plant.

8.6.2 Economic Analysis

The purpose of this economic analysis is to determine the production cost per unit enzyme product (USD/kg), estimate the selling price of the enzyme, and analyze the economic feasibility (payback time, net present value or NPV, and internal rate of return or IRR) of this enzyme production plant [37]. The capital expenditure (CAPEX) and the OPEX are automatically generated by SuperPro Designer, but the prices of raw materials need to be added manually. The equipment costs are estimated according to the SuperPro Designer's database as shown in Table 8.3.

According to Table 8.3, the most expensive equipment costs are the main fermenters at US$ 1,872,000. To achieve 12,500 l working volume and maximize the productivity, two 8000 l fermenters with a cost of US$ 936,000/unit are used simultaneously. The 1200 l inoculum bioreactor has the highest unit cost at US$ 1,299,000. This is in line with the heuristics of plant design in which most of the equipment costs should fall on the main bioreactor or fermenter where the core of the process happens [41]. Overall, the costs for upstream processing units are higher than the costs for downstream processing units. Compared to bio-based products that are used for pharmaceutical sector such as antibodies and vaccines, the separation and purification process for industrial enzymes requires less robustness and specificity. This is because the Quality Assurance standards for biopharmaceuticals are stricter to ensure their safety in human body, as opposed to industrial enzymes that are not used for human consumption [32].

Table 8.3 List of equipment purchasing costs.

Unit name	Size	Unit	Unit cost (USD)	Cost (USD)
Inoculum Bioreactor (BR-101)	1,200 l	1	1,299,000	1,299,000
Main Fermenter (FR-101)	8,000 l	2	936,000	1,872,000
Tray Dryer (TDR-101)	27.44 m^2	1	110,000	110,000
Grinder (GR-101)	376 kg/h	1	81,000	81,000
Storage Tank (V-101)	500 l	2	54,000	108,000
Plate and Frame Filter (PFF-101)	1.04 m^2	1	27,000	27,000
Ultrafilter (UF-101)	8.40 m^2	1	42,000	42,000
Spray Dryer (SDR-101)	200 l	1	109,000	109,000
Unlisted Equipment				986,000
Total				4,928,000

Source: Alifia (author).

The total CAPEX is US$ 31,216,000 and it consists of the working capital (the total OPEX during the first month) and the total direct fixed capital (DFC). DFC is split into total plant cost (TPC) and contractor's fee and contingency (CFC). TPC is divided into direct costs (TPDC) and indirect costs (TPIC). The main contributor to TPDC is the equipment purchasing costs. The result shows similarity with the case study on submerged fermentation of wheat chaff by Jovanović et al. [37] with slightly lower building cost and installation costs, which are due to the smaller fermenter size. In this chapter, the solid-state fermentation is not simulated due to its complexity in design and operation. The total CAPEX calculation is shown in Table 8.4.

The total operating costs (OPEX) consist of the costs for raw materials, labor, facilities, and the Quality Control/Quality Analysis/Laboratory. The first month of the total OPEX is defined as the working capital because it is needed as the initial capital investment to run the production plant. The annual costs of raw materials are defined from the SuperPro database and other trusted supplier's database and shown in Table 8.5.

The utility costs, labor costs, facility-dependent costs, and QC/QA/Lab costs are estimated by SuperPro with the values shown in the summary of OPEX in Table 8.6. The raw materials are rounded up to US$ 44,000. The main contributors to OPEX are the labor costs and facility costs. The raw materials are relatively much cheaper because the wheat chaff has a very low price.

The unit production costs can be obtained by dividing the annual OPEX by the annual enzyme production rate. With 47 kg of enzymes produced per batch and 61

Table 8.4 Total CAPEX calculation.

Item	Costs (USD)
Equipment Purchasing Costs	4,928,000
Installation	1,758,000
Process Piping and Instrumentation	3,696,000
Insulation and Electricity	641,000
Buildings and Yard Improvement	2,956,000
Auxiliary Facilities	1,971,000
Total Plant Direct Costs (TPDC)	**15,949,000**
Engineering	3,987,000
Construction	5,582,000
Total Plant Indirect Costs (TPIC)	**9,570,000**
Contractor's Fee and Contingency (CFC)	**3,828,000**
Total Direct Fixed Capital (DFC)	**29,347,000**
Total Working Capital	**1,869,000**
Total CAPEX	**31,216,000**

Source: Alifia (author).

Table 8.5 Annual costs of raw materials.

Bulk material	Unit cost (USD)	Annual amount (kg)	Annual cost (USD)	%
PDA	2	6,200	12,400	28.5
Ammonia	0.2	811	162	0.4
Sulfuric Acid	1.98	111	220	0.5
Sodium Hydroxide	44.08	443	19,527	44.8
Dipotassium Phosphate	1.2	446	535	1.2
Maltodextrin	0.5	1,612	806	1.9
Trichoderma reesei	10	363	3,630	8.3
Wheat Chaff	0.27	23,312	6,294	14.4
Water	0	73,618	0	0
Air	0	730,328	0	0
Ash	0	1,448	0	0
Total Raw Materials Cost			43,574	100

Source: Alifia (author).

Table 8.6 Summary of annual OPEX calculation.

Cost item	Cost (USD)	%
Raw Materials	44,000	0.28
Labor-Dependent	4,341,000	28.35
Facility-Dependent	5,530,000	36.11
Laboratory/QC/QA	651,000	4.25
Consumables	4,738,000	30.94
Utilities	11,000	0.07
Total OPEX	15,315,000	100

Source: Alifia (author).

batches operating in a year, 2,900 kg of enzymes are produced annually. The net unit production cost is US$ 5270/kg enzyme product. This is 125 times higher than the result in the experimental study by Jovanović et al. [37] because the production scale in this case study is 1000 times smaller. However, the net unit production cost is 15–50 times higher than the industrial enzymes production cost in general (US$ 100–300/kg). To assess whether this unit production cost is economically feasible or not, an analysis of the payback time, NPV, and IRR need to be conducted based on three scenarios: the average industrial grade cellulase enzyme bulk price (US$ 16/kg in average), pure cellulase enzyme from *Trichoderma* sp. (US$ 315 per 1 g vial in average), and the estimated bulk price of cellulase enzyme from *Trichoderma* sp. (US$ 8000/kg). Only cellulase enzyme prices are considered because it is too complex to model a multiproduct biochemical reaction in SuperPro Designer 12.0.

For the first scenario, the enzyme is sold at US$ 16/kg. This is a more realistic price range because the enzyme being produced is supposed to be used for the pulp and paper industry, therefore high purity enzymes is not necessary. According to this scenario, the annual revenue is US$ 46,000, the gross margin is −32,800%, the return on investment (ROI) is −39.92%, and the NPV at 7% interest is negative US$ 120,500,000, showing that the production plant is experiencing a huge loss to the point that the payback time and IRR cannot be estimated.

For the second scenario, the enzyme is sold at US$ 315,000/kg. Although this is not a realistic price range for industrial enzymes, the purity level of the product is possible to be adjusted to produce high purity cellulase enzymes by adding more downstream processing units. The annual revenue is US$ 914.6 million, the gross margin is 98.33%, the ROI is 1,737.55%, and the NPV at 7% interest is US$ 3.8 billion. As a result, the payback time is 0.06 years (22 days) and the IRR after taxes is 350%. This shows that this production plant is highly profitable, but this is not a realistic estimate of economic feasibility.

For the third scenario, the enzyme is sold at US$ 8,000/kg. This is a more realistic scenario, which gives annual revenue of US$ 23.2 million, the gross margin is 34.15%, the ROI is 24.18%, and the NPV at 7% interest is US$ 20.2 million. As a result, the payback time is four years and the IRR after taxes is 16.05%. Positive value of IRR shows that the production plant is profitable and within four years the capital investment will be returned by the revenue.

According to this economic analysis, this production plant is financially feasible when it is used to produce high-purity cellulase enzymes and sold at its bulk price. In-depth techno-economic analysis and sensitivity analysis with different production scales and enzyme prices maybe useful to assess how different factors contribute to the profitability and productivity of the enzyme production plant.

8.7 Challenges and Future Outlooks

The bleaching procedure on pulp seeks to remove lignin and, if lignin is still difficult to remove, then alter its structure so as to improve the physical and chemical qualities of the pulp (clearer and brighter hue) [42]. Massive use of dioxin, furan, and other chlorine-based chemicals generates extremely challenging environmental challenges. Other bleaching techniques may employ ozone, peroxide, or oxygen. However, the bleaching agent is less cost-effective, needs high pressure and temperature, and diminishes the paper's physical characteristics [43]. By replacing the bleaching agent with a bio-bleaching enzyme, it is possible to eliminate some of these problems.

Because bleaching enzymes are often produced by moderate thermophilic microorganisms, the majority of bio-bleaching enzymes routinely employed in the textile, pulp, and paper industries operate at temperatures below 60 °C [44]. High temperatures necessitates more energy to decrease the pulp temperature after delignification or increases the retention time prior to the bleaching procedure. In addition, the use of a single enzyme leads the bleaching procedure to be conducted in multiple phases and to take longer due to the application of varying temperatures. Since lignin removal is the ultimate objective, several characteristics must be met for efficient bleachings, such as high catalytic efficiency, thermal stability, and resistance to inhibition by by-products [45]. Using bio-bleaching enzyme in the form of "cocktails" is preferable to using bio-bleaching enzyme alone. It was believed that the performance of enzymes cocktails was superior to that of chemical bleaching treatments. Sharma et al. [43] compared the chemical bleaching procedure to the enzyme cocktail bleaching process (a mixture of xylanase, pectinase, amylase, protease, and lipase). With retention duration, temperature, and pH, it turns out that the performance of bio-bleaching enzyme is better with a paper brightness indication of approximately 84%.

It is known that enzymes generated by thermophilic and thermotolerant organisms have excellent thermal stability. Extremely sensitive to the structure, rigidity, flexibility, and adaption of membranes, proteins, and nucleic acids is a microbe's ability to withstand exposure to system temperatures that exceed the thermophilic organisms' optimal range [44]. The catalytic activity of intracellular enzymes increases according to the internal stiffness and flexibility of the cell's proteins. According to Scandurra et al. [46], thermostable proteins produced by thermophilic and thermotolerant microbes contain aromatic residues and hydrophobic amino acids that provide a "hydrophobic effect" to prevent interactions with water and protein folding with a low tendency for unfolding, thereby rendering the protein resistant to relatively high temperatures. The thermophilic and thermotolerant species from several kingdoms that produce bio-bleaching enzymes, such as Archaea (*Aquiflex pyrophilus, Thermocrinis rubber, Pyrococcus, Ferroglobus, Desulforococcus, Thermoproteus, Thermococcus,* and *Sulfolobus*), Asco- and Zygomycota (*Canariomyces, Acremonium, Corynascus, Rhizomucor, Thermomyces, Dactylomyces, Chaetomium, Corynascus, Melanocarpus,* and *Paecilomyces*), and yeast (*Candida bovina, Candida thermophilia. Torulopsis pinfolopesii, Candida parapsilosis, Candida sloofii,* and *Arxiozyma telluris*) [46].

Several researchers have made efforts to improve the heat stability of enzymes. This involves protein engineering, chemical modification, covalent immobilization, the development of reversible enzyme-inhibitor complexes, the addition of stabilizing agents, and the alteration of nucleic acids [47]. Ozawa et al. [48] enhance heat stability by site-directed mutation (one of the nucleic acid modification methods). Ozawa et al. [48] enhanced the heat stability of alpha-amylase from *Bacillus* sp. that produce enzymes with low to neutral acidity by substituting the 11 amino-terminal residues with thermophilic *Bacillus* amino acid residues. As a result, the thermal stability of the combined sequences of amino acids approaches 100%. Zhou and Wang [49] utilized site-directed mutagenesis (combinational mutations) and regional substitution techniques to boost the thermostability and alkalinity of the pectate lyase enzyme, therefore enhancing its thermostability and alkalinity. In the textile industry, pectate lyase is typically utilized for the ramie degumming process. Adding additives as a stabilizing agent is a simple and low-cost strategy. Nirmal and Laxman

[47] did a study on efforts to improve the thermal stability of a *Cornidiobolus brefeldianus* alkaline protease with high activity. Adding stabilizing chemicals (trehalose and sorbitol) increased the residual activity (as an indication of thermal stability) by a factor of up to fivefold. Thermostable bio-bleaching enzymes have different temperature ranges and optimal temperatures, such as xylanase, which has a temperature range of 45–105 °C, cellulase, which has a temperature range of 45–95 °C, and laccase, which has an ideal temperature of 70 °C [44].

The bio-bleaching enzyme's alkaline strength must also be addressed. Bio-bleaching enzymes are often very alkaline enzymes. Some bacteria that create bio-bleaching enzymes, such as *Bacillus licheniformis*, *Bacillus subtilis*, and *Bacillus amyloliquefaciens*, produce enzymes with a range of acidic to neutral regions, hence they do not qualify as bio-bleaching agents [48].

Several earlier investigations have demonstrated that thermostable enzymes and alkalinophiles produce superior bleaching results compared to enzymes that operate at mild temperatures. There are multiple recommendations for improving the quality of produced enzymes through technological progress. These suggestions include improving the quality of existing enzymes/proteins (through site-directed mutagenesis, random mutagenesis, and cold-active enzymes) and obtaining novel genes encoding specific target enzymes/proteins (through a sequence-dependent identification and function-activating approach). based identification), and quality improvement via microbial cell surface display [50].

8.8 Conclusions

Laccase, cellulase, and xylanase are types of bio-bleaching enzymes that can be produced using corncob, sugarcane bagasse, and wheat straw as a medium and an intermediate of organisms that manufacture these enzymes. Because three media have significant cellulose and hemicellulose content, they offer a strong potential for utilization as the primary carbon source. The recognized and mass-produced sources of bio-bleaching enzymes are mostly mesophilic species with high commercial potential. The production of bio-bleaching enzymes from agricultural waste begins with the upstream processes, which include the biomass pretreatment, the inoculum preparation, and the main fermentation. Then, it will proceed to downstream processes for the separation and purification process of enzyme product. From the case study of wheat chaff valorization for enzyme production, it is financially feasible to produce high-purity cellulase enzymes with submerged fermentation. In-depth techno-economic analysis and sensitivity analysis maybe useful to assess how different factors contribute to the profitability and productivity of the enzyme production plant. Some future challenges will be obtaining bio-bleaching enzymes with parameters that have high heat resistance and tolerance to mild alkalinity. The solutions to these issues are employing thermophilic microorganisms and genetic modification to engineer the enzyme-producing microbes. By using the correct microorganisms to produce bio-bleaching enzymes and integrating experimental and techno-economic simulation, the process of scaling up to industrial scale becomes more feasible and cost-effective.

References

1 Ravindran, R., Hassan, S., Williams, G., and Jaiswal, A. (2018). A review on bioconversion of agro-industrial wastes to industrially important enzymes. *Bioengineering* 5 (93): 1–20. https://doi.org/10.3390/bioengineering5040093.

2 Uçkun Kiran, E., Trzcinski, A.P., Ng, W.J., and Liu, Y. (2014). Enzyme production from food wastes using a biorefinery concept. *Waste Biomass Valorization* (6): 903–917. https://doi.org/10.1007/s12649-014-9311-x.

3 Sharma, A., Parul, M., and Bhardwaj, S. (2021). Just Agriculture. justagriculture. https://justagriculture.in/2021/january/publications-newsletter.html (accessed 25 November 2022).

4 Tiseo, I. (ed.) (2022). Pulp & paper market size globally 2021-2029. Statista. https://www.statista.com/statistics/1073451/global-market-value-pulp-and-paper (accessed 22 November 2022).

5 Liu, K., Wang, H., Liu, H. et al. (2020). Covid-19: challenges and perspectives for the pulp and paper industry worldwide. *BioResources* 15 (3): 4638–4641. https://bioresources.cnr.ncsu.edu/resources/covid-19-challenges-and-perspectives-for-the-pulp-and-paper-industry-worldwide.

6 Rullifank, K.F., Roefinal, M.E., Kostanti, M., and Sartika, L. (2020). Pulp and paper industry: an overview on pulping technologies, factors, and challenges. *IOP Conf. Ser. Mater. Sci. Eng.* 845 (012005): https://doi.org/10.1088/1757-899X/845/1/012005.

7 Yasmidi and Roosmini, D. (2008). Analysis of chlorinated organic compounds (AOX) concentration in aquatic around pulp and paper mill. *J. Selulosa* 43 (1): 29–38. http://dx.doi.org/10.25269/jsel.v43i01.165.

8 Gupta, G.K., Dixit, M., Kapoor, R.K., and Shukla, P. (2022). Xylanolytic enzymes in pulp and paper industry: new technologies and perspectives. *Mol. Biotechnol.* 64 (2): 130–143. https://doi.org/10.1007/s12033-021-00396-7.

9 Isak, E. (2017). Golden barley. Unsplash. https://unsplash.com/photos/rgYkWfIaWHs (accessed 10 November 2022).

10 Lü, X. and Guo, X. (2021). Chapter 2 – The need for biofuels in the context of energy consumption. In: *Advances in 2nd Generation of Bioethanol Production* (ed. X. Lu), 9–30. Duxford, Cambridgeshire: Woodhead Publishing https://doi.org/10.1016/B978-0-12-818862-0.00004-2.

11 Gandam, P.K., Chinta, M.L., Pabbathi, N.P. et al. (2022). Corncob-based biorefinery: a comprehensive review of pretreatment methodologies, and biorefinery platforms. *J. Energy Inst.* 101: 290–308. https://doi.org/10.1016/j.joei.2022.01.004.

12 Saleem Khan, T. and Mubeen, U. (2012). Wheat straw: a pragmatic overview. *Curr. Res. J. Biol. Sci.* 4 (6): 673–675. https://www.researchgate.net/publication/257645533_Wheat_Straw_A_Pragmatic_Overview.

13 Singha, P., Jawaid, M., Chandrasekar, M. et al. (2021). Sugarcane wastes into commercial products: processing methods, production optimization and challenges. *J. Cleaner Prod.* 328: 129453. https://doi.org/10.1016/j.jclepro.2021.129453.

14 Hernández, C., Escamilla-Alvarado, C., Sánchez, A. et al. (2019). Wheat straw, corn Stover, sugarcane, and agave biomasses: chemical properties, availability,

and cellulosic-bioethanol production potential in Mexico. *Biofuels, Bioprod. Biorefin.* 13 (5): 1143–1159. https://doi.org/10.1002/bbb.2017.

15 Chundawat, S.P.S., Beckham, G.T., Himmel, M.E., and Dale, B.E. (2011). Deconstruction of lignocellulosic biomass to fuels and chemicals. *Annu. Rev. Chem. Biomol. Eng.* 2: 121–145. https://doi.org/10.1146/annurev-chembioeng-061010-114205.

16 Blanch, H.W. and Clark, D.S. (1996). *Biochemical Engineering*. Taylor Francis, New York, Boca Raton: CRC Press.

17 El-Gendi, H., Saleh, A.K., Badierah, R. et al. (2021). A comprehensive insight into fungal enzymes: structure, classification, and their role in mankind's challenges. *J. Fungi* 8: 23–26. https://doi.org/10.3390/jof8010023.

18 Kalyani, D., Tiwari, M.K., Li, J. et al. (2015). A highly efficient recombinant laccase from the yeast *Yarrowia lipolytica* and its application in the hydrolysis of biomass. *PLoS One* 10 (3): https://doi.org/10.1371/journal.pone.0120156.

19 Nurudeen, O.O., Adetayo, O.M., Salaam, A. et al. (2015). Cellulase and biomass production from sorghum (*Sorghum guineense*) waste by *Trichoderma longibrachiatum* and *Aspergillus terreus*. *J. Microbiol. Res.* 5 (6): 169–174. https://doi.org/10.5923/j.microbiology.20150506.01.

20 Wang, Z., Ong, H.X., and Geng, A. (2012). Cellulase production and oil palm empty fruit bunch saccharification by a new isolate of *Trichoderma koningii* D-64. *Process Biochem.* 47 (11): 1564–1571. http://dx.doi.org/10.1016/j.procbio.2012.07.001.

21 Schindler, D. (2020). Genetic engineering and synthetic genomics in yeast to understand life and boost biotechnology. *Bioengineering* 7 (4): 1–18. https://doi.org/10.3390/bioengineering7040137.

22 Shariq, M. and Sohail, M. (2020). Production of cellulase and xylanase from *Candida tropicalis* (MK-118) on purified and crude substrates. *Pak. J. Bot.* 52 (1): 323–328. https://doi.org/10.30848/pjb2020-1%2814%29.

23 Li, Q., Pei, J., Zhao, L. et al. (2014). Overexpression and characterization of laccase from trametes versicolor in *Pichia pastoris*. *Appl. Biochem. Microbiol.* 50 (2): 140–147. https://doi.org/10.1134/S0003683814020124.

24 Miyazaki, K. (2005). A hyperthermophilic laccase from *Thermus thermophilus* HB27. *Extremophiles* 9 (6): 415–425. https://doi.org/10.1007/s00792-005-0458-z.

25 Chakdar, H., Kumar, M., Pandiyan, K. et al. (2016). Bacterial xylanases: biology to biotechnology. *3 Biotech* 6 (2): 150. https://doi.org/10.1007%2Fs13205-016-0457-z.

26 Lynd, L.R., Wimer, P.J., van Zyl, W.H., and Pretorius, I.S. (2022). Microbial cellulose utilization: fundamentals and biotechnology. *Microbiol. Mol. Biol. Rev.* 66 (3): 506–577. https://doi.org/10.1128/MMBR.66.3.506-577.2002.

27 Mhiri, S., Bouanane-Darenfed, A., Jemli, S. et al. (2020). A thermophilic and thermostable xylanase from *Caldicoprobacter algeriensis*: recombinant expression, characterization and application in paper biobleaching. *Int. J. Biol. Macromol.* 164: 808–817. https://doi.org/10.1016/j.ijbiomac.2020.07.162.

28 Walia, A., Guleria, S., Mehta, P. et al. (2017). Microbial xylanases and their industrial application in pulp and paper biobleaching: a review. *3 Biotech* 7 (1): 11. https://doi.org/10.1007/s13205-016-0584-6.

29 Paës, G., Berrin, J.-G., and Beaugrand, J. (2012). GH11 xylanases: structure/function/properties relationships and applications. *Biotechnol. Adv.* 30 (3): 564–592. https://doi.org/10.1016/j.biotechadv.2011.10.003.

30 Demuner, B.J., Pereira Junior, N., and Antunes, A.M.S. (2011). Technology prospecting on enzymes for the pulp and paper industry. *J. Technol. Manag. Innov.* 6 (3): 148–158. http://dx.doi.org/10.4067/S0718-27242011000300011.

31 Imran, M., Anwar, Z., Irshad, M. et al. (2016). Cellulase production from species of fungi and bacteria from agricultural wastes and its utilization in industry: a review. *Adv. Enzym. Res.* 04 (02): 44–55. http://dx.doi.org/10.4236/aer.2016.42005.

32 Headon, D.R. and Walsh, G. (1994). The industrial production of enzymes. *Biotechnol. Adv.* 12 (4): 635–646. https://doi.org/10.1016/0734-9750(94)90004-3.

33 Kuhad, R.C., Gupta, R., and Singh, A. (2011). Microbial cellulases and their industrial applications. *Enzym. Res.* 2011: 1–10. https://doi.org/10.4061/2011/280696.

34 Mate, D.M. and Alcalde, M. (2017). Laccase: a multi-purpose biocatalyst at the forefront of biotechnology. *Microb. Biotechnol.* 10 (6): 1457–1467. https://doi.org/10.1111/1751-7915.12422.

35 Crestini, C., Jurasek, L., and Argyropoulos, D.S. (2003). On the mechanism of the laccase–mediator system in the oxidation of lignin. *Chem. Eur. J.* 9 (21): 5371–5378. https://doi.org/10.1002/chem.200304818.

36 Yaver, D.S., Overjero, M.D.C., Xu, F. et al. (1999). Molecular characterization of laccase genes from the basidiomycetes *Coprinus cinereus* and heterologous expression of laccase Lcc1. *Appl. Environ. Microbiol.* 65 (11): 4943–4948. https://doi.org/10.1128/AEM.65.11.4943-4948.1999.

37 Jovanović, M., Vučurović, D., Dodić, S. et al. (2020). Simulation model comparison of submerged and solid-state hydrolytic enzymes production from wheat chaff. *Rom. Biotechnol. Lett.* 25 (5): 1938–1948. https://doi.org/10.25083/rbl/25.5/1938.1948.

38 Risdianto, H., Sofianti, E., Suhardi, S.H., and Setiadi, T. (2012). Optimisation of laccase production using white rot fungi and agriculture wastes in solid state fermentation. *ITB J. Eng. Sci.* 44 (2): 93–105. http://dx.doi.org/10.5614/itbj.eng.sci.2012.44.2.1.

39 Jourdier, E., Poughon, L., Larroche, C. et al. (2012). A new stoichiometric miniaturization strategy for screening of industrial microbial strains: application to cellulase hyper-producing *Trichoderma reesei* strains. *Microb. Cell Fact.* 11 (70): https://doi.org/10.1186/1475-2859-11-70.

40 Schutyser, M.A.I., Perdana, J., and Boom, R.M. (2012). Single droplet drying for optimal spray drying of enzymes and probiotics. *Trends Food Sci. Technol.* 27 (2): 73–82. https://doi.org/10.1016/j.tifs.2012.05.006.

41 Lewin, D.R., Seider, W.D., and Seader, J.D. (2002). Integrated process design instruction. *Comput. Chem. Eng.* 26 (2): 295–306. https://doi.org/10.1016/S0098-1354(01)00747-5.

42 Solomon, K.R. (1996). Chlorine in the bleaching of pulp and paper. *Pure Appl. Chem.* 68 (9): 1721–1730. https://doi.org/10.1351/pac199668091721.

43 Sharma, A., Balda, S., Gupta, N. et al. (2020). Enzyme cocktail: an opportunity for greener agro-pulp biobleaching in paper industry. *J. Cleaner Prod.* 271: 122573. https://doi.org/10.1016/j.jclepro.2020.122573.

44 Seibert, T., Thieme, N., and Benz, J.P. (2016). *Gene Expression Systems in Fungi: Advancements and Applications*, 59–96. Springer https://link.springer.com/book/10.1007/978-3-319-27951-0.

45 Lopes, A.M., Ferreira Filho, E.X., and Moreira, L.R.S. (2018). An update on enzymatic cocktails for lignocellulose breakdown. *J. Appl. Microbiol.* 125 (3): 632–645. https://doi.org/10.1111/jam.13923.

46 Scandurra, R., Consalvi, V., Chiaraluce, R. et al. (1998). Protein thermostability in extremophiles. *Biochimie* 80 (11): 933–941. https://doi.org/10.1016/s0300-9084(00)88890-2.

47 Nirmal, N.P. and Laxman, R.S. (2014). Enhanced thermostability of a fungal alkaline protease by different additives. *Enzym. Res.* 2014: 1–8. https://doi.org/10.1155/2014/109303.

48 Ozawa, T., Endo, K., Igarashi, K. et al. (2007). Improvement of the thermal stability of a calcium-free, alkaline alpha-amylase by site-directed mutagenesis. *J. Appl. Glycosci.* 54 (2): 77–83. https://doi.org/10.5458/jag.54.77.

49 Zhou, Z. and Wang, X. (2021). Rational design and structure-based engineering of alkaline pectate lyase from *paenibacillus* sp. 0602 to improve thermostability. *BMC Biotech.* 21 (32): 1–12. https://doi.org/10.1186/s12896-021-00693-8.

50 Baweja, M., Nain, L., Kawarabayasi, Y., and Shukla, P. (2016). Current technological improvements in enzymes toward their biotechnological applications. *Front. Microbiol.* 7: 1–13. https://doi.org/10.3389/fmicb.2016.00965.

9

Recovery of Industrially Useful Enzymes from Rubber Latex Processing By-products

Tan W. Kit[1,2], Yong Y. Seng[1,2], Siti N. Azlan[1,2], Nurulhuda Abdullah[3], and Fadzlie W. F. Wong[1,2]

[1] *Universiti Putra Malaysia, Department of Bioprocess Technology, Faculty of Biotechnology and Biomolecular Sciences, 43400 UPM, Serdang, Selangor Darul Ehsan, Malaysia*
[2] *Universiti Putra Malaysia, Bioprocessing and Biomanufacturing Research Complex, Faculty of Biotechnology and Biomolecular Sciences, 43400 UPM, Serdang, Selangor Darul Ehsan, Malaysia*
[3] *Malaysian Rubber Board, Technology and Engineering Division, 47000, Sungai Buloh, Selangor Darul Ehsan, Malaysia*

9.1 Introduction

Natural rubber latex (NRL) is produced naturally from rubber trees (*Hevea brasiliensis*) and can be collected via tapping activity around the tree bark. The structure of NRL is made up of a hydrophobic polyisoprene core, that is surrounded by hydrophilic proteins and phospholipids that stabilize the system [1]. The stability of NRL is influenced by surface charge, electrokinetic, and coagulation behavior, which ultimately influence the NRL properties [2]. The tapped NRL contains approximately 20–40% hydrocarbon molecules and 2–6% nonrubber components, such as proteins, lipids, carbohydrates, and other inorganic materials. The pH of unprocessed NRL is around pH 6.5–7.0, however, some NRL processing has set a standard for the processed NRL to be around pH 9.0 [3]. The increment of pH is due to the addition of ammonia for latex preservation, clarification, and longer shelf life. The ammonia also acts as an inhibitor that can stunt the growth of microorganisms, which can cause coagulation of NRL and result in a spoiled condition [4]. The addition of ammonia in the NRL system can also increase its stability by the increment in zeta potential value and steric repulsion [1]. Various rubber products have been developed from NRL processing, such as tyres, surgical instruments, gloves, balloons, catheters, foams, and rubber gloves due to the latex's high elasticity [5, 6]. Nevertheless, the protein content of NRL has given rise to a huge number of concerns from the consumer since some proteins in NRL can cause allergic reactions to the user [6]. Small factions of NR proteins will remain in the products as residual extractable proteins (EP) and most of proteins are eliminated when the latex undergoes certain processes, such as preservation, centrifugation, chemical addition, and manufacturing process for the production of intended

Chemical Substitutes from Agricultural and Industrial By-Products: Bioconversion, Bioprocessing, and Biorefining, First Edition. Edited by Suraini Abd-Aziz, Misri Gozan, Mohamad Faizal Ibrahim, and Lai-Yee Phang.
© 2024 WILEY-VCH GmbH. Published 2024 by WILEY-VCH GmbH.

Table 9.1 Allergens of the rubber tree *Hevea brasiliensis*.

Allergen	Biochemical name	Molecular weight (kDa)
Hev b1	Rubber elongation factor	14
Hev b2	β-1,3-Glucanase, glucan *endo*-1, 3-beta-glucosidase, and basic vacuolar isoform	35, 36.5, and 38
Hev b3	Small rubber particle protein	24
Hev b4	Lecithinase homolog	53–55
Hev b5	Acidic structural protein	16
Hev b6	Prohevein (hevein precursor)	20
Hev b7	Patatin-like protein	42
Hev b8	Profilin	15
Hev b9	Enolase	51
Hev b10	Superoxide dismutase	26
Hev b11	Chitinase class I	30
Hev b12	Nonspecific lipid transfer protein type 1 (nsLTP1)	9
Hev b13	Esterase	42
Hev b14	Hevamine	30
Hev b15	Serine protease inhibitor	7.5

products. Because of its solubility in the water, total EP can vary from product to product depending on the manufacturing protocols.

Ever since the identification of the first major NRL allergen, known as the rubber elongation factor (Hev b1), in 1993 by Czuppon et al. [7], to date, 15 native proteins of NRL have been identified (Table 9.1) and recognized as NRL allergens in the International Union of Immunological Societies (IUIS), and the proteins have been assigned official numbers in the nomenclature list of the International Nomenclature Committee of Allergens [8].

Of these 15 allergenic proteins, 6 allergens (Hev b1, Hev b2, Hev b3, Hev b5, Hev b6, and Hev b13) have been determined to be the major allergen that causes allergic in a specific risk group, such as spina bifida patients and healthcare workers. Studies show that these six allergens were found in 50% of the patients and were also found in almost all medical gloves examined [9]. On the other hand, Hev b6.01 (Prohevein) and Hev b6.02 (Hevein) are among the major NR allergens associated with "latex fruit syndrome." Prohevein is a precursor that yields two allergenic fragments; whereas Hevein is a major allergen with antifungal properties. Both of them exhibit cross-reactivity of chitinase.

Therefore, serious attention has been paid to reduce or eliminate allergenic proteins in gloves made of NRL. By far, the Malaysian glove manufacturers and NRL producers have managed to address and manage the protein allergy, hence the focus now is to seek other potentials of NRL that can bring economic values for future development and sustainability in the rubber sector. The most commonly used methods of protein removal from gloves are the leaching and deproteinization processes. Leaching is carried out for commercial natural rubber gloves by dipping into hot water, while deproteinization is for commercial high ammonia natural rubber latex (HANRL) with the addition of urea, sodium dodecyl sulfate (SDS), and acetone [10]. Nevertheless, problems arise when wastes are generated due to the allergenic protein removal from the gloves. At the same time, the C-serum and lutoid or bottom fractions, which contain allergenic proteins are being disposed of after the production of the natural rubber products. Hence, as an economical and environmental opportunity, these allergenic proteins, mainly the enzymes, can be recovered from the rubber processing by-products and recycled, to be used for other industrial applications. Of the allergenic proteins, chitinases and lysozymes are being targeted for valorization purposes as they are considered as the most abundant proteins in the NRL. Further, they also possess great potential for industrial applications due to their antibacterial and antifungal properties. Previously, several recovery methods have been done for extracting proteins from NRL and other plant latex for characterization purposes, mostly by precipitation methods and chromatographic methods [11].

In this chapter, firstly, the processing of NRL for the production of rubber products including the overview and structure of NRL, and preservation techniques is highlighted. Next, the general characteristics of lysozymes and chitinases from plants, animals, and bacteria are also presented. As the availability of a reliable and accurate qualitative and quantitative method for the enzymes is important before the development of recovery methods, the topic of conventional and alternative activity assays for lysozymes and chitinases is also covered. Next, the chapter presents the potential application of plant-derived lysozymes and chitinases in industry. These sections served as the background topics and are deemed important before exploring on the main sections: The developments in the recovery of the plant-derived lysozymes and chitinases (Section 9.5), and the potential strategy for recovering lysozymes and chitinases from NRL (Section 9.6). Taken together, this chapter provides insights on the potential strategy for the valorization of the by-products from NRL processing, and ultimately laying the ground for the transition to a circular bioeconomy.

9.2 Processing of Natural Rubber Latex for the Production of Rubber Products

9.2.1 NRL Overview

The rubber tree (*H. brasiliensis*), is an important industrial crop cultivated widely in Asia (especially Malaysia, Thailand, and Indonesia) for its ability to produce NRL [12, 13]. A rubber tree can strive for up to 25–30 years before it is deemed uneconomical for NRL production [14]. Asia is the largest supplier of NRL with the ability to supply approximately 97% of NRL worldwide [12, 15]. Malaysia, one of the largest cultivators of rubber trees has been cultivating rubber trees since the 1970s using the popularized rubber seed from India, which originates from the Valley of South America [14, 16]. The collection of NRL can be done via tapping activity from its bark that can be further processed into industrial goods, such as automotive parts, surgical instruments, and footwear. However, such process usually produces different types of by-products or wastes like rubber solids and serum especially after compounding of NRL. Rubber solids often being recycled by milling into fine granules, whereas serum is discarded as wastewater. The serum could be useful if its enzymes are recovered successfully rather than disposing them as waste.

9.2.2 NRL Structure

NRL is made up of approximately 94% rubber hydrocarbon of which 30–40% is polyisoprene polymer, which is hydrophobic by nature [13]. The presence of rubber particles in NRL can be generally classified into two categories: the large rubber particles (LRP) with a diameter of 0.40–0.75 µm and the small rubber particles (SRP) with a diameter of 0.08–0.20 µm [17, 18]. On the other hand, another 5–6% of nonrubber components are consisting of proteins (2–3%), carbohydrates, lipids, fatty acids, and inorganic materials, which are mostly hydrophobic [19]. As the nature of polyisoprene polymer is hydrophobic, the NRL particles are widely believed to form a core–shell. The core–shell is surrounded by hydrophilic materials (nonrubber particles) to form a barrier to support the stability of NRL particles [20, 21]. The presence of proteins also helps to fight against wounding, wood borers, and fungi attacks, although it might cause allergic reactions in humans with sensitive conditions [22].

9.2.3 NRL Preservation

Latex tends to coagulate right after collection, and therefore preservative agents like ammonia are added to NRL to maintain the stability of the latex [19]. The addition of ammonia can also help to denature some proteins due to its alkalinity, as the presence of proteins in NRL has been reported to cause an allergic reaction in some users [6, 19]. The increase in pH can inhibit microbial growth, which can cause coagulation and NRL spoilage during storing period. A long storage period will influence the NRL to increase its negative charge, thus, the addition of ammonia can reduce the instability of ions in NRL [19]. The percentage of ammonia added can influence the final product of NRL; 1–3% ammonia will create a low-ammoniated latex (LA latex), 4–5% ammonia for medium-ammoniated latex (MA latex), and 6–7% ammonia for creating a highly-ammoniated latex (HA latex). As ammonia is easily volatile, the addition and testing of ammonia content in the NRL sample should be continuous for longer preservation and storing time [4].

9.2.4 Deproteinization of NRL

The allergenic proteins can also be partially removed or denatured during centrifugation, leaching, and surfactants or other chemical treatments [19]. The presence of proteins in the latex contributed to its stability, and hence, the removal of proteins will cause an imbalance of ions as the hydrophilic layers are stripped off during the process [1]. The process of protein removal is called the deproteinization process and has been widely used due to its simplicity, rapidness, and inexpensiveness to reduce protein content by almost up to 99%. Generally, the deproteinization process can be done via physical, biological, or chemical methods [1, 2]. The chemical method is commonly used as chemicals, such as urea and surfactants are relatively easily available. As most of the protective layers of NRL are removed during the processing, surfactant add-on is vital for the colloidal stability enhancement, and the method is more versatile [1, 2]. Therefore, the addition of chemicals, such as surfactants can aid the latex to achieve a stable form as the addition of fatty acids soap can enhance latex stability greatly [2].

9.3 General Characteristics of Plant-derived Lysozymes and Chitinases

Figures 9.1 and 9.2 are the 3D models, which show the structural difference between lysozyme and chitinase [23]. Lysozyme is a common enzyme known to consist of 129 amino acids, cross-linked by four disulfide bridges. It consists of two main domains; the alpha domain (consists of alpha helices) and the beta domain (consists of beta sheets). The active site can be found within a cleft between these two domains [24].

Lysozymes can be classified into six types: chicken-type (c-type), goose-type (g-type), plant, bacteria, T4 phage (phage-type), and invertebrate (i-type) lysozyme. However, the molecular characteristics and activity differ between each of these sources. Their differences in amino acid sequence, active site location, and enzymatic activity (which is further dependent on pH, ionic strength, and the substrate)

Figure 9.1 Structure of lysozyme. Source: Adapted from Terwisscha et al. [23].

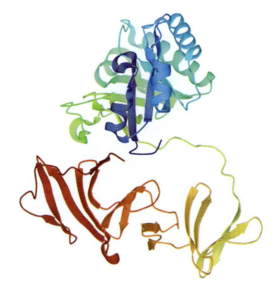

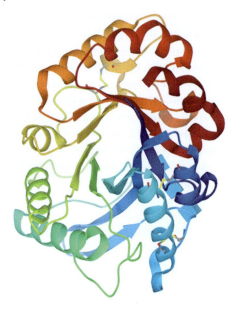

Figure 9.2 Structure of chitinase. Source: Adapted from Terwisscha et al. [23].

make them unique and identifiable through methods, such as chromatography and electrophoresis [24].

Furthermore, chitinase is another common enzyme produced by a variety of organisms, such as plants, animals, bacteria, and humans. It belongs to the glycosyl hydrolase family and has a molecular size from 20 to 90 kDa. Plant chitinases belong to two families, family 18 and family 19 [25]. Despite their similar amino acid sequence, these families differ in both activity and structure. Family 18 chitinases have an eightfold alpha/beta barrel structure, while family 19 chitinases have a bilobal structure, consisting primarily of alpha helices. Table 9.2 shows the structural differences between the chitinases produced by plants, yeast, fungi, insects, and mammals [26].

9.4 Conventional and Alternative Activity Assays for Lysozymes and Chitinases

Common assays for both lysozyme and chitinases are colorimetric assays, which are based on the observation of the appearance or disappearance of a colored compound. This method is particularly prevalent for chitinases and its sensitivity depends on the type of substrate used and the method of color detection. Radiometric assays are the most sensitive and are based on the formation of chitooligosaccharides (COS) from regenerated [3H] chitin [27]. It is more rapid and simpler than colorimetric assays but requires specialized laboratory equipment. Further, viscosimetric assays, which are based on the reduction of viscosity of glycol chitin, have also been developed to detect slight chitinase activity. The standard assay for detecting lysozyme activity is through the turbidity of a solution containing *Micrococcus lysodeikticus*. Activity is

Table 9.2 Structural differences of chitinase derived from plant, microbe, and animal.

Plants	Microbe		Animal	
	Yeast	Fungi	Insect	Mammalian
One catalytic domain	Has 4 domains; signal sequence, a catalytic domain, a serine/threonine-rich region, and a C-terminal chitin-binding domain	Has 5 domains; N-terminal signal peptide region, catalytic domain, chitin-binding domain, serine/threonine rich-region, and C-terminal extension region	Has 3 domains; catalytic, cysteine-rich chitin-binding, and serine/threonine-rich linker	Has a N-terminal catalytic domain with triose phosphate isomerase fold
Family 18 – eightfold alpha/beta barrel structure	—	—	Has beta/alpha barrel proteins	—
Family 19 – bilobal structure, high alpha-helical content. Same amino acid sequence as family 18	—	—	Beta sheets in parallel fashion	—

Table 9.3 Activity assays for lysozymes and chitinases.

Method	Enzyme source	Assay	Specificity	Sensitivity	References
Lysozymes					
CM-chitin-RBV (dye-labeled substrate) is soluble and stable in buffer systems. After the enzyme reaction is terminated, the substrate is precipitated and the absorbance is measured	Chicken egg white	Colorimetric	CM-chitin solution prepared from crustacean chitin, which is soluble and highly functional for chitinase and lysozyme activity assay	The colloidal chitin is highly sensitive substrate for quantitative colorimetric assays	[28]
Lysozyme form aggregates with Cys-Ala-Leu-Asn-Asn-capped gold nanoparticles (CALNN/pep-AuNP's). A specific lysozyme detection method was developed through this interaction	Human serum	Colorimetric	The aggregation of pep-AuNPs is not caused by electrostatic attraction but by the specific recognition of lysozyme	Response ranges 1–25 ng/ml	[29]
Rate of lysis of *Micrococcus lysodeikticus*.	*H. brasiliensis*	Turbidimetric	>75% of the total soluble protein in latex was found in the pellet fraction, and 25% identified as chitinases/lysozymes.	Lysozyme activity ranges from 1.8 to 79.6 units/mg $\times 10^{-3}$	[30]
Enzyme activity also measured as the rate of lysis of *M. lysodeikticus*. Clearing of a buffered suspension of *M. lysodeikticus* cells was successful	Fig	Turbidimetric	Fig lysozyme was 0.85× as active as egg-white lysozyme and 2.4× more active than papaya lysozyme, optimum at pH of 4.5	Lysis is sensitive to ionic strength. The maximum activity is found at an ionic strength of 0.02–0.03	[31]
M. lysodeikticus is re-suspended in phosphate buffer and assay is run for 5 min. The optical density is recorded at 30 s intervals at a wavelength of 450 nm	Hen egg white	Turbidimetric	Determined a slightly higher content of lysozyme enzyme (compared to the fluorescence-based method)	Range of detection is found to be 20–120 ng	[32]

Description	Source	Method	Notes	Ref
Total active lysozyme is measured using fluorescently quenched *M. lysodeikticus*	Hen egg white	Fluorimetric/fluorescence-based	An increase in the fluorescent signal is directly proportional to the amount of active lysozyme in the sample	Range of detection is found to be 2–150 ng	[32]

Chitinases

Description	Source	Method	Notes	Result	Ref
CM-chitin-RBV (dye-labeled substrate) is soluble and stable in buffer systems. After the enzyme reaction is terminated, the substrate is precipitated and the absorbance is measured	*Streptomyces*	Colorimetric	Colloidal chitin-RBV in quantitative assessment could not be adapted to microtiter plates	Colloidal-chitin-RBV substrate is approximately 200× less sensitive to activity compared to CM-chitin-RBV	[28]
A method to quantitatively assay chitinase activity using glycochitin as substrate. Based on the affinity of fluorescent brightener 28 with undigested glycochitin	Plants	Colorimetric	Reliable method that avoids hazardous chemicals. Automation and objective analysis make it highly reliable and replicable	Their reaction gel reveals the potential to detect as little as 12.5 µU/ml of chitinase	[33]
Determination of chitinase activity using a batch of glycol chitin. As the amount of chitinase increases, the faster the viscosity of glycol chitin is reduced	*Aspergillus niger* (fungus) and *Vibrio* spp. (bacteria)	Viscometric	Time-consuming to determine chitinase activity of numerous samples and required of specialized equipment. However, it is more specific to endochitinases only	A sensitive and effective procedure to detect chitinase activity as glucosamine is used as reference compound	[34]
[3H] chitin is prepared and the reaction with the enzyme is terminated by boiling. The radioactivity of the supernatant is determined to find activity	*H. brasiliensis*	Radiometric	Able to determine specific radioactivity activity of chitin	The reaction is monitored by spectrophotometer where chitinase activity ranges from 4.0 to 38.7 nkat/mg	[30]

measured as the rate of clearing of *M. lysodeikticus* cells. Other cited methods specific for lysozyme activity include fluorimetric assays. Table 9.3 details several methods reported in previous studies for both lysozymes and chitinases (respectively) as well as their sensitivity, where applicable.

9.5 Potential Application of Plant-derived Lysozymes and Chitinases

Based on the extensive research done on the uses of plant-derived lysozymes and chitinases (Table 9.4), plant-derived lysozymes can be potentially used in cleaning products (bactericide and biofilm removal). Meanwhile, for applications related to the food, agriculture, or pharmaceutical industry (which are common applications for egg white lysozyme), since the implications of the allergens on human health have not been explored fully, more research in this area must be done before making it commercially available.

Plant-derived chitinase has shown promising efficacy as a biocontrol agent, especially to mosquitoes, such as the *Aedes aegypti* mosquito species, and against a pathogenic fungus called powdery mildew (*Sphaerotheca humuli* Burrill). Another potential use of chitinase is to treat chitinous waste. The ability of chitinase to degrade chitin into a low molecular weight chitooligomers/COS opens up further opportunities, including seed and horticultural products and plant vaccines [41]. Plant-derived chitinase has a huge potential of being exploited for many more applications, however, the enzyme's stability, cost, activity within a temperature, and pH range still require ongoing research to be fully exploited for commercial uses.

9.6 Potential Strategy for Recovering Lysozymes and Chitinases from NRL

Table 9.5 shows different recovery and purification methods used in previous literatures. Recovery of plant-derived enzymes generally comes with multiple steps including pretreatment, such as ammonium sulfate precipitation or acid precipitation, followed by various types of chromatographic techniques. Rao and Gowda (2008) used chitin affinity chromatography to purify chitinase from tamarind and obtained a 64.6% yield [46]; the yield is generally very high compared to that of usually achieved by the combination of the traditional methods: precipitation and ion-exchange chromatography. Manikandan et al. (2015) presented a method of purifying lysozyme from cauliflower heads by using membrane filtration (Amicon YM3) and passing the filtrate through a Sephadex G100 column [45]. Overall, the aim of the purification studies was to characterize the enzymes and examine their activities, rather than for subsequent downstream application, except for the work of Nitsawang et al. (2006), which explored the use of alternative purification method, aqueous two-phase system (ATPS) to recover papain (which contains

Table 9.4 Potential applications of plant-derived lysozymes and chitinases.

Industry	Potential applications
Lysozymes	
Bactericide, disinfectants, and antiseptics (cleaning product industry) for Gram-negative bacteria	• Lysing activity of lysozyme can be employed for Gram-negative bacteria, such as *Escherichia coli*, *Pseudomonas aeruginosa* and *Azotobacter vinelandii*, when lysozyme at pH 7.5–9 is mixed with Versene (EDTA) as described by Ercan and Demirci (2016) [24] • Lysozyme can digest cell wall and cause disruption of cell wall, also known as apoptosis (sudden change in solute concentration inside and outside of the cell wall) and creating an osmotic shock in which lysozyme acts as cationic protein that can cause cell lysis via puncturing of cell wall [35]
Biofilm removal in nature (environmental science)	Biofouling of water pipes and marine vessels on ships where biofilm often grows on. *Staphylococcus aureus* is known to produce its own lytic transglycosylases called IsaA (immune-dominant Staphylococcal Antigen A) – a lysozyme-like enzyme that inhibits biofilm formation with the addition of tannic acid [36]
Increase biofilm removal efficiency of enzymatic cleaners	Enzymes such as protease, DNase I, amylase, and cellulase have shown to be effective in supporting biofilm removal. Lysozyme can be added to the testing to examine its efficacy. *S. aureus*, *P. aeruginosa*, and *Gardnerella vaginalis* biofilms were tested and have proven to be successful in reducing biofilm formation ability [37]
Chitinases	
Agriculture and cleaning products (antifungal in specific)	• Control of pathogenic fungi and plant diseases. The spraying of Chitinase E solution onto the powdery mildew spores (*Sphaerotheca humuli* Burrill) resulted in the complete disappearance of white spores, and no reappearance has been recorded for at least 2 wk [38]
Mosquito control insecticide	• In mosquito control and the spread of arbovirus disease – in arthropods, chitin is a key component of the skeleton and digestive lining tract. Chitinase attacks chitin, hence destroying these structures. The mortality of larvae has reached 100% within 48 h when exposed to 150 mg/l of chitinase [39]
Bactericide, biocontrol, and bioinsecticide	• Many plant-derived chitinases have lysozyme-like activity, which can cleave the peptidoglycan cell wall of bacteria (cleaves between C-1 of a *N*-acetylglucosamine and the C-4 of a *N*-acetylmuramate) [25]. • For example, Hevamine (family 18 glycosyl hydrolase), an endochitinase found in the latex of *H. brasiliensis* (a type of commercial rubber tree) [30]. It was found that 25% of the 75% of total soluble protein derived from the latex is from chitinases/lysozymes origin
Recycling and environmental protection	• Common source of chitin is the exoskeleton of shrimp shells, whereas chitosan is a biopolymer that is produced by the deacetylation of chitin. Chitinase can degrade chitin into low molecular weight chitooligomers/COS via the enzymatic hydrolysis of chitin and chitosan [40] • Better method compared to the conventional way of using chemicals, such as HCl and NaOH to decompose chitinous waste. However, high processing time for any pretreatment steps can be a challenge

Table 9.5 Recovery and purification methods for lysozymes and chitinases from various plant sources.

Source	Method(s)	Conditions	Yield/purity	Remarks	References
NRL (*H. brasiliensis*)	1. Ammonium sulfate precipitation 2. Size exclusion chromatography	1. Room temperature, overnight, 70% ammonium sulfate 2. Acetate buffer (pH 5)	5.2 mg of lysozyme yielded from 800 g of surgical gloves	Chromatography process is repeated for at least three times for lysozyme purification	[42]
	Acid precipitation using trichloroacetic acid (TCA) and phosphotungstic acid (PTA)	5% TCA and 0.2% PTA	Approximately 700 µg/ml of a wide range of proteins	N/A	[43]
	1. Anion exchange chromatography 2. Cation exchange chromatography	1. Pharmacia Mono Q 5/5 anion exchange column using pH 7.6 buffer 2. Pharmacia Mono S 5/5 cation exchange column using pH 6.5 buffer	56% recovery of chitinase and 79% recovery of lysozyme	Dialysis was done both before anion and cation exchange chromatography	[30]
Papaya latex (*Carica papaya*)	1. Ammonium sulfate precipitation 2. Sodium chloride precipitation	1. 4 °C, 30-min stirring, 11% ammonium sulfate 2. 4 °C, 1-h stirring, 20% sodium chloride	20% recovery of chitinase with 4.2-folds of purification	Papain contains considerable chitinase activity and hence is included in this table	[44]
Papaya latex (*C. papaya*)	Aqueous two-phase system (ATPS)	ATPS composed of 8% (w/w) PEG, 15% (w/w) $(NH_4)_2SO_4$, and papaya latex containing 20–40 mg protein/ml at pH 5	88% recovery and 100% purity were obtained	The ATPS system gave superior performance in terms of recovery and purification factor	[44]

Source	Method	Results	Notes	Ref
Ipomoea carnea latex	1. Ammonium sulfate precipitation 2. Hydrophobic interaction chromatography 3. Size exclusion chromatography	1. Ammonium sulfate saturation of 85% 2. Ether–Toyopearl 650S column 3. Phenyl–Toyopearl 650S column 4. Superdex S-200 column	1. 82.80% chitinase yield (1.13 purification fold) 2. 59.17% chitinase yield (1.70 purification fold) 3. 36.83% chitinase yield (2.90 purification fold) 4. 29.74% chitinase yield (3.95 purification fold)	N/A	[38]
Momordica charantia fruit	1. Turbidity assay for lysozyme activity 2. Ion-exchange chromatography	HPLC under Source Q and POROS 50 HS	41.2% lysozyme yield with 55.4 purification fold	Turbidity assay was used for lysozyme activity	[28]
Cauliflower (heads)	1. Membrane filtration 2. Sephadex G100 column	1. 200 g cauliflower head, crushed and obtain filtrate 2. 4 °C, 50 mM sodium acetate buffer	1.7% yield of lysozyme with purification fold 66.7	Purifying lysozyme using membrane filtration and Sephadex G100 chromatography, yields homogenized single band in SDS-PAGE	[45]
Tamarind	Single-step chitin bead affinity chromatography	1. Tamarind kernel powder extract was centrifuged at 8000 rpm for 30 min. 2. Equilibrate with 20 mM Tris–HCl buffer, pH 7.0	64.6% yield of chitinase with purification fold 1.64	N/A	[46]

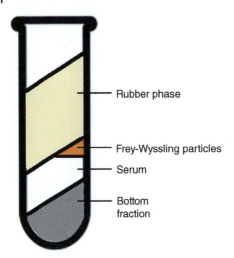

Figure 9.3 Schematic diagram of NRL fractions obtained upon centrifugation at 44,000 × g for one hour at 4 °C. Source: Drawn by Yong Yee Sheng.

significant chitinase activity) from papaya. The method proved to be more industrially friendly, i.e. generating high yield for specific applications (88% recovery and 100% purity), which was superior compared to the conventional two-step salt precipitation method [44].

Commercial rubber industries gather fresh NRL tapped from rubber tree *H. brasiliensis* and add ammonia to the NRL to prevent coagulation. The NRL is subsequently stored in an airtight environment under low temperatures during transportation. This is needed to minimize bacteria growth, as bacteria will produce lactic acid, which can cause coagulation of NRL, deteriorating the quality of NRL. The NRL requires to go through a compounding process, where chemicals, surfactants, minerals, and fillers will be added based on specific formulas in order to modify the mechanical properties of NRL to suit the final rubber product properties.

Further, some industries perform mechanical deproteinization by centrifuging the NRL before the compounding process to separate the protein-rich fraction from the NRL itself. The protein-rich fraction contains high concentration of lysozymes and chitinases and is often removed as by-products (waste). NRL are commonly separated into three different fractions by centrifugal force, although they can be further separated up to nine fractions [43]. The distinct three fractions are the upper phase (rubber phase), middle phase (C-serum), and bottom phase (Figure 9.3). The upper phase is a creamy white layer that contains different sizes of rubber particles that are surrounded by nonrubber particles [47]. The thinly-enclosed particles are negatively charged, playing important role in the colloidal stability of the NRL, and the presence of these allergenic proteins in this phase is described as acidic [43].

The middle phase, also known as the C-serum, refers to the aqueous medium containing the soluble substances from the cytoplasm [43, 47]. Apart from that, the middle phase has a higher protein content, which is twice the content of B-serum. On the other hand, the content of C-serum also includes another soluble suspension, such as inorganic salts and organic acids, and the presence of these allergenic proteins is also described as acidic [43].

Meanwhile, the bottom fraction of the NRL, is also known as B-serum or lutoid layer. This fraction comprised mainly lutoids, and also the presence of other minor organelles [43]. Lutoids make up 10–20% of the bottom layer and have a cationic protein presence [47]. This lutoid layer also contains the Frey–Wyssling complexes, a carotenoid-rich layer that gives its yellowish properties [46]. A range of 50–70% of the total bottom fraction is dominated by hevein, an anionic protein that is often associated with allergic reactions [43, 47]. B-serum is the only phase that has both the presence of acidic and basic allergenic protein [43].

Hence, the recovery scheme for the enzymes that can be anticipated should be initiated by centrifuging a fresh NRL using a high-speed centrifuge, specifically run for about one hour at 44,000 × g under 4 °C. Figure 9.3 shows the NRL fractions after high-speed centrifuge. Three fractions (rubber phase, C-serum, and bottom fraction) are formed, which can then be separated. The C-serum is removed from the tube and stored separately, whereas the bottom fraction is resuspended in a fresh sodium buffer because of its paste-like texture. These two fractions are reported to have the highest content of lysozymes and chitinases, thus further purification steps can be carried out to recover these enzymes [43]. Based on the literature, the possible primary recovery methods that could be applied for recovering the enzymes are chromatographic methods and alternative methods like ATPS. Depending on the potential application of the enzymes, it can be anticipated that high degree of purity is not generally required and the downstream processing is only required to enrich the enzyme fraction. Meanwhile, the rubber phase, which is easily hardened after the centrifugation, can be recovered as solid rubber. The solid rubber maybe used to produce block rubber, also known as technically specified rubber (TSR) by the process of crashing, cleaning and drying. The TSR is mainly used in the production of tyres [43].

9.7 Conclusions

The serum and bottom fractions of NRL, which are obtained from centrifugation process, are considered to be the by-products of NRL processing. These fractions contain major proteins, which are associated with allergens, but enzymes, such as lysozymes and chitinases possess economic value if successfully recovered and applied for industrial applications.

The potential strategy to recover the lysozymes and chitinases from the NRL can be initiated with reference to the existing industrial pretreatment process of NRL – by employing centrifugal force to obtain the C-serum and bottom fractions from the liquid NRL. Subsequently, the protein-rich fractions, which consist of targeted enzymes can be separated and subjected to further purification process. Chromatographic and alternative methods like ATPS could potentially be exploited for achieving the primary recovery of the enzymes from the separated fractions. However, the deproteinized NRL maybe added with modifiers (e.g. chemical, surfactant, and fillers) to replace the lost proteins and maintain the structural stability of the rubber fraction. Extensive research is required to produce NRL with minimal allergens, possessing superior physical properties and stability for rubber products manufacturing in different industries.

References

1 Singh, M., Esquena, J., Solans, C. et al. (2014). Influence of hydrophobically modified inulin (INUTEC NRA) on the stability of vulcanized natural rubber latex. *Colloids Surf. A Physicochem. Eng. Asp.* 451: 90–100.

2 Lim, H.M. and Misni, M. (2018). Effect of sodium dodecyl sulphate and polyoxyethylene dodecyl ether on the rheological behaviour and stability of natural rubber latex. *ASM Sci. J.* 11 (1): 44–55.

3 Lim, H.M. and Misran, M. (2013). Effect of sodium dodecyl sulphate, Brij 35 on natural rubber latex properties. *J. Rubber Res.* 16 (3): 162–168.

4 Riyajan, S.A., Santipanusopon, S., and Yai, H. (2010). Influence of ammonia concentration and storage period on properties field NR latex and skim coagulation. *KGK Kautsch. Gummi Kunstst.* 63 (6): 240–245.

5 Manroshan, S., Mustafa, A., Mok, K.L. et al. (2009). Comparison between sodium dodecyl sulfate and polyfructose surfactant systems in urea deproteinisation of natural rubber latex. *J. Rubber Res.* 12 (1): 1–11.

6 Filza, M.S., Qamarina, M.S.N., Nurulhuda, A., and Norhanifah, M.Y. (2018). Purified natural rubber latex: investigating effect of single anionic surfactant and centrifugation process to latex properties and addition of bio-filler towards latex film performance. *AIP Conf. Proc.* 1985 (1): 040008. AIP Publishing LLC.

7 Czuppon, A.B., Chen, Z., Rennert, S. et al. (1993). The rubber elongation factor of rubber trees (*Hevea brasiliensis*) is the major allergen in latex. *J. Allergy Clin. Immunol.* 92 (5): 690–697.

8 Allergen Nomenclature Allergen from *Hevea brasilliensis* natural rubber latex. http://allergen.org/search.php?allergensource=Hevea+brasiliensis (accessed 01 September 2022)

9 Yeang, H.Y. (2004). Natural rubber latex allergens: new developments. *Curr. Opin. Allergy Clin. Immunol.* 4 (2): 99–104.

10 Fukuhara, L., Miyano, K., Yamamoto, Y. et al. (2015). Removal of proteins from natural rubber latex and gloves. *KGK rubberpoint* 68: 24–29.

11 Duong-Ly, K.C. and Gabelli, S.B. (2014). Salting out of proteins using ammonium sulfate precipitation salting out of proteins using ammonium sulfate precipitation. *Methods Enzymol.* 541: 85–94.

12 Fox, J. and Castella, J.C. (2013). Expansion of rubber (*Hevea brasiliensis*) in Mainland Southeast Asia: what are the prospects for smallholders? *J. Peasant Stud.* 40 (1): 155–170.

13 Yu, L.J., Chi, C.W.Y., Tangavaloo, V. et al. (2019). Protein reduction of natural rubber films through leaching solvent. In: *MATEC Web of Conferences*, vol. 268, 01011. EDP Sciences.

14 Teoh, Y.P., Don, M.M., and Ujang, S. (2011). Assessment of the properties, utilization, and preservation of rubberwood (*Hevea brasiliensis*): a case study in Malaysia. *J. Wood Sci.* 57 (4): 255–266.

15 Hytönen, J., Nurmi, J., Kaakkurivaara, N., and Kaakkurivaara, T. (2019). Rubber tree (*Hevea brasiliensis*) biomass, nutrient content, and heating values in Southern Thailand. *Forests* 10 (8): 638.

16 Wei, Y., Zhang, H., Wu, L. et al. (2017). A review on characterization of molecular structure of natural rubber. *MOJ Polym. Sci.* 1 (6): 197–199.

17 Gomez, J.B. and Hamzah, S. (1989). Particle size distribution in Hevea latex-some observations on the electron microscopic method. *J. Nat. Rubber Res.* 4 (1): 204–211.

18 Xiang, Q., Xia, K., Dai, L. et al. (2012). Proteome analysis of the large and the small rubber particles of *Hevea brasiliensis* using 2D-DIGE. *Plant Physiol. Biochem.* 60 (3): 207–213.

19 Mahendra, I.P., Linh, M.K., Thang, N.N. et al. (2021). Protein removal from natural rubber latex with $Fe_3O_4@Al_2O_3$ nanoparticle. *J. Braz. Chem. Soc.* 32 (2): 320–328.

20 Gaboriaud, F., Gaudemaris, B., Rousseau, T. et al. (2012). Unravelling the nanometre-scale stimuli-responsive properties of natural rubber latex particles using atomic force microscopy. *Soft Matter* 8 (9): 2724–2729.

21 Rochette, C.N., Crassous, J.J., Drechsler, M. et al. (2013). Shell structure of natural rubber particles: evidence of chemical stratification by electrokinetics and cryo-TEM. *Langmuir* 29 (47): 14655–14665.

22 Wang, Y., Lomakin, A., Latypov, R.F., and Benedek, G.B. (2011). Phase separation in solutions of monoclonal antibodies and the effect of human serum albumin. *Proc. Natl. Acad. Sci.* 108 (40): 16606–16611.

23 Terwisscha, V.S., Kalk, K.H., Beintema, J.J., and Dijkstra, B.W. (1995). Crystal structures of hevamine, a plant defence protein with chitinase and lysozyme activity, and its complex with an inhibitor. *Structure* 2 (12): 1181–1189. RCSB Protein Data Bank. https://www.rcsb.org/structure/1HVQ.

24 Ercan, D. and Demirci, A. (2016). Recent advances for the production and recovery methods of lysozyme. *Crit. Rev. Biotechnol.* 36 (6): 1078–1088.

25 Santos, I.S., Da Cunha, M., Machado, O.L.T., and Gomes, V.M. (2004). A chitinase from *Adenanthera pavonina* L. seeds: purification, characterisation and immunolocalisation. *Plant Sci.* 167 (6): 1203–1210.

26 Rathore, A.S. and Gupta, R.D. (2015). Chitinases from bacteria to human: properties, applications, and future perspectives. *Enzyme Res.* 2015: 791907.

27 Sahu, B.B., Sahu, U., Barik, N.K. et al. (2017). Biorefinery products from shell fish processing waste: application of Chitin, Chitosan, Chitooligo saccharides and derivatives in organic agriculture. *Int. J. Fish. Aquat.* 2: 27–31.

28 Wirth, S.J. and Wolf, G.A. (1990). Dye-labelled substrates for the assay and detection of chitinase and lysozyme activity. *J. Microbiol. Methods* 12 (3–4): 197–205.

29 Huang, H., Zhang, Q., Luo, J., and Zhao, Y. (2012). Sensitive colorimetric detection of lysozyme in human serum using peptide-capped gold nanoparticles. *Anal. Methods* 4: 3874.

30 Martin, M.N. (1991). The latex of *Hevea brasiliensis* contains high levels of both chitinases and chitinases/lysozymes. *Plant Physiol.* 95 (2): 469–476.

31 Glazer, A.N., Barel, A.O., Howard, J.B., and Brown, D.M. (1969). Isolation and characterization of fig lysozyme. *J. Biol. Chem.* 244 (13): 3583–3589.

32 Ng, A., Heynen, M., Luensmann, D. et al. (2013). Optimization of a fluorescence-based lysozyme activity assay for contactlens studies. *Curr. Eye Res.* 38 (2): 252–259.

33 Velasquez, L. and Hammerschmidt, R. (2004). Development of a method for the detection and quantification of total chitinase activity by digital analysis. *J. Microbiol. Methods* 59 (1): 7–14.

34 Ohtakara, A. (1988). Viscosimetric assay for chitinase. In: *Biomass Part B: Lignin, Pectin, and Chitin*, Methods in Enzymology, 426–430. Elsevier.

35 Witholt, B., Heerikhuizen, H.V., and Leij, L. (1976). How does lysozyme penetrate through the bacterial outer membrane? *Biochim. Biophys. Acta, Nucleic Acids Protein Synth.* 443 (3): 534–544.

36 Sytwala, S., Günther, F., and Melzig, M.F. (2015). Lysozyme- and chitinase activity in latex bearing plants of genus *Euphorbia* – a contribution to plant defense mechanism. *Plant Physiol. Biochem.* 95: 35–40.

37 Hukić, M., Seljmo, D., Ramovic, A. et al. (2018). The effect of lysozyme on reducing biofilms by *Staphylococcus aureus*, *Pseudomonas aeruginosa*, and *Gardnerella vaginalis*: an *in vitro* examination. *Microb. Drug Resist.* 24 (4): 353–358.

38 Kumar, A., Kumar, V., Prakash, R. et al. (2010). Purification and characterization of a new chitinase from latex of *Ipomoea carnea*. *Process Biochem.* 45 (5): 675–681.

39 Mendonsa, E.S., Vartak, P.H., Rao, J.U., and Deshpande, M.V. (1996). An enzyme from *Myrothecium verrucaria* that degrades insect cuticles for biocontrol of *Aedes aegypti* mosquito. *Biotechnol. Lett.* 18 (4): 373–376.

40 Kasprzewska, A.N.N.A. (2003). Plant chitinases-regulation and function. *Cell. Mol. Biol. Lett.* 8 (3): 809–824.

41 Karasuda, S., Tanaka, S., Kajihara, H. et al. (2003). Plant chitinase as a possible biocontrol agent for use instead of chemical fungicides. *Biosci. Biotechnol. Biochem.* 67 (1): 221–224.

42 Yagami, T., Sato, M., Nakamura, A., and Shono, M. (1995). One of the rubber latex allergens is a lysozyme. *J. Allergy Clin. Immunol.* 96 (5): 677–686.

43 Yeang, H.Y., Arif, S.A.M., Yusof, F., and Sunderasan, E. (2002). Allergenic proteins of natural rubber latex. *Methods* 27 (1): 32–45.

44 Nitsawang, S., Hatti-Kaul, R., and Kanasawud, P. (2006). Purification of papain from *Carica papaya* latex: aqueous two-phase extraction versus two-step salt precipitation. *Enzym. Microb. Technol.* 39: 1103–1107.

45 Manikandan, M., Balasubramaniam, R., and Chun, S.-C. (2015). A single-step purification of cauliflower lysozyme and its dual role against bacterial and fungal plant pathogens. *Appl. Biochem. Biotechnol.* 177 (2): 556–566.

46 Rao, D.H. and Gowda, L.R. (2008). Abundant class III acidic chitinase homologue in tamarind (*Tamarindus indica*) seed serves as the major storage protein. *J. Agric. Food Chem.* 56 (6): 2175–2182.

47 Guerra, N.B., Pegorin, G.S.A., Boratto, M.H. et al. (2021). Biomedical applications of natural rubber latex from the rubber tree *Hevea brasiliensis*. *Mater. Sci. Eng. C* 126: 112126.

10

Sago Wastes as a Feedstock for Biosugar, Precursor for Chemical Substitutes

Mohd A. Jenol[1], Muhd N. Ahmad[2,3], Dayang S. A. Adeni[2], Micky Vincent[2], and Nurashikin Suhaili[2]

[1]*Universiti Putra Malaysia, Department of Bioprocess Technology, Faculty of Biotechnology and Biomolecular Sciences, Serdang, 43400, Selangor, Malaysia*
[2]*Universiti Malaysia Sarawak, Resource Biotechnology Programme, Faculty of Resource Science and Technology, Kota Samarahan, 94300, Sarawak, Malaysia*
[3]*i-CATS University College, Faculty of Agriculture and Applied Science, 93350, Sarawak, Malaysia*

10.1 Introduction

Sago or scientifically known as *Metroxylon sago* is one of the most important crops in Malaysia. To promote the growth of the sago sector, the Malaysian government has provided RM20 million in subsidies for sago palm plantations [1]. Sago palm plantation has unique adaptability against extreme growth conditions, which include variety of climate adaptability and wide range of pH (neutral to extremely acidic) soil conditions. According to Flores [2], sago palm has higher production capacity of starch as compared to cassava, corn, or rice, which accounted for 20–25 tonnes/ha/year. Sarawak has been acknowledged as one of the main sago starch exporters in the world, providing more than 47 thousand tonnes per year. Additionally, with the increasing demand, the value is expected to rise significantly, leading to the increment in waste production.

Generally, sago industry generates two types of wastes, which are in-plantation site and in-mill wastes. Sago fronds are the first-point waste generated in-plantation site of sago industry. It is generated during pruning and harvesting activities, which accounted for 7500 fronds discarded daily [3]. Meanwhile, in the processing mill, sago industry has generated another three main by-products, which are bark, hampas, and effluent. Sago bark resulted from the debarking process of sago palm trunk, which accounted for 5–15 tonnes daily [4]. Concurrently, sago hampas is a starchy lignocellulosic biomass produced upon completing the process of starch extraction. It is one of the largest solid biomass produced in the sago starch processing mill, which accounted for 150 tonnes per day [5]. Despite all the solid biomass generated, effluent is the largest by-product of sago starch processing mill, due to the huge amount of water used. The wastewater produced covers up to 97% of the total

Chemical Substitutes from Agricultural and Industrial By-Products: Bioconversion, Bioprocessing, and Biorefining, First Edition. Edited by Suraini Abd-Aziz, Misri Gozan, Mohamad Faizal Ibrahim, and Lai-Yee Phang.
© 2024 WILEY-VCH GmbH. Published 2024 by WILEY-VCH GmbH.

by-products generated, which accounted for 10^6 l of wastewater generated for every 1 tonne of sago starch produced [6].

Both solid and liquid wastes produced from sago industry has potential value to be converted into higher value-added products. Enormous efforts have been done exploring the utilization of each sago waste as a feedstock. Bioconversion of sago wastes, including sago bark [7], sago hampas [8], sago frond [3], and sago wastewater [9], into biosugars has been proficiently demonstrated by various works. This is due to the high concentration of starch and cellulosic content, which can be converted to wide range of pentoses and hexoses by the implication of enzymatic hydrolysis. This indicates the promising potential of sago wastes, which requires further efforts to explore their full potential.

Therefore, this chapter is aimed to bring light to the current development of each type of sago waste into value-added products, especially biosugars and other bioproducts. The challenges and future prospects of sago waste biorefinery are also discussed.

10.2 Current Status of Sago Starch Industry

10.2.1 Sago Palm Cultivation

Sago palm, a starch producer is a palm that can thrive well in lowland freshwater swamps and tropical rainforests, normally found in Southeast Asia including Papua New Guinea, Malaysia, and Indonesia [10]. The sago palm is characterized as soboliferous, grows in a cluster, produces a substantial amount of starch lump at the trunk, and is classified as a perennial plant. These characteristics of sago palm allow it to adapt to a variety of climate settings. It can grow on uncertain lands such as deep peat soils and areas with a high water table. The development of sago palm as a novel source of starch is crucial for increasing the output of global starch crops to meet the rising demand proportional to population growth. The innovative starch source can alleviate global hunger while overcoming crop production issues caused by climate change and global warming. According to Bintoro et al. [11], the extremely acidic and brown color of peat soil is the optimal setting for the mutually beneficial microorganism to develop with sago palms, which enables sago palms to thrive on riverbanks, near lakes, and in damp soil. Sago palms are well-adapted to marginal soils in Papua New Guinea, where most cash crops cannot grow. Sago palms can grow in a wide range of neutral to acidic soil conditions, from uplands to low-flooded places.

Sago palm has the remarkable ability to survive and thrive following a forest fire. The resilience of the sago palm against wildfires can lessen the time and expense required for replanting and help farmers recover from the devastation. The tropical climate in Malaysia is not only limited to severely hot and dry weather but is also vulnerable to monsoon seasons classified as Geo-Hazards in Malaysia that can inflict devastation owing to extended high rainfalls. Sago palm virtually resides in the freshwater of the swampy area, allowing it to be adept and withstand prolonged or severe floods [12]. The ability of the sago palm to withstand catastrophic natural disasters demonstrates its supremacy as a dependable resource for ensuring sustained food security and safety in the face of unfavorable climate change resulting from rapid worldwide urbanization. Thus, sago palm plantations must be expanded as a substitute for less drought- and flood-tolerant crops.

10.2.2 Sago Starch Industry in Malaysia

The sago palm industry is one of the primary contributors to Sarawak's Gross Domestic Product (GDP), mostly from the manufacturing of sago starch. Sarawak is one of the largest exporters of sago starch in the world, producing over 200 000 tonnes annually and exporting 47 000 tonnes to feed the worldwide starch supply. The export value of sago starch climbed from RM 135/tonne in 1970 to RM 2000/tonne in 2010 due to rising worldwide starch demand [13]. Primarily, sago starch is utilized for food production, especially traditional delicacies. In addition, sago starch is used as a raw material to create high fructose syrup and glucose as a substitute for imported cornstarch. Alongside tapioca, sago starch is utilized in the production of monosodium glutamate (MSG). As an extender for urea formaldehyde adhesive in the wood industry, an alternative to hydroxyethyl cellulose to control fluid loss in the petroleum industry, a stable and long-lasting biodegradable material for the production of plastic, and a low-cost substrate for fermentation industries, industrial grade sago starch has a wide range of applications [12].

Starch is produced primarily from maize, cassava, potato, and rice, supplemented by sago starch by less than 0.05% of the world's starch need. With a 24 MT/ha/year yield, sago palm has a greater land use efficiency than rice (6 MT/ha/year), maize (5.5 MT/ha/year), wheat (5 MT/ha/year), and potato (2.5 MT/ha/year) [12]. Significantly, the global demand for starch is expected to increase at a pace of 7.7% per year, which is equivalent to 3.85 million tonnes per year. Malaysia was estimated to produce 2.4 million tonnes of sago starch with a gross income of RM 4.8 billion by 2020 by allocating 250 000 ha of land for sago palm plantation to meet the worldwide demand for starch due to the high land productivity and efficiency of the sago palm [13]. The current plantation area for sago palm is only 32, 328.5 ha with a production capacity of 131, 249.2 MT in 2020. For the last 10 years, sago palm plantation area undergoes 47% reduction responsible for 30% decrease in sago starch productivity [14].

10.2.3 Sago Starch Processing in Sarawak

There are eight sago plants in operation in Sarawak as of 2003, seven of which are located in the Mukah – Dalat regions and one in the Igan – Sibu region [15]. Each harvested sago palm from the storage farm is first cut into approximately one-meter-long per log. These logs are sent to slicers, which separate the pith and bark. Logs of sago are delivered to the factory by boat through the river or, more recently, by lorries when roads are accessible.

Debarking of sago logs was accomplished manually, with an automatic debarking machine, or with continuous feeding in rasping. The bark accounted for around 20% of the weight of each log. For a medium-sized sago mill that produces approximately 12 tonnes of dry sago starch per day, the daily average consumption of sago logs will be approximately 600 logs, resulting in a daily total of 15.6 tonnes of sago bark waste [16]. Sago barks are often sun-dried and stacked for use as firewood or flooring material in sago mills.

Each of the 1 m long debarked sago logs is fed by conveyor belt into the high-speed mechanical rasper with chrome nails set on a drum to pulverize sago logs into fine pith. Following processing, the fine pith and water mixture is now a white slurry containing around 20–25% starch. The resulting starch slurry is forced through a succession of centrifugal sieves to separate coarse fibers and then condensed in holding tanks. A water-saving and water-recycling tank that also recovers leftover starch. The resultant slurry is pumped into a dehydration tank, which further concentrates the starch between 30% and 35% moisture [16].

A rotating vacuum drum dryer is utilized to dehydrate the starch in the dehydration tank. Starch solid from the rotary vacuum drum dryer is fed into a cyclone dryer, which is heated with a diesel–coal mixture to dry the slurry into starch powder. The starch bag is made of woven polythene with a thick, transparent plastic lining to reduce the effects of moisture. Each bag weighs around 50 kg and multiple sacks are transported together. Modern starch extraction process companies, especially in Sarawak imply advanced mechanical procedures for the entire process, aimed at shorter processing time.

10.3 Sago Wastes Biomass

10.3.1 Sago Wastewater

Sago wastewater is slurry waste generated by a sago processing plant. It is estimated that approximately 10^6 l of wastewater are generated for every 1 tonne of sago starch produced [6]. According to several studies, an average sago plant is reported to produce approximately 7 tons of effluent daily [16, 17]. The bulk of the wastewater is liquid (up to 97%), while the solid portion (approximately 5%), known as sago hampas makes up the remaining waste [18]. This liquid waste is highly rich in organic materials. According to Vincent et al. [18], when stored at ambient conditions, sago effluent usually undergoes oxidation that results in color changes and the release of obnoxious odor.

In recent years, sago wastewater volume generated from sago mills has increased tremendously as demands for sago starch escalate. As typical sago starch extraction plants are small-scale, no proper waste treatment facilities such as efficient ponding systems, anaerobic digestors, and bioreactors are available. Therefore, the common practice is usually to discharge the sago wastewater directly into the nearby waterways. This practice is very controversial as sago effluent contains sufficient nutrients to support algal growth and would cause eutrophication if discharged directly into the waterways, causing water pollution that negatively affects the surrounding aquatic environments [18].

Currently, several studies are being conducted to address the issues of improper waste management. One example is the use of an up-flow packed bed digester for the anaerobic fermentation of sago wastewater to exploit the high carbon-to-nitrogen ratio [18]. Huang [19] suggested that the voluminous sago plant waste stream to be used as a cheap and cost-effective culture medium to cultivate algal cells to produce single-cell protein. Awg–Adeni et al. [20] added that the solid portion of sago wastewater (sago hampas) can potentially be used as substrate to produce enzymes (such as amylases, cellulases, and proteases) and reducing sugars for microbes, manufacture of particleboard and growth medium for mushroom production. Sago hampas is suitable for this application due to the high cellulose, hemicellulose, and lignin content of 23.0%, 9.2%, and 4.0% w/w, respectively [20].

The reducing sugars produced from sago wastewater can also be used as feedstock in the production of bioethanol and other high-value chemicals [9]. Another more recent study also reported the use of sago wastewater as a substrate for aquaculture feed production due to the high carbon-to-nitrogen ratio (105 : 0.12) content [18]. According to this study, submerged fermentation (SmF) was used to biologically convert the complex substrates in sago effluent for the production of high protein fungal biomass (HPFB) using *Rhizopus oligosporus,* while at the same time reducing the starch content, as well as the BOD and COD levels. This promising treatment was shown to demonstrate the reduction of starch content, BOD, and COD by up to 96.70%, 89.81%, and 78.30%, respectively [18].

10.3.2 Sago Bark

Sago bark is the hard encasing layer of the sago palm trunk with a thickness of about 2 cm [4]. In a sago starch processing plant, sago bark is a major solid waste product. According to Izaan et al. [21], a typical sago starch extraction process generates 0.5 tonnes of sago bark waste per tonne of sago flour produced. The bark accounts for about 17–25% of the total logs processed [9], which is approximately 5–15 tonnes in weight, daily [4]. Proximate analysis of the sago bark reported moisture and ash content of 9.00% and 2.09% w/w, respectively. Further ultimate analysis of sago bark estimated the carbon content at 45.10%, hydrogen at 6.31%, and oxygen at 48.54% [22]. Although the carbon content of sago bark is high, a larger portion is deposited in the trunk as pith biomass, which contains 60–70% starch [23].

Sago bark is an example of lignocellulosic biomass. Unlike sago palm pith, sago bark is largely made of cellulose, hemicellulose, and lignin. Cellulose in sago bark is a major structural component of the cell wall, which provides mechanical strength and chemical stability. The chemical structure of cellulose is shown in Figure 10.1. Cellulose is a linear biopolymer consisting of D-glucose subunits linked together by β-(1,4)-glycosidic bonds in intricate arrangement with the hydroxyl groups oriented to form strong bonds, creating a semicrystalline structure that is very rigid and difficult to break via enzymatic hydrolysis. The cellulose chains are then embedded in the core of the sago bark biomass [22].

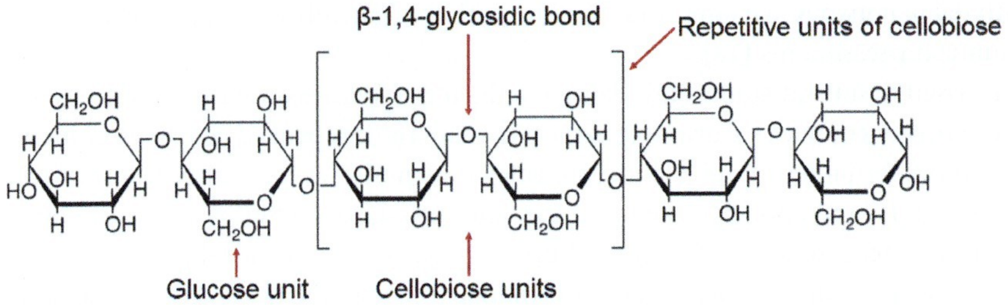

Figure 10.1 The chemical structure of cellulose.

Hemicellulose is the second most abundant heteropolysaccharide in sago bark. Hemicellulose functions as a cross-linking agent in the cell wall that binds structural and nonstructural polysaccharides through a variety of covalent and noncovalent interactions. In short, hemicellulose provides structural backbone to plant cell walls. Unlike cellulose, hemicellulose is a branched structure that consists of several 5-C sugars (xylose and arabinose) as well as 6-C sugars (glucose, galactose, and mannose) [24]. The major sugars in hemicellulose are xylose and glucose. Depending on the plant species, hemicellulose may also contain acetylated sugars, such as glucuronic, galacturonic, and methylgalacturonic acids [24].

Lignin in sago bark is strongly associated with the hemicelluloses in the cell walls of sago pith. The main block of the lignin is the phenyl–propane unit. Lignin is a heterogeneous polymer comprises complex non-condensed syringyl units and polyphenolic substances, such as p-coumaryl, conifery, and sinapyl alcohols [24]. These compounds are held together by different kinds of linkages, but most commonly by ether bonds. Lignin, which is present in the cellular wall of sago bark acts like glue by filling the gap between and around the cellulose and hemicellulose. It is insoluble in water and provides rigidity, making lignin degradation very hard. More importantly, this complex biopolymer confers microbial resistance and oxidative stress [24].

Approximately, 85% of the sago bark waste is unutilized after the sago extraction process [25]. Most of the sago bark is discarded into the environment for natural degradation [4]. Sago processing plants are also reported to incinerate sago bark waste for power generation. However, the voluminous amount of discarded sago bark provides many opportunities for this unwanted waste to be utilized for many downstream applications. Sago bark is used as a platform around the sago plant and as footpaths for houses. In the North Luwu communities, it is used as a replacement for boards and temporary bridges [4]. Sago bark ash from the incinerator units can also be added to soil to improve texture, soil nutrient availability, and crop productivity [25]. According to Kasi et al. [26], sago bark waste is a suitable culture medium to support fungal growth, especially basidiomycetes fungi, because of its high cellulosic and organic content. Because the sago bark contains lignin as one of its main components, it is also a suitable alternative for fuel and building materials [4]. Currently, sago bark is processed to produce sago plywood, wall tiles, and particleboards, which have potential as a building material. Sago bark can potentially be used as a material replacement for advanced industrial purposes such as lightweight concrete bricks that meet the requirements of SNI 03-0349-1989 for minimum compressive strength required [21]. Izaan et al. [21] further added that sago bark-based building materials are also economical to produce. The same study also reported that the physical properties of the sago bark composite board also meet the requirements of SNI 03-2105-2006. Another application of sago bark includes as a filler for fly ash bricks that possess the compressive strength that meets several ASTM requirements [21].

10.3.3 Sago Hampas

Sago hampas also known as sago pith residue is a starchy lignocellulosic by-product left behind upon completing the sago starch extraction process. It was estimated that 1 ton of sago hampas is discharged for every 1 ton of sago starch produced and 20 tons of sago effluent will concomitantly discarded directly into nearby streams. As of 2015, a total of about 1500 tons/day of refined starch were produced from about 10 modern sago mills in operation in Sarawak [5]. Hence, it is approximately 1500 tons of sago hampas are available on daily basis in Sarawak itself. The current waste management conditions potentially deteriorate the rivers and the environment. In rivers, the oxygen content will eventually decrease due to the high consumption of dissolved oxygen for degradation of the wastes via microbial activity, hence a severe drop in water quality able to endanger aquatic lives [27].

The chemical composition of raw sago hampas was stated in Table 10.1. All the sago starch content shown in the table is comparable despite the pulverization method of the mills. The amount of residual starch that is still trapped in the hampas depends on the quality of the extraction processes implemented by the sago mills.

According to Alias et al. [29], sago hampas has a highly beneficial advantage to be used for the fermentation feedstock as it contains high carbohydrate composition with low lignin and extractives content. Apart from that, sago hampas have been used as animal feed, compost for mushroom culture, substrate for producing confectioners' syrup, and for particleboard manufacture. Sago hampas are also applied as a cheap carbon source for the production of α-amylase and cellulase [30]. More importantly, there is no pretreatment required before the saccharification process due to the low lignin content in sago hampas [8].

Table 10.1 Chemical composition of raw sago hampas.

Starch (%)	Cellulose (%)	Hemicellulose (%)	Lignin (%)	Reference(s)
49.5	26.0	14.5	7.5	[8]
58.0	23.5	8.2	6.3	[17]
58.0	21.0	13.4	5.4	[28]
56.0	20.7	11.2	3.1	[29]

Table 10.2 Lignocellulosic composition of sago fronds at different growth stages.

Growth stage of sago frond	Hemicellulose (%)	Cellulose (%)	Lignin (%)
Adolescent	15.83 ± 0.10	41.43 ± 0.42	6.06 ± 0.60
Parent	18.32 ± 0.27	10.76 ± 0.23	38.82 ± 0.36
Pruned	17.62 ± 0.25	8.02 ± 0.08	40.63 ± 0.23

Source: Ahmad [31]/Universiti Malaysia Sarawak.

10.3.4 Sago Frond

Sago fronds are leafy waste, disposed of after harvesting sago palms for starch extraction [19]. It was estimated about 500 palms are harvested daily in Mukah, Sarawak, and at about 15 fronds/palm, at least 7500 fronds are discarded, each day [3]. Sago frond is a potential waste, but also a valuable resource if processed to produce sugars or other by-products. It is a lignocellulosic material that comprises primarily cellulose, hemicellulose, and lignin and if left on the plantation floor, these fronds can potentially cause serious environmental issues due to their slow degradability rate. Currently, some of these fronds are used to make crude path layers to roll sago logs out from the farm to the nearest road or river [19]. The remaining fronds are left to degrade in sago palm estates, which potentially pose fire hazards in the dry season, concomitantly accommodating various pests that endanger the livelihood of the sago farmers. It was observed that adolescent sago fronds have deep green color skin and leave with white and compact pith. Parent sago fronds have a combination of brown and dark green color of the skin and leaves. The white part of the pith is rather compact but the brown section is relatively loose due to dryness. Pruned sago fronds have brownish skin and pith, the latter being loose and sagging. The lignocellulosic composition of sago frond samples is shown in Table 10.2.

Analyses conducted conclude that enzymatic hydrolysis of cellulosic fiber in sago frond using Celluclast 1.51 produced cellobiose as the main sugar (16.16%) followed by glucose (9.17%). The adolescent stage sago frond is the best source for the production of sugars due to the high cellulose content (41.43%) [3]. On the other hand, the sago frond sap naturally contains glucose (28–68 g/l) without going through pretreatment process or involving addition of enzymes [19]. Sago frond sap also contains high amount of minerals (Ca, K, Mn, Mg, and P), which are useful in ethanol and kojic acid fermentation [19]. Utilization of sugars from sago frond fiber and sap for producing value-added products creates alternative biomass sources, hence concomitantly minimizing wastage from sago palm plantations.

10.4 Bioconversion of Sago Wastes into Biosugars and its Derivative Precursors

The sago palm is a very versatile agricultural commodity as a source of biomaterials, especially carbohydrates, because of its starch-rich lignocellulosic properties.

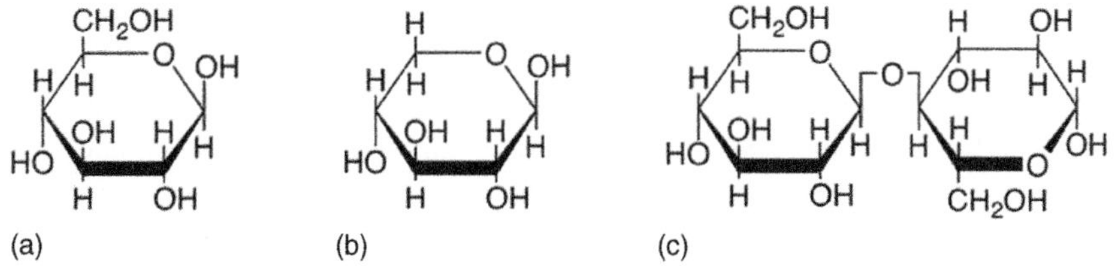

Figure 10.2 The chemical structure of (a) Glucose, (b) Xylose, (c) Cellobiose.

Sago starch has been recognized as a new sugar source and cheaper carbohydrate alternative to replace sugarcane sugar [32]. According to Toselong et al. [33], sago carbohydrate levels are almost equal to carbohydrate levels found in other starchy crops, such as rice flour, cassava, and potatoes. Globally, sago starch is well known as a basic material for various kinds of industrial applications, such as in food products, paper, adhesives, textiles, and cosmeceuticals [33, 34]. As an industrial raw material for the food industry, sago starch can be used to produce sweeteners such as glucose, fructose syrup, liquid sugar, and flavorings. While in the nonfood industries, sago starch can be derivatized to produce bioenergy, biofuels, and bioproducts. Even sago starch processing residues such as sago effluent (sago hampas and sago effluent hydrolysate) can be processed to produce sugars. Previous studies have reported that the cellulose fraction of sago hampas can yield up to 89% glucose, while the hemicelluloses contain xylose and glucose, along with small amounts of arabinose, galactose, mannose, and uronic acids [3].

Glucose (Figure 10.2a), also known as dextrose, is a polyhydroxyl aldohexose monosaccharide. Glucose is odorless and colorless and is usually found as white crystalline or granular powder [24]. It is the most abundant monosaccharide in sago starch after complete hydrolysis. Glucose is produced from sago via amylolytic hydrolysis of the starch. Glucose can also be produced from dried sago fronds at 9–11% w/w yields [3].

Glucose may exist in two forms, which are D-glucose and L-glucose. However, only the D-glucose confirmation is found in nature, while the latter is produced synthetically in small amounts artificially.

Xylose (Figure 10.2b) is an aldopentose that consists of five carbon atoms and an aldehyde functional group. Xylose is a common sugar in lignocellulosic biomass. In sago palm, xylose is the second major monosaccharide sugar after glucose, making it a promising carbon source [35]. In similarity to glucose, the dextrorotary form of xylose (D-xylose) is the naturally occurring form of the sugar, while the levorotary form (L-xylose] is only available when synthesized in the laboratory [36]. Xylose contributes to the color and flavor of food as it is often associated with flavonoids and oligosaccharides. This 5-C sugar is not readily processed by human metabolic machinery, making it widely used as a diabetic sweetener in food items and beverages [36].

Enzymatic hydrolysis of physically treated sago palm residues can also be used for the production of cellobiose (Figure 10.2c). Cellobiose is a disaccharide, formed from 2 β-glucose molecules linked by β-glucosidic bonds. A recent report by

Dewayani et al. [32] documented that cellobiose can be produced at 12% w/w from fresh sago leaves and sago frond (as dried powder) using a cellulase enzyme complex (containing cellobiohydrolases, endoglucanases, and β-glucosidases). Another related study further reported cellobiose production at 16–18% [3]. Production of cellobiose is potentially very profitable as cellobiose is significantly more expensive (about US$ 2000/kg) than glucose, maltose, and sucrose. Although cellobiose is not as common as table sugar, it has several prebiotic properties, especially for the fermentation of dairy products associated with bifidobacterial [6].

10.5 Bioprocessing Sago Wastes Fermentable Sugar for Chemicals Substitute

10.5.1 L-Lactic Acid

Acknowledge as one of the most versatile organic acids, lactic acid is applicable to be utilized in various industries such as food, cosmetic, pharmaceutical, and bioplastic. The global lactic acid revenue is forecast to be worth US$ 3.7 billion in 2020 and expected to increase up to US$ 8.7 billion in 2025 [37]. The unfortunate event of the COVID-19 pandemic hit the bioplastic manufacturing industry at the right spot to escalate the global demand for lactic acid for the production of consumable food packaging and personal protection equipment (PPE) due to protective effect of lactic acid against pathogenic microorganisms and biodegradable properties.

Lactic acid conventionally produces through chemical synthesis; however, due to the potential toxicity threat of D-lactic acid subcomponent, demands the production of pure L-lactic acid the endogenous compound that effectively produces via biotechnological fermentation of homofermentative lactic acid bacteria. Initially, food-based crops utilized as major resources to produce lactic acids such as sugarcane, maize, and cassava, which later also include sago starch. Enzymatic hydrolysis of sago starch superficially recovers 100% of sugar in form of glucose that later use as feedstock to produce pure L-lactic acid using *Lactococcus lactis* IO-1 strain [38]. However, overexploitation of food-based crop to produce chemical substitute and biofuel raise an ethical concern that case may compromise global food security and safety, especially for third world countries [31]. Hence, agricultural by-products and waste are proposed as ideal alternative raw materials to synthesize L-lactic acid.

In sago industry, frond is suggested as an elegant alternative resource to produce multiple chemical substitutes and biofuel at the same time providing the ultimate solution to sago plantation due to long maturation period and nonexistence of alternate commodities while growing. Sago frond possessed lignocellulosic components and sap that can be utilized as substrate to produce L-lactic acid through fermentation [39]. In the meantime, sago frond sap simply can be extracted by using a roller presser machine, which possesses mixture of glucose, xylose, and residual starch. Boiling is sufficient to sterilize the sap and with the aid of thermophilic enzymes can simultaneously convert residual starch into glucose to maximize the carbon source in the substrate to produce L-lactic acid [39].

Presterilized sago frond sap was then formulated with sufficient amount of nitrogen sources, salt, and neutralizing agent to enhance the fermentation process and stimulate the growth of the strain under minimum control and maintenance of batch fermentation system. However, the excessive carbon source of sago frond sap is not favorable toward *L. lactis* growth performance due to high osmotic pressure caused by high sugar concentration also prevents the strain to convert sugar into lactic acid efficiently. Commonly, *L. lactis* stain performs best at low sugar concentrations without pH control in lactic acid yield and productivity. Hence, sago frond sap needs to be diluted to the optimum concentration. Fermentation on the optimized condition of sago frond sap as substrate manages to increase the ideal sugar concentration for *L. lactis* IO-1 up to 75% (35 g/l) at the best yield, efficiency, and productivity of L-lactic acid [39]. Production of lactic acid through fermentation of sago frond sap is not only cost-effective, practical, and sustainable but also can avoid the ethical conundrum of using food-based crops to produce chemical substitutes or biofuel.

10.5.2 Antimicrobial and Prebiotic Sugar (Cellobiose) from Sago Frond

Partial hydrolysis is conducted by cellulase enzyme originating from *Trichoderma reesei* on sago frond's cellulose to produce mixture of cellobiose and glucose. Cellobiose categorized as disaccharide builds up from two glucose molecules bound together by β-glycosidic bond. Despite being an excellent substrate to produce L-lactic acid, cellobiose exhibits a unique beneficial effect by providing concurrently prebiotic and antibacterial properties. Due to lack of cellulase enzyme, cellobiose is unable to utilized by human body, which offers beneficial effects to probiotic microflora throughout the digestive system due to noncompetitive carbon source against human cells and digestive acid tolerant [40].

According to Awg–Adeni et al. [3], the capability of *L. lactis IO-1* to utilize cellobiose allowed it to grow better in the substrate solution that contains mixture of cellobiose and glucose. Regardless of being an ideal substrate for the production of L-lactic acid, cellobiose from sago frond act as slow-release carbon source that allows *L. lactis* IO-1 to consume energy efficiently, resulting in complete sugar consumption at the end of the fermentation.

On the other hand, as reported by Ahmad [31], complex structure of cellobiose also inhibits the growth of *Staphylococcus aureus*, *Escherichia coli*, and *Salmonella typhi*, which are associated with foodborne disease. Those pathogenic bacteria are naturally unable to metabolize cellobiose. Once cellobiose attaches and remains at the membrane receptor, it can prevent the cell to absorb consumable carbohydrates leading the cell to starve to death. Meanwhile, study on the fermentability of sago frond sugar to bioethanol shows the incapability of *Saccharomyces cerevisiae* to metabolize cellobiose and also shows the potential of the respected disaccharide to offer inhibition effect on yeast or other genera of fungi. Preliminary study on the antifungal properties of cellobiose shows promising results where *Candida albicans* was selected as the target pathogen. Brown sago frond sugar made up of cellobiose and glucose also possessed significant amounts of phenolic and flavonoid compounds shows higher antifungal activity than purified sago frond sugar [31].

10.5.3 Sago Frond Silage

Residual sago frond fiber from the extraction of sap and sago leaves can be used as raw material to produce silage. The ideal mixture of sago frond fiber and leaves is achieved at 50 : 50 ratios, which provide a balance composition of moisture content, total water-soluble sugar, protein, digestible fiber, and pH value. The *Lactococcus lactis* IO-1 recovered from post-fermentation of L-lactic acid by using sago frond sap repurpose as an inoculant for the manufacturing of sago frond silage. The viability of *L. lactis* IO-1 to produce L-lactic acid from sago frond sap suggests the potential of the respected strain to provide probiotic effect during ensiling process by producing substantial amount of lactic acid to improve the performance of the process and preserve the silage for better shelf life. Hence, inoculation of *L. lactis* IO-1 manages to minimize up to 50% of the protein degradation rate during the ensiling process and preserve the sago frond silage for up to 24 months without the addition of antibiotics and antifungals [39].

Inoculation of *L. lactis* IO-1 increased the edibility of sago frond silage by reducing the neutral detergent fiber (NDF) content of the silage. *L. lactis* IO-1 can conduct partial hydrolysis to metabolize the reducing end (xylan) of hemicellulose structure into simple, reducing sugar, such as xylose [41]. Consequently, xylose can be utilized by *L. lactis* IO-1 as the source of carbohydrates for lactic acid production. Yet, hemicellulose possesses a major component in NDF. Substantial degradation of hemicellulose by inoculation of *L. lactis* IO-1 directly reduced the NDF content of the silage. Thus, the application of *L. lactis* IO-1 as an additive is expected to improve the *in vitro* edibility of the sago frond silage. These results strongly imply that amendment with *L. lactis* IO-1 can increase the amount of feed intake by the ruminant.

Ironically, an extensive acidic condition in the sago frond silage due to inoculation of *L. lactis* IO-1 activates the lactic acid degrading bacteria leading to high concentration of acetic acid by the end of the ensiling process. A preliminary study on the diversity of lactic acid bacteria shows the existence of an indigenous strain of heterofermentative lactic acid bacteria in the sago frond that may be responsible for the lactic acid degradation phenomenal. This situation was similarly reported by Filva [42], where inoculation of *Lactobacillus buchneri* resulted in high concentration of lactic acid at the beginning of the ensiling process and end up with substantial amount of acetic acid. High concentration of acetic acid in the silage will increase the aerobic stability against yeast and mold since acetic acid has stronger protective effect against fungi compared to lactic acid. In addition, acetic acid has lower acidity compared to lactic acid leading to increase in pH value of the silage suggesting the respected situation also the natural mechanism to stabilize the acidic condition to maintain the viability of lactic acid bacteria in the silage [39].

With the production of silage from sago fronds, farmers can offer alternative sources of raw material for the production of animal feed that is locally available for the production of livestock in the meantime can generate alternative commodities for the sago plantation while waiting for the sago palm to be harvestable.

10.5.4 Enzymes

One of the potential bioproducts that can be generated from microbial fermentation using the waste stream generated from the sago bioethanol production is laccases. Laccases are polyphenol oxidases that have manifold applications, particularly in effluent treatment, bioremediation, textile, and petrochemical industries. One of the bottlenecks of industrial laccase production is the high cost of enzyme production. The use of sustainable feedstock for industrial laccase production is, therefore, crucial in producing the enzyme economically for wide applications.

Having glycerol as the main by-product of bioethanol fermentation has made the bioethanol waste stream an attractive feedstock for growing microorganisms that can metabolize glycerol for producing target products. The presence of glycerol as the by-product of sago bioethanol fermentation has increased the economic viability of the sago bioethanol waste stream. Several works have reported the use of waste glycerol from biofuel industry as feedstock to produce some value-added bioproducts. To the best of our knowledge, the use of glycerol-based stillage generated after sago bioethanol fermentation is still scarcely reported.

In recent work, Mamat et al. [43] have demonstrated the feasibility of laccase production by *P. pastoris* GS115 using sago bioethanol liquid waste (SBLW) as the feedstock. The work demonstrated that the use of 40% (v/v) SBLW as the fermentation medium gave the highest laccase activity, which is equivalent to 73% of the enzyme activity attained in fermentations employing standard synthetic medium. Further supplementation of 40% (v/v) SBLW with 1% (w/v) yeast extract was found to increase the maximum biomass and laccase activity by 1.2-fold and 1.5-fold, respectively from that produced using standard synthetic medium.

In a subsequent research, Mamat et al. [44] reported the decolorization percentage of laccases produced using the optimal SBLW medium, which is 68.6%. The value represents 91% of the decolorizing performance of laccases produced from fermentations using standard synthetic medium. Interestingly, the decoloring performance of laccases produced using SBLW medium was achieved under nonoptimal decoloring conditions and in the absence of mediators. It is envisaged that higher decolorizing performance can be achieved under optimal conditions. This suggests that the laccases were successfully produced as a functional protein with promising decoloring ability as that of those produced using synthetic medium. Furthermore, the finding also indicates that the enzyme has a promising direction to be used for bioremediation application.

In general, the aforementioned preliminary studies suggest the promising feasibility of SBLW as an alternative glycerol-based fermentation feedstock for producing laccases, an important biocatalyst that has diverse industrial applications. Valorization of waste stream from sago bioethanol fermentation is seen as a sustainable effort in reducing the waste that will be otherwise causing severe environmental pollution. Moreover, this will also create an economic advantage for future sago biorefinery industry through the creation of potential on-site product stream. The proof-of-concept of SBLW as a fermentation substrate can be further extended to other bioproduction in increasing its utility and flexibility in the near future.

10.5.5 Kojic Acid and its Derivatives

Having residual starch that can be further converted to fermentable sugars, sago fiber can serve as a potential substrate for microbial fermentation to produce various bioproducts. One of the potential products that can be produced from sago fiber is kojic acid. Kojic acid is an organic acid that has diverse industrial applications, such as in food, cosmetics, and medicine.

The conventional method of producing kojic acid, which is via chemical synthesis, is normally deemed as complex and costly. Hence, there is a need to transition to alternative strategies. Several published works have reported the production of kojic acid via fermentation [45]. However, one of the challenges of the industrial production of kojic acid is the high cost of synthetic media.

Utilization of starchy agricultural waste such as sago fiber as fermentation feedstock for the production of kojic acid offers a low-cost and sustainable solution to the aforementioned challenge. The feasibility of kojic acid production from starchy substrates such as cornstarch and sago starch via SmF has been reported by Mohamad et al. [46]. One of the limitations of employing SmF as a mode of fermentation for starchy substrates is that the substrate needs to be first hydrolyzed to release fermentable sugars that can be metabolized by the microorganisms during the fermentation. The pretreatment of the starchy substrates is laborious and costly, and this may contribute to the complexity of the kojic acid production via SmF.

In a later work, Spencer et al. [47] reported the feasibility of kojic acid production from sago fiber by *Aspergillus flavus* Link 44–1 via solid-state fermentation (SSF). In contrast to SmF, SSF allows direct use of agricultural waste as a fermentation substrate, hence eliminating the need for substrate hydrolysis before the fermentation. The optimization of kojic acid production from sago fiber using Central Composite Design of Response Surface Methodology was conducted by Spencer et al. [47]. The work revealed that the data was found best to be represented by a quadratic model. Two factors namely inoculum density of 30% (v/w) and incubation time (18 days) were identified as the significant factors for kojic acid production. The highest kojic acid production achieved under the optimal conditions was found to be two times higher than the titer produced under the nonoptimized conditions. In summary, the prior works revealed the feasibility of transforming sago fiber into a potential bioproduct, such as kojic acid.

Sie et al. [48] demonstrated the feasibility of transforming kojic acid produced from sago fiber into kojic acid derivatives. Kojic acid has been widely used as a precursor for producing various types of derivatives that have improved properties than kojic acid. The derivatives can be used in various industrial applications including as antibacterial and antioxidant compounds and for dye-sensitized solar cells. As reported by Sie et al. [48], kojic acid produced from sago fiber by *A. flavus* Link 44-1 was chemically incorporated with chalcones and azobenzene. The results showed that kojic ester-bearing chalcone derivatives exhibited active inhibition against *Staphylococcus aureus*. In the case of kojic ester-bearing azobenzene derivatives, a moderate antibacterial activity against *Escherichia coli* was observed. The biological activities demonstrated by kojic acid derivatives in this study can be associated with the presence of C=C and N=N reactive moieties in the molecules [48].

In summary, the forenamed studies suggest another potential of sago fiber as a feedstock for producing kojic acid as well as kojic acid derivatives that have diverse industrial applications. In either case, adoption of sago fiber as the feedstock offers a sustainable and environmentally friendly solution to the conventional production of kojic acid and its derivatives.

10.6 Challenges and Prospect of Sago Wastes Biorefinery

10.6.1 Challenges

In sago processing industry, it is well-known constraints related to the processing line of the starch extraction process. These include (i) the implementation of the conventional processing method, which uses the simple technique with simple equipment; (ii) the low yield in productivity; and (iii) transportation. These situations have contributed to the slow development not only for the main product but for the by-products as well. This situation is mainly due to the isolated location of the processing mill and sago plantation, which contributed to the aforementioned limitation. The initiative measure should be taken, especially for the high technology processing machineries to improve the efficiency in sago starch processing industries. Apart from that, due to limitations in the integration of waste utilization, the possibility of water pollution in the river is higher. This is because most of the sago starch processing mill uses the river as their main transportation for the sago logs. The concern about the toxic component contained in the sago log eventually inhibited the microorganisms, which is vital in the degradation of organic matter in wastewater. In addition, the process of sago starch requires a lot of water. Therefore, a massive amount of wastewater is generated on a daily basis. The lack of constituent knowledge of physicochemical properties has made it hard to utilize in higher chain of value-added production.

On the other hand, the limitation faced by the sago starch industry is low expansion and development. This limitation is mainly due to the several reasons [5], including (i) the apathy of sago plantation caused by the long juvenile phase, suitable land as well as fluctuation of the market; and (ii) inconsistent supply has contributed to the disinclined favor of the customer. This matter should be highlighted to improve in order to ensure the sustainability of sago starch industry However, the limited knowledge related to the food and nonfood application of sago starch in comparison to other industrial starches can be found.

10.6.2 Future Direction of Sago Waste Utilization

Sago wastes are considered an underutilized and understudied type of biomass as compared to other industrially vital starches, which are cassava, maize, and potato

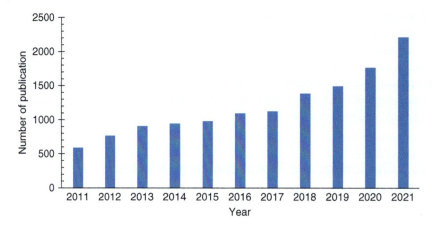

Figure 10.3 Number of publications in each year with keyword "sago waste". Source: Adapted from Google Scholar.

starches Figure 10.3 illustrates the number of publications with the keyword of "sago wastes" in Google Scholar for each year, since 2011. Based on the figure, the number of publications is deemed to be increased throughout the years, indicating that the efforts toward the utilization of sago waste have great potential value as substrate. Thus, it's a great opportunity for sago waste to be a key in the biotransformation of various industries. Genetic modification is one of the promising tools to be utilized on the sago wastes, especially starch component. The genetic mapping of the starch component of sago wastes will provide further understanding of the properties as well as the structure of starch from various origins. The transformation thru genetic modification will expand the potential application of sago waste. This situation is due to the better understanding of functionalities of sago wastes.

As for the starch, the "new" physical modification applied to the current industrially important starches, such as potato and maize starch can be implemented in the sago starch processing technique. These include electric fields, high pressure, plasma, and ultrasound treatment [49] aimed to treat the starch as clean label ingredient. Apart from that, sago wastes contained high amount of fiber, which is beneficial in various applications, especially in construction industries. According to Venkatesan et al. [50], the application of sago waste with other types of biomass has resulted in energy-saving wall structure equipped with heat transfer properties. A wide range of applications of sago waste can be further elevated by the accumulated understanding of the structure and physicochemical properties of this biomass, which is further beneficial to various applications and industries.

Nanotechnology has become the trend of interest among the new research era. The utilization of biomass-assisted polymer received much attention due to the performances of sago starch-based nanocomposite were greatly improved as compared to the neat polymer.

10.7 Conclusions

Sago wastes offer a new opportunity as an alternative resource for the production of biosugar in which several chemicals can be derived from its precursor. The various compositions of polymeric compounds found in sago wastes could induce the valorization of bioproducts, hence providing another option for biorefinery concept from agriculturally based biomass. Therefore, it can be deduced that the conversion of sago wastes into useful products is a promising approach to mitigate environmental pollution while concomitantly reducing the dependency on food resources. To be competitive, a wide range of sago waste applications can be further explored by understanding the structure and physiochemical properties of the biomass, which will be beneficial for industrial applications.

References

1. Embas, D.U. (2017). Sarawak Is the World Largest Sago Exporter. https://dayakdaily.com/sarawak-is-worlds-largest-sago-exporter (accessed on January 2023).
2. Flores, D.M. (2008). The versatile sago (*Metroxylon sagu* Rottb.) and its green potential for Mindanao. *Banwa* 5 (1): 8–17.
3. Awg-Adeni, D.S., Ahmad, M.N., and Bujang, K.B. (2018). Maximising production of prebiotic sugar (cellobiose) from sago frond. *Malays. Appl Biol.* 47 (1): 89–96.
4. Tenriawaru, E.P., Supu, I., and Cambaba, S. (2018). Physical properties of sago bark. *IOP Conf. Ser. Earth Environ. Sci.* 187: 012007. https://iopscience.iop.org/article/10.1088/1755-1315/187/1/012007.
5. Jong, F.S. (2018). An overview of sago industry development, 1980s–2015. In: *Sago Palm*, 75–89. Singapore: Springer Singapore http://link.springer.com/10.1007/978-981-10-5269-9_6.
6. Bujang, K.B. (2018). Production, purification, and health benefits of sago sugar. In: *Sago Palm*, 299–307. Singapore: Springer.
7. Ethaib, S., Omar, R., Mazlina, M.K.S. et al. (2016). Microwave-assisted dilute acid pretreatment and enzymatic hydrolysis of sago palm bark. *BioResources* 11 (3): 5687–5702.
8. Jenol, M.A., Ibrahim, M.F., Yee, P.L. et al. (2014). Sago biomass as a sustainable source for biohydrogen production by *clostridium butyricum* A1. *Bioresources* 9 (1): 1007–1026.
9. Hung, H.C., Adeni, D.S.A., Johnny, Q., and Vincent, M. (2018). Production of bioethanol from sago hampas via simultaneous saccharification and fermentation (SSF). *Nusant. Biosci.* 10 (4): 240–245. https://smujo.id/nb/article/view/3270.
10. Johnson, D. (2018). Distribution of sago making in the old world. In: *Sago-76: Papers of the First International Sago Symposium* (ed. K. Tan), 65–75. Kuala Lumpur, Malaysia: Kemajuan Kanji Sdn. Bhd.

11 Bintoro, M.H., Iqbal Nurulhaq, M., Pratama, A.J. et al. (2018). Growing area of sago palm and its environment. In: *Sago Palm*, 17–29. Singapore: Springer.

12 Bujang, K.B. and Ahmad, F.B. (1999). Production and properties of sago starch in Malaysia. In: *Sustainable Small-scale Sago Starch Extraction and Utilization: Guidelines for the Sago Industry*, 82–94. Nakhon Si Thammarat, Thailand: Kasertsart Agric and Agro-Indust Prod Dev Institute (KAPI) and FAO.

13 Ministry of Plantation Industries and Commodities (2020). Dasar Komoditi Negara. https://www.pustakasarawak.com/eknowbase/attachments/1596527434.pdf (accessed on December 2020).

14 Department of Agriculture Malaysia (2020). Perangkaan Tanaman Industri 2020. www.doa.gov.my/index/resources/aktiviti_sumber/sumber_awam/maklumat_pertanian/perangkaan_tanaman/perangkaan_tnmn_industri_2019.pdf (accessed on December 2020).

15 Manan, D.M.A., Chie, R., and Rumie, A.M. (2003). Sago starch technology: the Sarawak experience. *CRAUN Bull.*

16 Sikem, N.D. (2017). *Physio-Chemical Analyses of Raw and Purified Sago Sugars*. Universiti Malaysia Sarawak.

17 Vincent, M., Senawi, B.R.A., Esut, E. et al. (2015). Sequential saccharification and simultaneous fermentation (SSSF) of sago hampas for the production of bioethanol. *Sains Malays.* 44 (6): 899–904.

18 Vincent, M., Junaidi, F., Bilung, L.M. et al. (2020). Simultaneous reclamation of sago starch processing effluent water and *Rhizopus oligosporus* cultivation at different pH conditions. *J. Water Environ. Technol.* 18 (4): 254–263. https://www.jstage.jst.go.jp/article/jwet/18/4/18_19-152/_article.

19 Matnin, N., Awang Adeni, D.S., Ahmad, M.N., and Suhaili, N. (2021). The characteristics of sago frond sap from two selected growth stages; Angkat Punggung and Upong Muda palms. *Malays. J. Sci.* 40 (3): 43–53. https://mjs.um.edu.my/index.php/MJS/article/view/28862.

20 Awg-Adeni, D.S., Bujang, K.B., Hassan, M.A., and Abd-Aziz, S. (2013). Recovery of glucose from residual starch of sago hampas for bioethanol production. *Biomed. Res. Int.* 2013.

21 Hadi Izaan, I., Suraya Hani, A., Norhayati, A.W. et al. (2022). Preliminary study of sago fine waste as a sand replacement material for cement brick. *IOP Conf. Ser. Earth Environ. Sci.* 1022 (1): 012052. https://iopscience.iop.org/article/10.1088/1755-1315/1022/1/012052.

22 Erabee, I.K., Ahsan, A., Daud, N.N.N. et al. (2017). Manufacture of low-cost activated carbon using sago palm bark and date pits by physiochemical activation. *BioResources* 12 (1): 1916–1923.

23 Pasolon, Y.B. (2015). Environment, growth and biomass production of sago palm (*Metroxylon sagu* Rottb.): a case study from Halmahera, Papua and Kendari. *Int. J. Sustain. Trop. Agric. Sci.* 2 (1): 243093.

24 Huang, C.H. (2018). *Production of Liquid Biofuel from Fresh Sago Effluent*. Universiti Malaysia Sarawak.

25 Anwar, Z., Gulfraz, M., and Irshad, M. (2014). Agro-industrial lignocellulosic biomass a key to unlock the future bio-energy: a brief review. *J. Radiat. Res. Appl. Sci.* 7 (2): 163–173.

26 Kasi, P.D., Tenriawaru, E.P., Cambaba, S., and Triana, B. (2021). The abundance and diversity of Basidiomycetes fungi in sago bark waste. *IOP Conf. Ser. Earth Environ. Sci.* 739 (1): 012063. https://iopscience.iop.org/article/10.1088/1755-1315/739/1/012063.

27 Lai, J.C., Rahman, W.A.W.A., and Toh, W.Y. (2013). Characterisation of sago pith waste and its composites. *Ind. Crops Prod.* 45: 319–326. https://linkinghub.elsevier.com/retrieve/pii/S0926669013000149.

28 Jenol, M.A., Ibrahim, M.F., Kamal Bahrin, E. et al. (2019). Direct bioelectricity generation from sago hampas by *Clostridium beijerinckii* SR1 using microbial fuel cell. *Molecules* 24 (13): 2397. https://www.mdpi.com/1420-3049/24/13/2397.

29 Alias, N.H., Abd-Aziz, S., Yee Phang, L., and Ibrahim, M.F. (2021). Enzymatic saccharification with sequential-substrate feeding and sequential-enzymes loading to enhance fermentable sugar production from sago hampas. *Processes* 9 (3): 535.

30 Khan, F. and Husaini, A. (2006). Enhancing a-amylase and cellulase in vivo enzyme expressions on sago pith residue using *Bacilllus amyloliquefaciens* UMAS 1002. *Biotechnology* 5 (3): 391–403.

31 Ahmad, M.N. (2017). Antibacterial and prebiotic properties of cellobiose from sago frond and application as substrate for the production of l-lactic acid. [Kota Samarahan]: Universiti Malaysia Sarawak.

32 Dewayani, W., Arum, R.H., and Septianti, E. (2022). Potential of sago products supporting local food security in South Sulawesi. In: *IOP Conference Series: Earth and Environmental Science*, 012114. IOP Publishing.

33 Toselong, M.A., Bulkis, S., and Jamil, M.H. (2018). Sago agribusiness development as sustainable local food. *Int. J. Sci. Res.* 7 (1): 1900–1901.

34 Murod, M., Kusmana, C., Bintoro, M.H., and Hilmi, E. (2019). Strategy of sago management sustainability to support food security in regency of Meranti Islands, Riau Province. *Indnes. Adv. Agric. Bot.* 11 (1): 1–20.

35 Ochoa-Chacón, A., Martinez, A., Poggi-Varaldo, H.M. et al. (2022). Xylose metabolism in bioethanol production: *Saccharomyces cerevisiae* vs non-*Saccharomyces* yeasts. *Bioenergy Res.* 15 (2): 905–923. https://link.springer.com/10.1007/s12155-021-10340-x.

36 Lee, S.M., Jellison, T., and Alper, H.S. (2012). Directed evolution of xylose isomerase for improved xylose catabolism and fermentation in the yeast *Saccharomyces cerevisiae*. *Appl. Environ. Microbiol.* 78 (16): 5708–5716. https://journals.asm.org/doi/10.1128/AEM.01419-12.

37 Global View Research. Lactic acid market size, share and trend analysis report by raw material (corn, sugarcane), by application (industrial, food and beverages, polylactic acid), by region, and segment forecasts, 2019–2025. https://www.grandviewresearch.com/industry-analysis/lactic-acid-and-poly-lactic-acid-market (accessed on January 2021).

38 Bujang, K.B. and Jobli, S. (2002). Effect of Glucose Feed Concentration on Continuous Lactate Production from Sago Starch. In: *International Symposium of Tropical Natural Resources and Green Chemistry Strategy*. Grand Cube, Osaka.

39 Ahmad, M.N., Awang Adeni, D.S., Suhaili, N., and Bujang, K. (2022). Optimisation of pre-harvest sago frond sap for the production of l-lactic acid using *Lactococcus lactis* IO-1. *Biocatal. Agri. Biotechnol.* 43: 102435.

40 Slavin, J. (2013). Fiber and prebiotics: mechanisms and health benefits. *Nutrients* 5 (4): 1417–1435. http://www.mdpi.com/2072-6643/5/4/1417.

41 Ohara, H., Owaki, M., and Sonomoto, K. (2006). Xylooligosaccharide fermentation with *Leuconostoc lactis*. *J. Biosci. Bioeng.* 101 (5): 415–420. https://linkinghub.elsevier.com/retrieve/pii/S138917230670603X.

42 Filya, I. (2003). The effect of *Lactobacillus buchneri* and *Lactobacillus plantarum* on the fermentation, aerobic stability, and ruminal degradability of low dry matter corn and sorghum silages. *J. Dairy Sci.* 86 (11): 3575–3581.

43 Mamat, F.W., Suhaili, N., Ngieng, N.S. et al. (2021). Feasibility of sago bioethanol liquid waste as a feedstock for laccase production in recombinant *Pichia pastoris*. *Res. J. Biotechnol.* 16 (4): 172–179.

44 Mamat, F.W., Suhaili, N., Ngieng, N.S. et al. (2021). Intensification of recombinant laccase production from sago bioethanol liquid waste and evaluation of the enzyme for synthetic dye decolourisation. *Arab Gulf J. Sci. Res.* 39 (3): 209–220.

45 Shakibaie, M., Ameri, A., Ghazanfarian, R. et al. (2018). Statistical optimization of kojic acid production by a UV-induced mutant strain of *Aspergillus terreus*. *Braz. J. Microbiol.* 49 (4): 865–871. https://linkinghub.elsevier.com/retrieve/pii/S1517838217309450.

46 Mohamad, R., Arbakariya, A., Hassan, M.A. et al. (2002). Importance of carbon source feeding and pH control strategies for maximum kojic acid production from sago starch by *Aspergillus flavus*. *J. Biosci. Bioeng.* 94 (2): 99–105.

47 Spencer, A.M., Suhaili, N., Husaini, A. et al. (2019). Optimising production of kojic acid from sago fibre by solid-state fermentation using response surface methodology. *Borneo. J. Resour. Sci. Technol.* 9 (2): 94–100.

48 Sie, C.Z.W., Ngaini, Z., Suhaili, N., and Madiahlagan, E. (2018). Synthesis of kojic ester derivatives as potential antibacterial agent. *J. Chem.* 2018: 1–7.

49 Zhu, F. (2018). Modifications of starch by electric field based techniques. *Trends Food Sci. Technol.* 75: 158–169. https://linkinghub.elsevier.com/retrieve/pii/S0924224417307628.

50 Venkatesan, M., Raja, M., Sivalaksmi, S. et al. (2022). Experimental study of thermal performance on waste in-filled building wall construction. *Int. J. Thermophys.* 43 (10): 156. https://link.springer.com/10.1007/s10765-022-03082-1.

11

Biofertilizer and Other Chemical Substitutes from Sugarcane By-products

Is Fatimah, Ganjar Fadillah, Tatang S. Julianto, Rudy Syahputra, and Habibi Hidayat

Universitas Islam Indonesia, Kampus Terpadu UII, Chemistry Department, Jl. Kaliurang Km 14, Sleman, Yogyakarta, 55584, Indonesia

11.1 Introduction

Saccharum officinarum, or sugarcane, is a tropical and subtropical plant widely used in industry, especially for sugar production. However, in addition to producing food products such as sugar, the sugarcane industry produces less environmentally friendly by-products. For example, in the process of making sugar from sugarcane, the essential ingredients of sugarcane must be carried out in a series of processes that include physical treatment (filtering and flashing) and chemical treatments such as (bleaching and polymerization). These treatments aim to remove some nonsugar components, colloids or particles, and colors to produce high-purity sugar. The by-products from the sugar industrial process can be seen in Figure 11.1. From the process, it is about 25–20% sugarcane bagasse (SCB) and 3.4% press mud cake (PMC) are left as by-products. Moreover, from the distillation in the alcohol production, about 12–15 l of wastewater called spent wash (SW) is also produced for a liter of alcohol.

However, some by-products from sugar production include molasses, filter cake, vinasse, yeast, and other by-products. Each of these by-products has different characteristics, so the management of these products must be appropriate so that they can be reused and have a high value [1, 2].

As early mentioned, these by-products from the sugarcane industry have different physical and chemical characteristics because they go through various stages of the pretreatment process. Bagasse is the most waste generated from the sugarcane industry. Bagasse is obtained from the extraction process of sugarcane juice with a chemical composition of almost 40% cellulose and 24.4% hemicellulose. This by-product is categorized as biomass waste and has the potential to be developed as a source of bioethanol (energy recovery). In addition, the high cellulose content of this by-product causes this product to be widely used as a raw material for paper, pulp, and fermentation in the production of enzymes and microorganism substrates. This characteristic is different from other by-products, such as press mud,

Chemical Substitutes from Agricultural and Industrial By-Products: Bioconversion, Bioprocessing, and Biorefining,
First Edition. Edited by Suraini Abd-Aziz, Misri Gozan, Mohamad Faizal Ibrahim, and Lai-Yee Phang.
© 2024 WILEY-VCH GmbH. Published 2024 by WILEY-VCH GmbH.

Figure 11.1 Scheme of sugarcane industrial process. Source: Fatimah et al. (author).

which only contains 11% cellulose and is high in protein content of up to 15%. Press mud is produced from the carbonization and sulfitation processes. Physically, this by-product has the characteristics of chewy, soft, and brown. Based on the literature, the different chemical compositions of various by-products produced by the sugarcane industry are presented in Figure 11.2 [3].

One of the potential by-products of sugarcane industrial process is SCB (SCB). This by-product is cheap, widely available, and primarily utilized for fuel in sugar

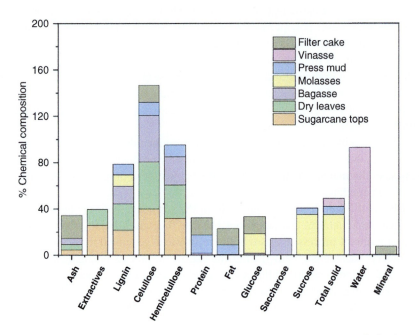

Figure 11.2 The chemical composition of sugarcane by-product. Source: Fatimah et al. (author).

processing. As the SCB is the result of the extraction process of sugarcane juice, SCB contains fibers that consist of approximately 40–50 μl cellulose and 25–35% hemicellulose. The remainder includes lignin, waxes, etc. Cellulose has a crystalline structure (approximately 50–90% crystalline, depending on the cellulose source), whereas hemicellulose is an amorphous structure containing xylose, glucose, etc. Cellulose closely resembles natural linear polymers containing anhydrous glucose units linked by glycosidic β 1,4 linkages. C-2 and C-3 have secondary –OH groups and the primary –OH is at C-6, thus containing three hydroxyl groups with different reactivities. These hydroxyl groups contribute to the formation of strong intermolecular and intramolecular hydrogen bonds. These cellulosic polymers are distributed in fibrils surrounded by hemicellulose and lignin. Lignin acts as an adhesive between cellulose and hemicellulose, helping the material gain stiffness. With the high percentage compared to other by-products, SCB placed as the highest potency. The production of SCB from several countries can be seen in Figure 11.3.

The amount suggests that the utilization of SCB to be more effectively support the sugarcane industry is an important strategy. Furthermore, for several purposes, the pretreatment process determines the main characteristics of this by-product.

Vinasse is the final residue from the sugarcane refining process, which is fermented to produce ethanol. Physically, this by-product has a dark brown color due to the oxidation process of oxygen. This vinasse is produced from the distillation process, so the by-product is categorized as a liquid by-product that contains almost 90% water. Until now, this by-product has received enough attention because it can produce an average of nearly 15 l to get 1 l of ethanol. Vinasse contains several different organic compounds, such as ethanol, lactic acid, acetic acid, and phenolic [4]. In addition, physicochemical, the by-product of vinasse has a high chemical oxygen

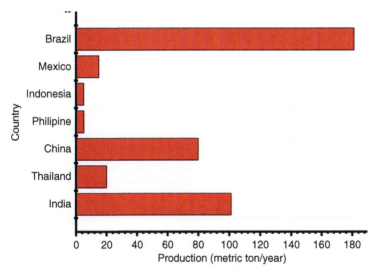

Figure 11.3 Sugarcane bagasse production annually for several countries. Source: Fatimah et al. (author).

demand (COD) value and a low pH, so the reuse of this by-product cannot be carried out directly because it can cause soil acidification and increase soil salinity [5].

Sugarcane molasses contains a lot of total solids and sucrose, up to more than 50%. In contrast, other compositions consist of several compounds with smaller structures, such as organic acids, proteins, oligosaccharides, and several compounds containing nitrogen functional groups. Because this compound has a high amount of sucrose and can be fermented, this by-product is suitable for feedstock for fermentation [6].

Generally, the by-products produced from the sugar industry process contain high amounts of sucrose, so the direction of the utilization of this by-product is mostly directed to energy recovery and biofertilizer. While the physical characteristics, this by-product has a slightly acidic color and pH due to the oxidation process and high heating during the sugar production process. Based on the studies above, each sugar industry's by-product has different characteristics. This difference in physiochemical characteristics is influenced by the treatment process carried out during the procedure for sugar production. Furthermore, knowing the physicochemical properties of the by-products of the sugar industry is very important because it will determine the processing technology and reuse of these by-products.

11.2 Sugarcane By-products Conversion into Biofertilizer

Using biofertilizers in agricultural land provides several advantages, significantly reducing chemical fertilizer's and pesticide's adverse environmental effects. In addition, this type of biofertilizer can improve the quality of the food produced. The conversion of sugarcane waste into biofertilizer can be part of the circular economy in the whole process. SCB and vinasse are the potential raw materials for this purpose. Microbial cultivation is vital for good quality and high-efficiency biofertilizers. This process can be achieved by carrying out solid-state fermentation (SSF) or submerged fermentation (SmF) processes [7, 8]. If viewed from an environmental perspective, the SSF process is considered more profitable than the SmF process because the SSF method can use agro-industry waste as a source of energy and carbon for the growth of microorganisms. One of the potential wastes to be developed as a cheap source of carbon and energy is SCB. The use of bagasse as a substrate in the SSF process has been extensively studied [9]. Additional doses of several nutrients play an essential role in the growth of microorganisms, such as sucrose phosphate and $(NH_4)_2SO_4$. These results proved that optimizing the composition could increase mineral solubility and reduce production costs.

The use of SCB ash in combination with chicken manure and sewage sludge as soil fertilizers have been reported. The chemical compositions of SCB ash were strongly modified by thermal coprocessing (co-gasification and co-combustion) of bagasse pellets and their composition of chicken manure or sewage sludge. Compared to the non-fertilized control plants, the fertility was significantly increased by the addition of SCB ash. The availability of phosphorus (P) from the fertilizer composite is a crucial factor in fertilizer efficiency. According to earlier studies, the extractability and/or plant availability of P from $AlPO_4$ and Ca-based phosphates is as follows: $Ca(Na,K)PO_4 > CaK_2P_2O_7 > Ca_9M(PO4)_7 > AlPO_4$. Further analyses of the released P phases and detailed elucidation of whitlockite structure in SCB ash concluded that the increased Ca, Mg, Na, K, Fe, and Al are of great importance for phosphate formation in the composite and governs the solubility [10]. A similar result suggests the capability of SCB to release P and K fertilizer, especially for acid soil. The capability of SCB ash as a liming material was reported by the combination of SCB with two types of soil: Haplic Arenosol and Haplic Cambisol. The pH of soil increased from 5.1 to 5.9 in a Haplic Arenosol and from 5.3 to 6.0 in a Haplic Cambisol. These pH changes by the ash were like those amended with calcitic and dolomitic limestone. The combination gave soil extractable P and K increasing by 254% and 869% in the Haplic Cambisol, about 183% and 208% in the Haplic Arenosol. It is reflected by an increasing corn yield of 32% in the Haplic Cambisol and 11% in the Haplic Arenosol relative to the unamended treatments [11]. In line with the influence of soil acidity, SCB ashes were good sources of nutrients to enhance wheat yield in acidic soil [12].

The enhancement of phosphorus solubility and stability in the soil can be obtained by composting SCB. Over the past decades, effective inoculation by fungi and bacteria has been applied in practice. For example, *Aspergillus niger* and *Trichoderma viride* strains were used together as a fungal activator in the presence or absence of farmyard manure (FM) for composting bagasse enriched with rock phosphate. A faster decomposition was represented to get the most suitable conditions for phosphate solubilization.

Sugarcane vinasse is another biofertilizer source with a high amount as its production reaches up to 10–15 l per ethanol produced [2]. Vinasse is the primary liquid by-product of the conversion process with a composition consisting of water up to 93%, organic matter (6%), and the remainder composed of several nutrients such as calcium (Ca), magnesium (Mg), nitrogen (N), and potassium (K). The vinasse's chemical composition depends on the soil's characteristics, harvest time, sugarcane variety, and, specifically, the technology processes used in ethanol production. These factors significantly affect the nutritional content, especially mineral contents, biological oxygen demand (BOD), and other organic compounds, such as alcohol, glycerol, and organic acids. Therefore, the composition of sugarcane vinasse paradise can be developed as a biofertilizer. However, before processing vinasse as a biofertilizer, the vinasse biodigestion process is still needed to reduce its organic content and degrade existing organic compounds to become more straightforward so that some nutrients can dissolve when applied to a biofertilizer. The sugarcane vinasse pretreatment process before being used as a biofertilizer is proven to increase and improve the composition of the vinasse. Anaerobic biodigestion is a processing method and can potentially be used to process sugarcane vinasse. This process consists of the biodegradation of organic matter in vinasse waste and nutrient supply. The vinasse biodigestion process has been shown to increase the composition of total nitrogen and minerals, which play an essential role in the characteristics of the biofertilizer as shown in Figure 11.4. Using vinasse as a fertilizer has several advantages and differentiators from other fertilizers. The vinasse can

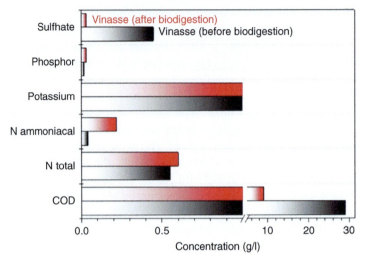

Figure 11.4 The different nutrient and mineral compositions of sugarcane vinasse after and before the biodigestion pretreatment process. Source: Adapted from Bettani et al. [13].

provide nutrient compounds that can function in chemical recovery from the soil on the surface and in the subsurface soil. Improvement of nutrient composition can be achieved by mixing vinasse sugarcane with other ingredients, such as microalgal biomass. The mixing process has been proven to increase nutrients, especially nitrogen composition, by up to 70%. In addition, the mixing process can also increase fertilizer use efficiency because the nutrient release process occurs in a slow-releasing process. The production of biofertilizers from vinasse sugarcane involves several crucial steps, such as (a) vinasse solid–liquid separation for isolation of solid vinasse from liquid waste produced, (b) vinasse fermentation with degrading enzymes and microorganisms, (c) vinasse mixing with ammonium sulfate, ammonium carbonate, calcium–magnesium phosphate, potassium sulfate, humic acid for maxing biofertilizer, (d) vinasse crushing, (e) vinasse organic fertilizer granulation, (f) vinasse organic drying and cooling, (g) vinasse organic fertilizer packing.

11.3 Sugarcane Bagasse as Raw Material for Soil Improver: Phenol Degradation

Bioremediation is valued as eco-friendly, noninvasive, and cheaper than conventional methods. In terms of green chemistry application, the entrapment of microbial cells into a stable carrier is a well-known strategy and widely used in bioremediation a strategy to sustain the activity of enzymes or active sites for remediation. Besides many solids, some research works reported that the immobilized fungi and bacteria in SCB-based material govern the enzyme by the stability and activity improvement. As the SCB showed the capability to stabilize and enhance the pH of the soil, SCB has the potential to simulate bacterial activity that in turn increases the decomposition rate of the humus layer and N mineralization. The immobilization of bacteria into SCB prevents the cells from leaking into the environment but may limit the exchange of nutrients and metabolites. In principle, the enzyme contained in the microorganism that is capable to convert organic contaminant is applied in a stabilized form. With regards to the great C/N ratio in bagasse, the composting with controlled conditions for optimized enzyme activity so the composted material can also act to enrich soil characteristics. The SCB itself shows capability as an adsorbent for some organic pollutants, so the additional functionality of microbial immobilization.

SCB-supported *Bacillus pumilus* HZ-2 was successfully prepared and exhibited capability for the bioremediation of mesotrione-contaminated soils [14]. It was over 75% of mesotrione in soil was degraded within 14 days. The wide range of pH (5–8) and temperature (25–35 °C) are the advantages characteristic for the operational condition of biodegradation, especially for mesotrione at a low concentration range (5–20 mg/kg). By a similar concept of immobilization, ACB-immobilized *Bacillus badius* ABP6 demonstrated 85.32% of atrazine biodegradation [15].

The growth of fungi or bacteria in the SCB compost is a crucial step of immobilization. Theoretically, there are five main techniques of immobilization: adsorption, binding on a surface (electrostatic or covalent), flocculation (natural or artificial), entrapment, and encapsulation. The most important parameter in the entrapment of microorganisms is the ratio of the size of the pores of the carrier to the size of the cells. From many studies, the entrapment of microbial into SCB is also directed by the potency of hydrophilic and hydrophobic bonding of the surface functional group as well as its porosity. The structure of SCB also gives some advantages related to suitability for gas evolution during cell growth and providing sufficient carbon and energy for cell growth.

Some investigations and innovations remarked on the applicability of SCB not only as organic fertilizer but also to improve soil characteristics. The combination of SCB with phosphate and nitrogen enrichment has been developed by the inoculated bacteria [16]. From previous research [16], the interacted sugarcane with biofertilizer produced from P and K rocks supplemented with elemental sulfur inoculated with *Acidithiobacillus* bacteria and mixed with earthworm compost enriched in N by improved the effectiveness in nutrients uptake in the soil, especially total N, ammonium-N, P and K. In a larger scale, the sugarcane combination in vermicompost showed the successful increase in the crop yield and soil fertility [17, 18]. In addition, the SCB-immobilized bacteria that exhibited biodegradation of polluted soil can be the potency of waste to be the eco-friendly composite to support the environment. Among the composting microorganisms, bacteria, fungi, and actinomycetes constitute the major active groups. Bacteria are also the most diverse group of compost organisms, using a broad range of enzymes to chemically degrade a variety of organic matter. The high surface/volume ratio of bacteria allows a rapid transfer of soluble substrates into the cell. The ubiquitous genus *Bacillus* is often found in an environment with the capacity to produce spores for producing extracellular polysaccharide hydrolyzing enzymes.

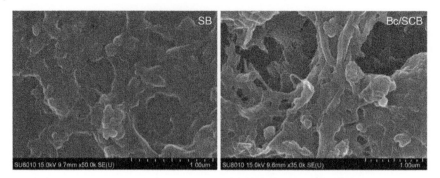

Figure 11.5 SEM images of Bc/SCB and SB. Source: Fatimah et al. (author).

The *Bacillus cereus*-immobilized sugarcane bagasse (Bc/SCB) has been prepared for phenol biodegradation application. The background of research was the capability of 67.76% phenol degradation within 168 hours of incubation by *B. cereus*, and on another side, SCB showed high affinity in phenol removal by adsorption mechanism. The physicochemical features of the material were studied by such instrumental analyses consisting scanning electron microscopy (SEM), transmission electron microscopy (TEM), and surface area analysis, meanwhile, the phenol removal was conducted in a batch system. Determination of the phenol removal was performed based on high-performance liquid chromatography analysis.

Physicochemical characterization of Bc/SCB in comparison to the SCB sample was first studied by SEM analysis, with the images presented in Figure 11.5.

As it can be seen from the images, the porous structures appeared in both materials as the characteristics of biomass materials with a cellulosic component. By comparing both samples, a relatively maintained porous structure of SB also appeared in Bc/SCB sample without any significant difference. This means that *B. cereus* immobilization does not give crucial changes to the SB chemical and physical composition. Furthermore, the gas sorption analysis was employed to elucidate the porous structure in more detail. The N_2 adsorption plots are presented in Figure 11.6.

Similar to that was obtained by SEM images, from the isotherm plots, there is no change of adsorption type or classification. Both samples represent a type I isotherm plot corresponding to the microporous structure in materials. Based on Brunair–Emmet–Teller (BET) equation, the BET-specific surface area of SCB and Bc/SCB are 2.85 and 2.48 m^2/g, respectively. Using the pore size distribution analysis, the compared pore distribution of SB and Bc/SCB are depicted in Figure 11.6 It is observed that some dominant pores are identified in Bc/SCB, while the pore size in the SB sample is dominantly ranging from 2 to 8 Å. The data suggests that the immobilized bacteria formed aggregates to create some new holes in the surface. This assumption is proven by the TEM analysis with the images presented in Figure 11.7.

The TEM image of Bc/SCB presents a more compact surface compared to the SB sample implying the appropriateness with the possibility of bacteria creating

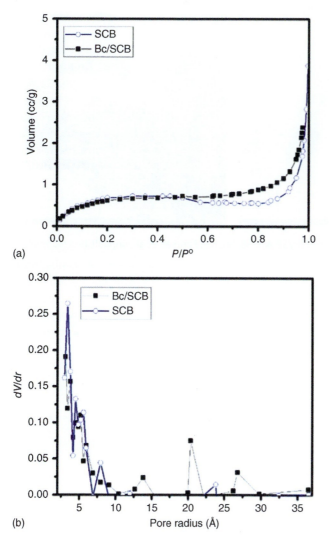

Figure 11.6 (a) Gas adsorption isotherm plots of materials; (b) Pore size distribution of Bc/SCB and SCB. Source: Fatimah et al. (author).

aggregates on the SB surface. The experiment of phenol biodegradation was performed for phenol solution with a concentration of 10 mg/l. After it was incubated for 24 hours, the concentration of phenol was evaluated by HPLC analysis. The comparison of the initial and treated phenol solution is presented in Figure 11.8.

The occurrence of phenol biodegradation was observed after the phenol solution was treated by using Bc/SCB. The phenol removal was identified by the disappearance of the phenol peak, meaning that the removal is 100%. In addition, small peaks are identified with a retention time of 2.69 and 3.05 minutes. These small peaks are probably associated with biodegradation products that are potentially synthesized

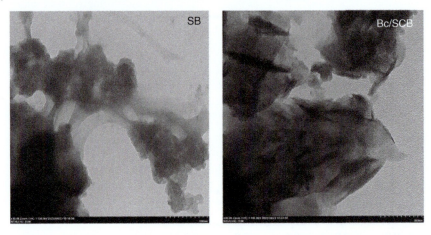

Figure 11.7 TEM images of Bc/SB and SB. Source: Fatimah et al., unpublished material.

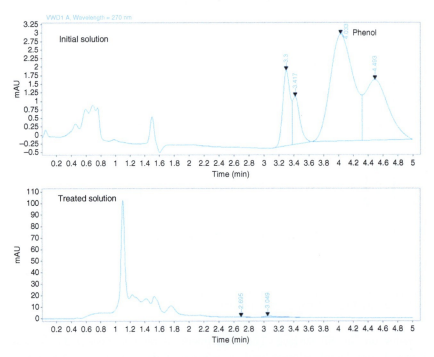

Figure 11.8 HPLC analysis results of initial and treated phenol solution. Source: Fatimah et al. (author).

by the enzyme of *B. cereus*. Referring to the 95% biodegradation efficiency of phenol with a concentration ranging from 100 to 300 mg/l by *B. cereus* IrC2 after 24 hours of incubation, the result from this work is acceptable. It means that immobilization does not influence much the capability of phenol removal. Generally speaking, from this experiment, Bc/SB expressed the biodegradability toward phenol, and further detailed kinetics and mechanism.

11.4 Sugarcane By-products Conversion into Chemical

Over the years, many value-added products have been generated from SCB across the diverse fields of energy, construction, food, textile, cosmetics, and many others. SCB serves as a substrate for the production of enzymes, such as xylanases. As lignocellulosic material, there are some possible products of the conversion as presented by the scheme in Figure 11.9.

Lignin can be extracted from SCB with an alkaline process. Usually, small pieces of dry SCB are mixed with a solution containing 40 wt% NaOH. The hydrothermal reaction is carried out at 90 °C for two to four hours to get separate liquid and solid parts, where the liquid part is called black liquor containing mainly lignin. Some intensifications of the lignin extraction were performed by the addition of various alcohol and solvents and by the use of the solvothermal method [19, 20]. Lignin is mainly an amorphous tridimensional polymer of three primary units: coniferyl (3-methoxy 4-hydroxycinnamyl), sinapyl (3,5-dimethoxy 4-hydroxycinnamyl), and p-coumaryl (4-hydroxycinnamyl) alcohols joined by ether and C–C linkages. These three monolignols are also known as guaiacyl (G), syringyl (S), and p-coumaryl alcohol (H) units (Figure 11.10).

Lignin is convertible into some chemicals such as vanillin, vanillic acid, lignin monomers, benzene-toluene-xylene (BTX), dimethyl ether (DME), cyclohexane, etc. Figure 11.11 represents the summary of lignin conversion.

Industrialization of vanillin production from lignin has been carried out since the 1950s and until the 1990s, it is the main route of worldwide vanillin production. The conversion is by aerobic oxidation of sodium lignosulfonate (waste from sulfite pulping) in the presence of NaOH. Furthermore, the progression for the selectivity toward vanillin and related compounds were obtained by alkaline nitrobenzene oxidation (AN oxidation). Considering that nitrobenzene is toxic, nitrobenzene-free methods for vanillin production are currently being developed. Utilizations of Bu_4N^+ and OH^- from various sources [21] and also bioconversion mechanism by

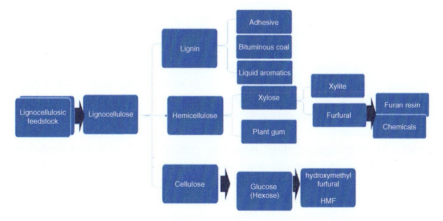

Figure 11.9 Scheme of lignocellulosic conversion. Source: Fatimah et al. (author).

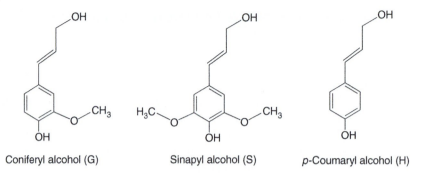

Figure 11.10 Chemical structure of conifer alcohol (G), synapyl alcohol (S), p-coumarin, and coumarin alcohol (H). Source: Li et al. [19]; Pereira et al. [20].

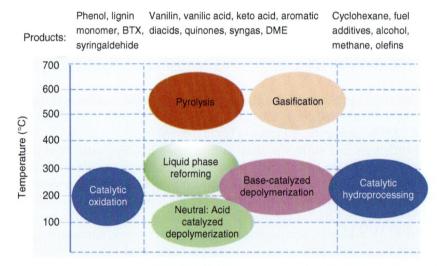

Figure 11.11 Various products from lignin conversion. Source: Fatimah et al. (author).

using *Bacillus* sp. are other routes that have been reported [22]. Other bioconversion of ligninolytic systems using fungi and bacteria are applied to produce phenolic compounds, such as vanillic acid, ferulic acid, and syringil alcohol. Catalytic hydro-processing including hydrothermal liquefaction (HTL) is another baseline technology of lignin depolymerization. The process usually performed at high pressure (5–28 MPa) and moderate temperature (200–400 °C) produces low-oxygen liquid bio-oil and/or phenolics (e.g. syringol, vanillin, guaiacol, as well as phenolic trimers to oligomers).

Similar to other lignocellulosic materials, cellulose in SCB is surrounded by a matrix of lignin and hemicelluloses that is organized into fibrils. Cellulose is characterized by the presence of three hydroxyl groups (secondary OH at the C-2 and C-3 and primary OH at the C-6 position) with different reactivity, and has strong intermolecular and intramolecular hydrogen bonds. Cellulose extraction generally involves acid, basic, or oxidative treatments. These include the acid-induced destruction process and hydrolysis via the cleavage of glycosidic bonds. Some pretreatments to the fibers by using mechanical and chemical treatments were applied with various purities and effectivities. Mechanical treatments including physical procedures such as mechanical grinding and milling, ultrasound irradiation, and microwave irradiation are optional choices to enhance the purity of cellulose by enhancing accessible separation by disrupting the lignocellulosic structure. In addition, the chemical treatment methods including the use of hydrogen peroxide, NaOH, acetic acid and $Na_2B_4O_7 \cdot 10H_2O$, HNO_3, and, ethanol was reported [19].

The transformation of cellulose into glucose and fructose has been carried out by mineral acids in a homogeneous system, heterogeneous catalysts, and enzymes. Some mineral acids such as sulfuric acid (H_2SO_4), phosphoric acid (H_3PO_4), hydrochloric acid (HCl), nitric acid (HNO_3), and perchloric acid ($HClO_4$) have been previously utilized to hydrolyze cellulose into glucose [23]. Even though it is low cost and fast, homogeneous catalysis has drawbacks related to the acid residue, which cannot be recoverable, corrosive, and requires thorough separation. By using a green and sustainable chemistry perspective, solid acid catalysts such as acid immobilized in activated carbon or zeolite were applied. A slower reaction is expressed by these catalysts, but the reusability and recoverability of the solid are important aspects of an eco-friendly process [24]. In addition, heteropoly acids having the combination of hydrogen cations such as Broensted acid and reduction–oxidation capabilities received much attention for the high conversion of cellulose to glucose with potential reusability. Within this scheme, $H_3PW_{12}O_{40}$ was reported to show a high yield and selectivity of glucose with 50.5% and 92.3%, respectively [25]. Enzyme hydrolysis has been carried out by cellulases, a complex mixture of enzymes that convert cellulose into fermentable sugars. The high yield obtained under mild conditions is the characteristic of enzyme hydrolysis, but it has some limitations to apply including high cost, needs tight control conditions, and a tricky separation process [26]. A similar perspective for the catalysis process of hydrolysis is expressed for the conversion of hemicellulose into xylitol [27–29].

From the extracted cellulose, glucose and levulinic acid (LA) are the important chemicals for further value-added chemicals production, as shown in Figure 11.12.

Acid-catalyzed hydrolysis of cellulose is a classic way to breakdown cellulose into glucose, which historically started in 1883 when the method of dissolving and hydrolyzing cotton cellulose with concentrated sulfuric acid was invented. Furthermore, from 1937 to 1948, there are three classified acid hydrolysis of cellulose, namely the Bergius–Rheinau process, the Peoria process, and the Hokkaido process were deployed at large scale [30]. In principle, the sulfate ions and hydronium ions form electron donor–acceptor complexes within the system leading to the break of the intra and intermolecular hydrogen bonds followed by the swelling and separation of the cellulose molecular chain. However, in the progression of the green chemistry perspective, the use of mineral acid leads to environmental problems such as nonreusable catalysts that generate liquid waste and super-acidity that cause corrosion. The replacement of liquid acid with solid acids for cellulose

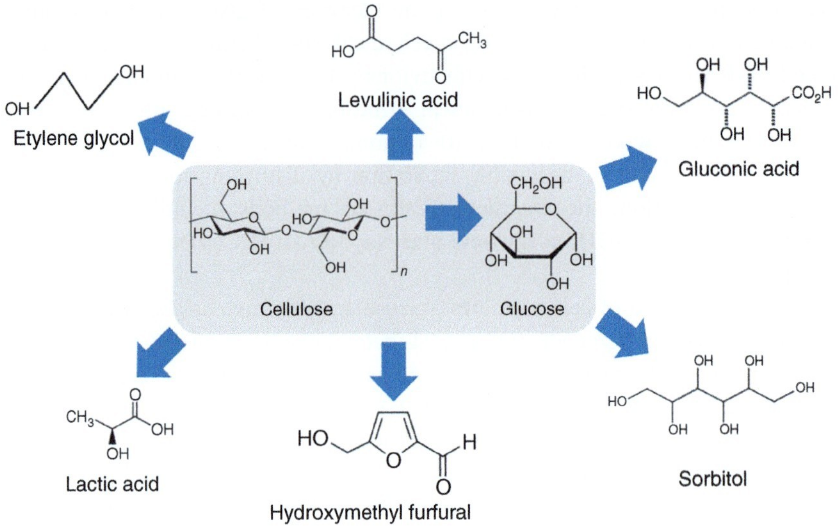

Figure 11.12 Chemical conversion of glucose to valuable chemicals. Source: Adapted from Liu et al. [14].

conversion into LA was attempted. Such TiO_2, ZrO_2, and Sn-based catalysts were reported as effective catalysts for the conversion [31–33].

The formation of ethylene glycol from sugar has been reported since 1935. Glucose can be converted into glycol aldehyde via a retro-aldol reaction, which is further transformed into ethylene glycol by hydrogenation. It was only since the discovery of the Nickel-promoted Tungsten Carbide catalyst, for the direct catalytic conversion of cellulose to ethylene glycol, that the selective formation of ethylene glycol was conceivable. The simpler conversion of glucose was represented in the formation of gluconic acid. A strong oxidant of H_2O_2 has been reported to be a reactant for gluconic acid conversion. Under the optimal conditions of 70 minutes at the reaction temperature of 80 °C a yield of 79.02% gluconic acid [34]. By applying excessive oxidation catalysts such as Pd/C and gold nanoparticles, high yields of about 98% and 92% were obtained, respectively [35, 36]. Ethylene glycol, LA, and gluconic acid are potential chemicals in industrial applications. LA is widely used in plasticizers, resin, animal feed, coating, and as an antifreeze usage, meanwhile gluconic acid, besides its use in plastics, solvent, and additive of paint and ink, is also popular in food processing industries as a flavoring agent and preservative.

Lactic acid is one of the most promising chemicals from the conversion of cellulose, as its demand is growing about 5% during 2018–2023. Lactic acid and its salt have been widely used in food and beverages, and for the polymer industry, it is the raw material for many products. Currently, the production of lactic acid is predominantly by fermentation of sugar. Enzyme-catalyzed saccharification and fermentation using a cellulase preparation derived from a mutant of *Penicillium janthinellum* (EU1) and *Lactobacillus delbrueckii* mutant were reported as an effective method of lactic acid production in a batch experiment. It is found that

L. delbrueckii mutant Uc-3 was capable to give a high yield (0.9 g lactic acid per g cellobiose) [37]. The progression of catalytic conversion was reported by using yttrium (Y), modified siliceous Beta zeolites catalyst. One-pot synthesis of LA from cellulose gave a yield of 49.2% within 30 minutes under the condition of 220 °C and 2 MPa N_2 [38].

11.5 Sugarcane By-product as Material for Biocomposites

Biocomposites are generally natural fiber-reinforced biopolymer-based composites with many applications and purposes. The development of biocomposites is being studied extensively, and cellulosic materials are a better option for being used as reinforcements in composite. Some published research utilized SCB in biocomposite formation with a variety of polymer matrices like polypropylene (PP), polyethylene (PE), polylactic acid (PLA), and polystyrene. The findings were that SCB shows a potential opportunity to substitute and replace synthetic fibers such as fiberglass and aramid, which are commonly used as polymer reinforcement. The improvement of mechanical performance is reflected by such parameters of tensile, flexural strength, and Young's modulus at the optimum composition, usually at the mass percentage of 10–40%.

Table 11.1 Some SCB-based biocomposites for various applications.

Biocomposite	Description	Application	Reference
SCB–gluten composite	The composites exhibited good thermal stability until 250 °C and similar flammability with gypsum tiles, and moderate acoustic insulation and heat transmission. Biocomposite showed maximum water absorbed was 62% (w/w)	Replacement of gypsum tile	[40]
Polypropylene–sugarcane bagasse	SCB addition increases the crystallization temperature and thermal stability of the biocomposites at 7 and 39 °C, respectively	Bioplastic	[41]
PLA/SCB	The biocomposite demonstrated strength and stability that is not altered by fiber size	Bioplastic	[42]
Polypyrolle/SCB	The composites were characterized to have surface and thermal stability. The biocomposites were the potential to be applied as adsorbents for 2,4-Dicholorophenoxyacetic acid (2,4-D)	Adsorbent	[43]
Polystyrene/SCB	SCB showed capability as reinforcement for high-impact polystyrene (HIPS) composites. The increasing tensile strength (17%), tensile modulus (96%), and flexural modulus (34%) was obtained by reinforcement until 30% of SCB	Manufacture fiber	[44]

SCB expressed the capability to contribute to thermal stability, biodegradability, and improvement of some features in biocomposite formations. Biocomposite panels prepared from SCB and polyvinyl acetate exhibited characteristic resistance to thermal conductivity according to the physical appearance and properties standard. The composite formation with recycled PE demonstrated the improved rigidity and hardness values of the composites. In principle, the availability of functional groups including hydroxyl groups in cellobiose monomers of SCB facilitates the strong hydrogen bonds with other polymers. However, like many other cellulosic materials, SCB is lack uniformity and incompatibility with a hydrophobic polymer. Numerous surface modification techniques were applied, for example, alkaline treatment, and surfactant treatment to the fibers. Another strategy is to homogenize the fiber's particle morphology by selecting a nanocellulose structure. Nanocellulose-derived SCB has lightweight, higher strength, and stiffness [39]. Table 11.1 presents some SCB-based biocomposites for various applications.

11.6 Future Perspective of Sugarcane By-products Conversion in the Sugarcane Industrial Cycle

The sugarcane industry has been a pioneer and continues to be developed as an energy biorefinery in the future. In general, the sugarcane agro-industry has traditionally produced various products such as ethanol and sugar with multiple by-products such as vinasse, molasses, straw, bagasse, and yeast. Several new processes have been developed and implemented to recycle these by-products into new products. Some of the main by-products produced by the sugarcane industry and various products obtained from these by-products are summarized in Figure 11.13.

Although this by-product has been widely developed into new products, this can also be developed into supporting materials such as organic nutrients for sugarcane replantations. Dotaniya et al. (2016) studied the reuse of industrial sugarcane by-products for monitoring soil health in sugarcane production [17]. They converted bagasse into solid press mud (SPM) and applied it directly to sugarcane fields. The results indicated that SPM could increase sugarcane yield

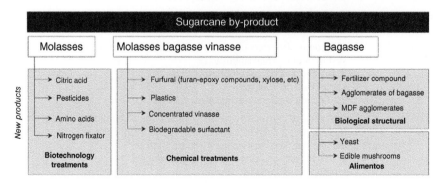

Figure 11.13 The valuable product recovery chart of sugarcane by-products with different technologies. Source: Adapted from Dotaniya et al. [17].

and the weight of sugarcane produced. In addition, the application of SPM can also improve the physical condition of the soil by increasing macrospores and reducing bulk density to increase the sugarcane yield. Biomass by-products are waste that is widely produced and has the potential to be developed in the recycling of the sugarcane industry. In addition, sugarcane residue as raw material for conversion can continue to be developed through the biorefinery process. The process of converting biomass into several high-value products such as bioenergy, biofuels, and biomaterials is possible by involving several main stages such as biochemical processes (enzymatic reaction), conversion (thermochemical processes), chemical process (acid hydrolysis), and physical function (distillation). However, the conversion of by-products such as biomass into high-value products is closely related to the conversion rates of polymers into their monosaccharides. For example, previous studies reported that the enzymatic conversion process increased conversion values at room pressure and temperature ranging from 40 to 45 °C. However, the relatively long processing time and high cost of enzymes became obstacles in the conversion process of these by-products [45]. Therefore, several current studies have studied various inexpensive and effective enzymes in converting sugarcane residues into bioproducts, such as cellulase, amylase, xylanase, and a mixture of several other enzymes [2].

11.7 Future Usage and Applications of Sugarcane By-products

Biotechnology currently makes it possible to develop new sugarcane varieties that not only focus on the amount of sugarcane production but can produce sugarcane plants and their residues with specific characteristics, such as high fiber, sugar, and biomass content. Then, it can be used to support the bioenergy and biofertilizer productivity of plants. Nevertheless, developing an integrated system of sugarcane processing by-products is still an important and exciting topic. For example, waste biomass from sugarcane is converted into biofertilizer in favor of recycled source material. From an agriculture and agro-industrial perspective, biofertilizer is better choice with respect to the consideration in minimizing environmental impact and assuring profitability and economic benefits to industrial enterprises via the circular economy concept.

However, this system still encounters several obstacles, such as the high cost of enzymes and processes that make it less economically attractive. Therefore, one of the main challenges of integral use of sugarcane residue is using financially viable straw, especially in preprocessing and selecting the best biorefinery exploitation. Several studies have shown positive progress in converting sugarcane industry by-products into value-added products. Nevertheless, one of the biggest obstacles in biorefinery are the economic feasibility of pretreatment, the development of cost-effective, and efficient conversion processes for biomass feedstocks.

Besides biochemical conversion, chemical conversion in producing furfural compounds is mostly carried out in the conversion of by-products from the sugarcane industry. *Furfural* is a chemical that has durable characteristics and can be converted into other compounds, such as furfuryl alcohol and furan. These compounds have been widely used in various industries, such as industrial agrochemicals, pharmaceuticals, biofertilizers, and petrochemicals [46]. The hydro-processing and catalyst process plays a vital role in the process of converting this furfural into several high-value products. The conversion of biomass derived from sugarcane residue as a source material into biochar is currently becoming a potential topic to be developed. This sugarcane residue can be converted into biochar and returned to the land to increase the absorption of carbon nutrients and improve the soil quality. The process of converting sugarcane residue into biochar can be carried out by chemical and biochemical processes involving carbonization and microbial mineralization techniques [47].

The optimization of by-products from the sugarcane industry is an important part of the sustainable development goals, either from green chemistry or green technology perspective. The productivity of the sugarcane industry depends on several factors such as management and monitoring, integration of sugarcane plantation and industrialization with the environment, nutrition of sugarcane and fertilizers, control of pests and defenses (biological control agents, impact of agrochemicals and use of pheromones), and harvesting (recovery of straw for biorefinery) [48].

11.8 Conclusions

In conclusion, sugarcane by-products demonstrated potential usage in many sectors of applications. Many applications have been expressed by such schemes of sugarcane by-products into biofertilizers and chemicals. Biofertilizer is important for future generations in agricultural development as there are many proofs of P, K, and N releasing effectiveness, as well as the adaptable feature, especially for soil enhancement. In addition, SCB was reported as effective matrix for immobilizing bacteria, especially for organic and toxic degradation in soil systems. Referring to the dominancy of lignocellulosic in SCB, various products could be obtained from either chemical or biochemical conversions. As parts of these conversions for example, furfural and glucose are the important platforms for many important chemicals for various petrochemical productions.

Optimizing the conversion of sugarcane by-products into fertilizer, various chemicals, and biocomposites could be an important strategy for circular economy improvement on sugarcane industrial activity. Furthermore, the progression of the process involves various kinds of technology at each stage. Therefore, the technology integration system in converting sugarcane and its by-products into high-value materials is still fundamental and attractive.

References

1 Lima, C.M.G., Benoso, P., Pierezan, M.D. et al. (2022). A state-of-the-art review of the chemical composition of sugarcane spirits and current advances in quality control. *J. Food Compos. Anal.* 106: 104338. https://doi.org/10.1016/j.jfca.2021.104338.

2 Guerra, S.P.S., Denadai, M.S., Saad, A.L.M. et al. (2019). *Sugarcane: Biorefinery, Technology, and Perspectives*. Academic Press https://doi.org/10.1016/B978-0-12-814236-3.00003-2.

3 Singh, S.P., Jawaid, M., Chandrasekar, M. et al. (2021). Sugarcane wastes into commercial products: processing methods, production optimization and challenges. *J. Cleaner Prod.* 328: 129453.

4 Rajagopal, V., Paramjit, S.M., Suresh, K.P. et al. (2014). Significance of vinasses waste management in agriculture and environmental quality – review. *Afr. J. Agric. Res.* 9: 2862–2873. https://doi.org/10.5897/ajar2014.8819.

5 Ahmed, O., Sulieman, A.M.E., and Elhardallou, S.B. (2013). Chemical and microbiological characteristics of vinasse, a by-product from ethanol industry article in. *Am. J. Biochem. Biotechnol.* 2013: 80–83. https://doi.org/10.5923/j.ajb.20130303.03.

6 Carioca, J.O.B. and Leal, M.R.L.V. (2011). Ethanol production from sugar-based feedstocks. In: *Comprehensive Biotechnology*, 2e (ed. M. Moo-Young), 27–35. Academic Press.

7 Mendes, G.O., Dias, C.S., Silva, I.R. et al. (2013). Fungal rock phosphate solubilization using sugarcane bagasse. *World J. Microbiol. Antimicrob.* 29: 43–50.

8 Freitas, J.V., Bilatto, S., Squinca, P. et al. (2021). Sugarcane biorefineries: potential opportunities towards shifting from wastes to products. *Ind. Crops Prod.* 172: 114057.

9 Mendes, G.D.O., Silva, N.M.R.M., Anastácio, N.B.V.T.C. et al. (2015). Optimization of *Aspergillus niger* rock phosphate solubilization in solid-state fermentation and use of the resulting product as a P fertilizer. *Microb. Biotechnol.* 930–939.

10 Dombinov, V., Herzel, H., Meiller, M. et al. (2022). Sugarcane bagasse ash as fertilizer for soybeans: effects of added residues on ash composition, mineralogy, phosphorus extractability and plant availability. *Front. Plant Sci.* 13: 1–13. https://doi.org/10.3389/fpls.2022.1041924.

11 Pita, V., Vasconcelos, E., Cabral, F., and Ribeiro, H.M. (2012). Effect of ash from sugarcane bagasse and wood co-combustion on corn growth and soil properties. *Arch. Agron. Soil Sci.* 58: 17–20. https://doi.org/10.1080/03650340.2012.698000.

12 Gonfa, A., Bedadi, B., and Argaw, A. (2018). Effect of bagasse ash and filter cake amendments on wheat (*Triticum turgidum* L. var. *durum*) yield and yield components in nitisol. *Int. J. Recycl. Org. Waste Agric.* 7: 231–240. https://doi.org/10.1007/s40093-018-0209-7.

13 Bettani, S.R., de Oliveira Ragazzo, G., Leal Santos, N. et al. (2019). Sugarcane vinasse and microalgal biomass in the production of pectin particles as an alternative soil fertilizer. *Carbohydr. Polym.* 203: 322–330. https://doi.org/10.1016/j.carbpol.2018.09.041.

14 Liu, J., Chen, S., Ding, J. et al. (2015). Sugarcane bagasse as support for immobilization of *Bacillus pumilus* HZ-2 and its use in bioremediation of mesotrione-contaminated soils. *Appl. Microbiol. Biotechnol.* 99: 10839–10851. https://doi.org/10.1007/s00253-015-6935-0.

15 Khatoon, H. and Rai, J.P.N. (2018). Sugarcane-bagasse as immobilizing support for *Bacillus badius* ABP6 and its use in biodegradation of atrazine. *Environ. Ecol.* 36: 446–456.

16 Chandran, M., Manisha, A., and Subhasini, A. (2014). Production of phospate biofertilizer using lignocellulosic waste as carrier material. *Asian J. Chem.* 26: 6745–6750.

17 Dotaniya, M.L., Datta, S.C., Biswas, D.R. et al. (2016). Use of sugarcane industrial by-products for improving sugarcane productivity and soil health. *Int. J. Recycl. Org. Waste Agric.* 5: 185–194. https://doi.org/10.1007/s40093-016-0132-8.

18 Ansari, A.A. and Jaikishun, S. (2011). Vermicomposting of sugarcane bagasse and rice straw and its impact on the cultivation of *Phaseolus vulgaris* L. in Guyana, South America. *Int. J. Agric. Technol.* 7: 225–234.

19 Li, C., Zhao, X., Wang, A. et al. (2015). Catalytic transformation of lignin for the production of chemicals and fuels. *Chem. Rev.* 115: 11559–11624. https://doi.org/10.1021/acs.chemrev.5b00155.

20 Pereira, A.A., Martins, G.F., Antunes, P.A. et al. (2007). Lignin from sugarcane bagasse: extraction, fabrication of nanostructured films, and application. *Langmuir* 23: 6652–6659. https://doi.org/10.1021/la063582s.

21 Maeda, M., Hosoya, T., Yoshioka, K. et al. (2018). Vanillin production from native softwood lignin in the presence of tetrabutylammonium ion. *J. Wood Sci.* 64: 810–815. https://doi.org/10.1007/s10086-018-1766-0.

22 Kaur, H., Pavithra, P.V., Das, S., and Kavitha, M. (2021). Bio-conversion of lignin extracted from sugarcane bagasse and coconut husk to vanillin by *Bacillus* sp. *Int. J. Curr. Res. Rev.* 13: 118–123. https://doi.org/10.31782/IJCRR.2021.13716.

23 Guleria, A., Kumari, G., and Saravanamurugan, S. (2019). *Cellulose Valorization to Potential Platform Chemicals*. Elsevier B.V https://doi.org/10.1016/B978-0-444-64307-0.00017-2.

24 Shrotri, A., Kobayashi, H., and Fukuoka, A. (2018). Cellulose depolymerization over heterogeneous catalysts. *Acc. Chem. Res.* 51: 761–768. https://doi.org/10.1021/acs.accounts.7b00614.

25 Tian, J., Wang, J., Zhao, S. et al. (2010). Hydrolysis of cellulose by the heteropoly acid $H_3PW_{12}O_{40}$. *Cellulose* 17: 587–594. https://doi.org/10.1007/s10570-009-9391-0.

26 Spano, L.A., Medeiros, J., and Mandels, M. (1976). Enzymatic hydrolysis of cellulosic wastes to glucose. *Resour. Recovery Conserv.* 1: 279–294. https://doi.org/10.1016/0304-3967(76)90039-1.

27 Lavarack, B.P., Griffin, G.J., and Rodman, D. (2002). The acid hydrolysis of sugarcane bagasse hemicellulose to produce xylose, arabinose, glucose and other products. *Biomass Bioenergy* 23: 367–380. https://doi.org/10.1016/S0961-9534(02)00066-1.

28 Abdul Manaf, S.F., Md Jahim, J., Harun, S., and Luthfi, A.A.I. (2018). Fractionation of oil palm fronds (OPF) hemicellulose using dilute nitric acid for fermentative production of xylitol. *Ind. Crops Prod.* 115: 6–15. https://doi.org/10.1016/j.indcrop.2018.01.067.

29 Yi, G. and Zhang, Y. (2012). One-pot selective conversion of hemicellulose (Xylan) to xylitol under mild conditions, *ChemSusChem* 5: 1383–1387. https://doi.org/10.1002/cssc.201200290.

30 Chang, J.K.W., Duret, X., Berberi, V. et al. (2018). Two-step thermochemical cellulose hydrolysis with partial neutralization for glucose production. *Front. Chem.* 6: 1–11. https://doi.org/10.3389/fchem.2018.00117.

31 Lanziano, C.S., Rodriguez, F., Rabelo, S.C. et al. (2014). Catalytic conversion of glucose using TiO_2 catalysts. *Chem. Eng. Trans.* 37: 589–594. https://doi.org/10.3303/CET1437099.

32 Joshi, S.S., Zodge, A.D., Pandare, K.V., and Kulkarni, B.D. (2014). Efficient conversion of cellulose to levulinic acid by hydrothermal treatment using zirconium dioxide as a recyclable solid acid catalyst. *Ind. Eng. Chem. Res.* 53: 18796–18805. https://doi.org/10.1021/ie5011838.

33 Yan, W., Hoekman, S.K., Broch, A., and Corronella, C.J. (2014). Effect of hydrothermal carbonization reaction parameters on the properties of hydrochar and pellets, Environ.Prog.Sustainable. *Energy.* 33: 676–680. https://doi.org/10.1002/ep.11974.

34 Mao, Y.M. (2017). Preparation of gluconic acid by oxidation of glucose with hydrogen peroxide. *J. Food Process. Preserv.* 41: 1–5. https://doi.org/10.1111/jfpp.12742.

35 Liu, A., Huang, Z., and Wang, X. (2018). Efficient oxidation of glucose into gluconic acid catalyzed by oxygen-rich carbon supported pd under room temperature and atmospheric pressure. *Catal. Lett.* 148: 2019–2029. https://doi.org/10.1007/s10562-018-2409-1.

36 Qi, P., Chen, S., Chen, J. et al. (2015). Catalysis and reactivation of ordered mesoporous carbon-supported gold nanoparticles for the base-free oxidation of glucose to gluconic acid. *ACS Catal.* 5: 2659–2670. https://doi.org/10.1021/cs502093b.

37 Adsul, M.G., Varma, A.J., and Gokhale, D.V. (2007). Lactic acid production from waste sugarcane bagasse derived cellulose. *Green Chem.* 9: 58–62. https://doi.org/10.1039/b605839f.

38 Ye, J., Chen, C., Zheng, Y. et al. (2021). Efficient conversion of cellulose to lactic acid over yttrium modified siliceous Beta zeolites. *Appl. Catal., A* 619: 118133. https://doi.org/10.1016/j.apcata.2021.118133.

39 Kumar, A., Negi, Y.S., Bhardwaj, N.K., and Choudhary, V. (2013). Synthesis and characterization of cellulose nanocrystals/PVA based bionanocomposite. *Adv. Mater. Lett.* 4: 626–631. https://doi.org/10.5185/amlett.2012.12482.

40 Guna, V., Ilangovan, M., Hu, C. et al. (2019). Valorization of sugarcane bagasse by developing completely biodegradable composites for industrial applications. *Ind. Crops Prod.* 131: 25–31. https://doi.org/10.1016/j.indcrop.2019.01.011.

41 Mohomane, S., Linganiso, L., Songca, S.P. et al. (2019). Comparison of alkali treated sugarcane bagasse and softwood cellulose/polypropylene composites. *Plast. Rubber Compos.* 48: 401–409. https://doi.org/10.1080/14658011.2019.1639027.

42 Khoo, R.Z. and Chow, W.S. (2017). Mechanical and thermal properties of poly(lactic acid)/sugarcane bagasse fiber green composites. *J. Thermoplast. Compos. Mater.* 30: 1091–1102. https://doi.org/10.1177/0892705715616857.

43 Khan, M.M., Khan, A., Bhatti, H.N. et al. (2021). Composite of polypyrrole with sugarcane bagasse cellulosic biomass and adsorption efficiency for 2,4-dicholrophonxy acetic acid in column mode. *J. Mater. Res. Technol.* 15: 2016–2025. https://doi.org/10.1016/j.jmrt.2021.09.028.

44 Benini, K., Voorwald, H., and Cioffi, M. (2017). Manufacturing and characterization of high impact polystyrene (HIPS) reinforced with treated sugarcane bagasse. *J. Res. Updat. Polym. Sci.* 6: 2–11. https://doi.org/10.6000/1929-5995.2017.06.01.1.

45 de Lucas, R.C., de Oliveira, T.B., Lima, M.S. et al. (2021). The profile secretion of *Aspergillus clavatus*: different pre-treatments of sugarcane bagasse distinctly induces holocellulases for the lignocellulosic biomass conversion into sugar. *Renew. Energy* 165: 748–757. https://doi.org/10.1016/j.renene.2020.11.072.

46 Ntimbani, R.N., Farzad, S., and Görgens, J.F. (2021). Techno-economic assessment of one-stage furfural and cellulosic ethanol co-production from sugarcane bagasse and harvest residues feedstock mixture. *Ind. Crops Prod.* 162: https://doi.org/10.1016/j.indcrop.2021.113272.

47 Jeong, C.Y., Dodla, S.K., and Wang, J.J. (2016). Fundamental and molecular composition characteristics of biochars produced from sugarcane and rice crop residues and by-products. *Chemosphere* 142: 4–13. https://doi.org/10.1016/j.chemosphere.2015.05.084.

48 De Matos, M., Santos, F., and Eichler, P. (2019). *Sugarcane World Scenario*. Elsevier Inc. https://doi.org/10.1016/B978-0-12-814236-3.00001-9.

12

Cocoa Butter Substitute from Tengkawang (*Shorea stenoptera*)

Muhammad A. Darmawan[1,2], *Suraini Abd-Aziz*[3], *and Misri Gozan*[1]

[1] *Universitas Indonesia, Faculty of Engineering, Department of Chemical Engineering, Depok, 16424, West Java, Indonesia*
[2] *Research Center for Process and Manufacturing Industry Technology, Research Organization for Energy and Manufacture, National Research and Innovation Agency, South Tangerang, 15314, Indonesia*
[3] *Universiti Putra Malaysia, Faculty of Biotechnology and Biomolecular Sciences, Department of Bioprocess Technology, Serdang, 43400, Selangor, Malaysia*

12.1 Introduction

Tengkawang (*Shorea* sp.) belongs to the Dipterocarpaceae family, as summarized in the taxonomy in Table 12.1. The Dipterocarpaceae family consists of 13 genera that include about 300 species. The island of Borneo is the origin of the Tengkawang tree. Tengkawang trees in Indonesia are primarily scattered in Kalimantan and a small part of Sumatra. Tengkawang is one of the flora that grows in the forests of West Kalimantan and has been cultivated since 1881. The distribution of tengkawang is influenced by soil type, climatic conditions, and altitude [1]. Tengkawang grows in tropical rain forests with rainfall types A and B and on latosol soil, red–yellow podzolic, and yellow podzolic at an altitude of up to 1300 m above sea level [2]. The forest types most occupied by *Dipterocarp* species are lowland forests, hills, riverbanks, and coastal forests.

In contrast, the forest types with extreme conditions where the soil is poor in nutrients and drainage will affect the number of species that can grow in these conditions [1]. Tengkawang belongs to the genus *Shorea* or *Meranti*, which has good economic value and is a vegetable oil producer. One of the meranti, which is endemic to West Kalimantan is the red *meranti* (*Shorea stenoptera*), which in the local language is called Tengkawang *Tungkul* or illipe nut/Borneo tallow nut in English. Besides *Shorea stenoptera*, there are several types of tengkawang, namely *S. pinanga, S. mecisopteryx,* and *S. macrophylla*. The distribution of *S. stenoptera* is in West Kalimantan (Kapuas Valley), Central Kalimantan (Muara Teweh), Sarawak, and Sabah [3].

In general, tengkawang tree blooms in September – October and the fruit ripens in January – March [2]. Flowering patterns of *Dipterocarp* species in the forest do not occur yearly but have irregular time intervals with varying intensity, where flowering is sometimes abundant. It usually bears heavy fruit after a long dry season, and

Chemical Substitutes from Agricultural and Industrial By-Products: Bioconversion, Bioprocessing, and Biorefining,
First Edition. Edited by Suraini Abd-Aziz, Misri Gozan, Mohamad Faizal Ibrahim, and Lai-Yee Phang.
© 2024 WILEY-VCH GmbH. Published 2024 by WILEY-VCH GmbH.

Table 12.1 Taxonomy.

Kingdom	Plantae
Subkingdom	Tracheobionta
Super division	Spermatophyta
Division	Magnoliophyta
Class	Magnoliopsida
Subclass	Dilleenidae
Ordo	Malvales
Family	Dipterocarpaceae
Genus	Shorea
Species	Shorea stenoptera

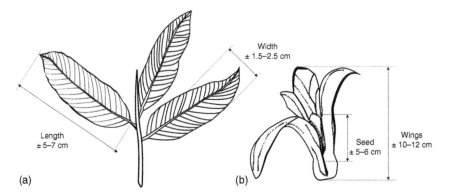

Figure 12.1 The leave (a), and seed (b) of Tengkawang (*S. stenoptera*). Source: Illustration and drawn by Gozan.

once every seven years, there is usually a big harvest because the fruit is abundant. Tengkawang *S. stenoptera* has pink or red flowers with 15 stamens, 3 long and 2 short seeds. The long seed wings measure 3.5–7.5 × 1.5–2.5 cm, while the short seed wings measure 2.0–5.5 × 0.5–1.0 cm. Tengkawang seeds have a size of 40–60 × 30–35 mm and have short hair [3]. The leaves and seeds of tengkawang (*S. stenoptera*) can be seen in Figure 12.1a,b.

Tengkawang seeds are known as Borneo tallow and green butter because of their distinctive yellowish-green color due to the presence of beta-carotene and chlorophyll [4], while the fat is known as illipe butter or illipe tallow. Tengkawang fat or tengkawang butter (TB) in importing countries is widely used as a raw material for making candles, soaps, and chocolate mixtures. Another use of tengkawang fat is as a mixture of cosmetic ingredients, cooking oil ingredients, drugs, and lubricants or oils [5–7]. Tengkawang oil is reported to have the same or equivalent properties as cocoa butter (CB) [8, 9]. The fat content of tengkawang seeds varies depending on the type and quality, generally ranging from 43% to 61%, while the fat content of tengkawang reaches 50–70% [10]. Tengkawang fat is a mixture of triglycerides, namely esters of glycerol and several unsaturated fatty acids. The fatty acid composition of tengkawang fat is dominated by stearic acid, oleic acid, and palmitic acid, with levels of 42–44%, 31–33%, and 19–21%, respectively [11]. Tengkawang fat and cocoa fat have almost the same saponification number, iodine number (IN), melting point, and refractive index.

12.2 Composition and Characteristics of Tengkawang Butter

12.2.1 Fatty Acids Profile of Tengkawang Butter

TB contains fatty acids, which are dominated by stearic acid (18 : 0), oleic acid (18 : 1), and palmitic acid (16 : 0) [9]. The fatty acid composition of tengkawang differs based on the type and regional origin of the tengkawang tree. TB producing areas in Kalimantan originate from Bengkayang, Nanga Yen, Sintang, Kapuas Hulu, Sahan, and several areas in West Kalimantan, Indonesia. Besides Indonesia, tengkawang also grows in several areas in Sarawak, Malaysia. Figure 12.2 shows the fatty acid composition of tengkawang from the Sintang, Nanga Yen, Kapuas Hulu, and Bengkayang areas.

The composition of stearic acid, oleic acid, and palmitic acid in tengkawang ranges from 40% to 46%, 30% to 32%, and 16% to 24%, respectively. In general, the composition of tengkawang fat is similar to brown fat, which has the dominant fatty acids: stearic acid, oleic acid, and palmitic acid. Therefore, tengkawang fat can be used as cocoa butter equivalent (CBE). CBE is a vegetable butter with a composition, properties, and characteristics close to CB [12]. The compositions of stearic acid, oleic acid, and palmitic acid of CB and CBE were 33–36%, 32–35%, and 25–27%, respectively [12, 13]. Gunstone [8] reported that the percentages of palmitic, stearic, and oleic acids in tengkawang fat make it a valuable component of CBE blends and allow it

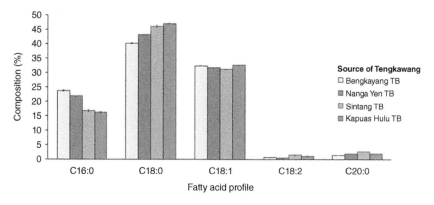

Figure 12.2 Tengkawang butter fatty acid profile from Bengkayang, Nanga Yen, Sintang, and Kapuas Hulu, Borneo, Indonesia. Source: Darmawan et al. [9]/Taylor & Francis/Licensed under CC BY 4.0.

Table 12.2 Fatty acid profile of various types of fat.

Fat resource	C12:0	C14:0	C16:0	C18:0	C18:1	C18:2	C18:3	C20:0	References
S. pinanga	—	0.030	11.780	1.560	42.790	22.040	—	—	[14]
S. mecisopteryx	—	0.040	14.510	0.800	31.280	27.050	—	—	[14]
Cocoa butter	—	—	25.600	36.600	32.700	2.800	—	—	[15]
CBE	—	—	27.000	33.000	35.000	3.000	—	2.000	[13]
Shea butter	0.090	0.020	3.560	43.500	44.470	6.110	0.150	1.440	[16]
Nanga Yen TB	0.009	0.044	19.710	44.267	31.894	0.526	0.135	2.182	[11]
Sintang TB	0.150	0.080	20.220	42.600	31.290	1.010	0.450	3.530	[11]

TB = tengkawang butter.

to be used directly without further processing. [8]. Table 12.2 shows the fatty acid composition of various types of fat. Based on the data in Table 12.2 tengkawang fat has similarities with CBE and CB.

12.2.2 Quality Parameters of Tengkawang Butter

Some physicochemical parameters determine the quality of tengkawang fat: fatty acid, acid number, peroxide number (PN), IN, saponification number, and melting point. Table 12.3 shows the physicochemical properties of tengkawang fat from several areas in Kalimantan.

The melting point is one of the parameters considered to get the quality of fat following food fat. The melting point is related to the amount of saturated and

Table 12.3 Physicochemical properties of tengkawang butter from various areas.

Chemical properties	Unit	Cocoa standard [17]	Tengkawang standard [18]	Tengkawang source				
				Bengkayang [9]	Nanga Yen [9]	Sintang [9]	Kapuas Hulu [9]	Sahan [19]
Melting point	°C	35–37	35–39	36.8	36.7	36.6	36.7	29
Water content	%	7	5	0.23	0.25	0.11	0.08	—
Acidity	%	1.5	3.5	6.88	9.68	10.71	15.94	5.01
Peroxide number	meq O_2/kg	10	10	0.41	0.56	4.33	8.27	—
Iodine number	I_2/100 g	33–42	25–38	21.72	29.33	32.46	31.38	20.46
Saponification number	mg KOH/g	188–198	189–200	200.99	201.88	198.74	196.51	423.35

unsaturated fatty acids. The greater the saturated fat content, the greater the melting point. Vice versa, the greater the content of unsaturated fat, the lower the melting point. Fats used as food have a melting point in the body temperature range of 36–37 °C [20]. Because it absorbs heat from the mouth walls, solid fat at ambient temperature and melts below body temperature will create a smooth, cooling feeling in the mouth [21]. According to the thermal data in that study, TB starts to melt in the mouth if consumed because its melting point is slightly lower than that of the human body (37 °C).

Acidity is a measurement that is equal to the free fatty acids (FFA) content of fat. Its high acidity rating shows the traditional TB's high FFA content. Due to the hydrolysis process, the conventional fat extraction method, which includes employing high-temperature steaming and hot air drying, can raise the FFA value [11]. However, there are currently no additional purifying methods that can enhance the quality of TB. To enhance the quality of traditional TB, a treatment procedure following industrial standards is required during the extraction and purifying process.

An indication for keeping track of the oxidation of oils and fats is the peroxide value (PV)/number (PN). A high PV indicates an increase in the hydroperoxide composition or the breakdown of fats/oils. The PV and the rancidity of the fat/oil are correlated. To prevent the rancidity of the fat, the PV should typically not exceed 10–20 meq/kg [22]. TB is more tolerant to oxidation processes that lead to rancidity if the PV is low. However, tengkawang still has a high acidity value. There is a link between rancidity and the quantity of FFA that may go through the oxidation process that results in rancidity.

The relative level of unsaturation in fats and oils is measured by iodine value (IV)/number (IN) reactions with halogens. The IV aids in determining the quality of fats/oils since unsaturation affects the melting point and oxidative stability [23]. Using an acid–base titration procedure, the saponification number determines the number of ester linkages in fats and oils. The saponification number is determined by the molecular weight and concentration of fatty acids in the fat/oil [24]. Therefore, a fat or oil's average relative molecular mass is determined using a sufficient saponification number.

12.2.3 Solid Fat Content of Tengkawang Butter

Solid fat content (SFC) is one of the quality parameters of fats and oils. The SFC values of various fats and oils are shown in Table 12.4. SFC in fats and oils at a specific temperature is the ratio of the magnetic signal of protons in the solid phase to the liquid phase [28]. SFCs in fats and oils are responsible for several characteristics of fatty foods, such as physical properties, organoleptic properties, and spreadability, as well as affecting the plasticity of oil/fat products [29, 30]. SFC affects the structure and sensory of vegetable fats related to composition, unsaturation content, and fatty acid chain length [31, 32]. The higher the composition of saturated fatty acids (SFA) and long-chain fatty acids, the greater the value of SFC [31, 32]. The value of SFC at a temperature of 20–25 °C determines product stability and fat/oil resistance

Table 12.4 Solid fat content of various fat/oil sources.

Fat/Oil source	Solid fat content (%)						References
	15 °C	20 °C	25 °C	30 °C	35 °C	40 °C	
Cocoa butter	—	78.11	72.95	55.53	0.00	—	[25]
Cocoa butter equivalent	—	74.37	62.51	45.73	2.66	—	[25]
Shea butter	87.15	79.13	71.15	67.15	34.51	0.60	[26]
Palm stearin/Canola oil	—	23.10	—	11.60	—	4.20	[27]
Palm stearin/Olive oil	—	22.00	—	13.00	—	4.50	[27]
Bengkayang TB	84.97	70.55	24.82	0.46	0.03	0.00	[9]
Nanga Yen TB	87.12	69.70	26.40	2.71	0.97	0.00	[9]
Sintang TB	75.48	58.92	15.45	0.38	0.03	0.00	[9]
Kapuas Hulu TB	89.84	72.68	35.97	5.89	3.11	0.00	[9]

at room temperature; a value of not less than 10% is significant for preventing the oxidation process [33]. SFC values between 35 and 37 °C indicate the thickness and flavor release properties of fat in the mouth. Butter without a mouth waxy feel must have an SFC value below 3.5% at 33.3 °C and entirely melt at body temperature [34]. Table 12.4 compares SFC values at several temperatures of various fats and oils. As seen in Table 12.4, the SFC value of TB at a temperature of 35 °C is close to the content of CB and CBE. Therefore, TB has great potential to become CBE, both pure and mixed.

12.2.4 Thermal Properties of Tengkawang Butter

The thermal properties of TB are determined by the fatty acid composition of triglycerides (TGA). Table 12.5 shows the thermal properties of various types of fats.

Table 12.5 Thermal properties of potential cocoa butter.

Fat resources	Temperature (°C)		Enthalpy (J/g)	References
	Onset	Offset		
Cocoa butter	25.97	37.70	116.20	[16]
Dark chocolate	12.54	32.32	121.52	[35]
Cocoa butter equivalent	26.20	34.60	30.70	[35]
Shea butter	7.13	17.82	58.93	[36]
Nanga Yen TB	37.70	48.40	119.34	[11]
Sintang TB	39.10	48.40	98.99	[11]
Bengkayang TB	33.10	37.80	113.80	[9]
Kapuas Hulu TB	35.30	38.90	153.50	[9]

The greater the saturated fat content, the higher the onset and offset temperatures will be because it requires more energy to melt these components. Conversely, if the unsaturated fat content is greater, the onset and offset temperatures will be lower because unsaturated fatty acids melt at a lower temperature than SFA. Tengkawang fat, dominated by SFA (stearic acid and palmitic acid) and large chain unsaturated fatty acids (oleic acid), makes the onset and offset temperature values close to human body temperature. Therefore, based on the melting point and SFC value, which is close to 0% at a temperature of 35 °C, it makes TB one of the components of CBE.

12.3 Traditional Treatment Process

Tengkawang fat processing is traditionally done by high-temperature steaming and smoking. Processing of tengkawang fat from Nanga Yen uses heating at high temperatures (100–150 °C) while tengkawang from Sintang uses a smoking process [11]. One of the tools used in the traditional processing of tengkawang fat is the "*apit*." The steamed tengkawang seeds will be processed with 4–5 kg of "*apit*" to produce 0.75–1 kg of tengkawang fat [19]. Figure 12.3 shows a clamping device that comes from the village of Sahan.

Tengkawang fat produced using "*apit*" is about 25–30% of the total weight of the tengkawang raw material [19]. The produced Tengkawang fat is then stored in bamboo with a diameter of 4 cm and a length of 30–40 cm. Occasionally, Tengkawang fat is also traditionally stored in 1 cm bamboo. The traditional treatment process of tengkawang has several weaknesses according to the quality of TB. TB produced by

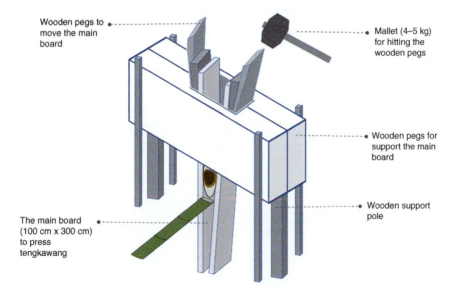

Figure 12.3 Apit instrument for traditional tengkawang process from Sahan village. Source: Illustration and drawn by Misri Gozan and Untung Nugroho Herwanto.

the traditional process has a high PV and acid number [11, 19]. The high values of PN and acid number make TB rancid easily, so the shelf life and quality become lower. Therefore, extraction and purification processes are needed to improve the quality of TB with industry standards.

12.4 Extraction and Purification Process of Tengkawang

The processing of tengkawang fat into products used as CBE is extraction and purification [14]. Figure 12.4 shows a diagram of the process of extracting tengkawang fat from tengkawang fruit. In the extraction process, the tengkawang fruit is separated into kernel, calyx, and shell. Calyx and shell from tengkawang are sources of lignocellulose that can be used as animal feed or produce fine chemicals through the lignocellulose valorization process. Tengkawang kernels are then dried using an oven or tray dryer in a modern process and sunlight in a traditional process to remove water content, thereby reducing the risk of hydrolysis of fat after extraction. The dried tengkawang kernels are then mashed to facilitate the extraction process. The extraction process can be used chemically and physically. The physical process uses hot or cold presses, while the chemical process uses chemical solvents such as

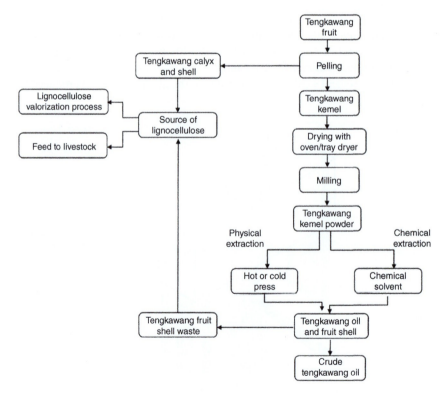

Figure 12.4 Process diagram of tengkawang processing. Source: Illustration and drawn by Darmawan, MA and Untung Nugroho Herwanto.

n-hexane, toluene, and benzene. The remaining skin from the tengkawang kernel can be used as animal feed or as a source of lignocellulose. The extracted tengkawang fat is crude TB.

The bright greenish-yellow color produced from a study [14] indicated that the color of the tengkawang fat was the same as that of commercial tengkawang fat. However, complete data on the quality of Tengawang were not obtained from the results of this study.

12.4.1 Chemical Purification of Tengkawang

The process of purification of tengkawang fat can be carried out in two ways: the chemical and physical processes. Figure 12.5 shows a diagram of the purification process from tengkawang fat. The chemical process uses four stages: degumming, neutralization, bleaching, and deodorization. In contrast, the physical process used three stages: degumming, bleaching, and deodorization. In the chemical degumming process, phosphoric acid is used to separate impurities, metal ions, and gums containing proteins, phosphatides, carbohydrates, and resins without reducing the content of FFA in fats [37]. The degumming process uses phosphoric acid with a concentration of 20%, as much as 1% (w/w), which is added to liquid tengkawang fat for 30 minutes [11]. Tengkawang fat that has gone through the degumming process is then neutralized with base or other reagents to separate FFA to produce soap stock. Sodium hydroxide reduces FFA, dyes, and impurities in sap and mucus in tengkawang fat [37]. 5–10% (w/w) sodium hydroxide was added to

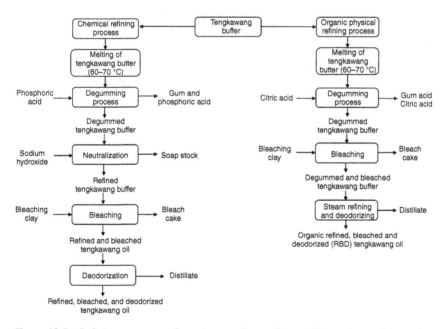

Figure 12.5 Refining processes of tengkawang butter. Source: Illustration and drawn by Darmawan, MA and Untung Nugroho Herwanto.

the tengkawang fat and stirred for 30 minutes to reduce FFA and produce soap stock [11]. The bleaching process is carried out to reduce residual impurities, such as phosphatides, FFA, and gums, and remove dyes in fats and oils [38]. The bleaching process uses adsorbents, such as activated bentonite, activated carbon, activated zeolite, activated bleaching earth, and activated alumina [4, 39]. The adsorbent surface will cover the oil's color, colloid suspensions (gums and resins), and oil breakdown by-products like peroxides [37]. The final step in the chemical refining process is deodorization to remove components that cause unpleasant odors and flavors in oil and butter. The fundamental step in deodorization is distilling hot steam with oil or butter under either vacuum or ambient pressure.

12.4.2 Physical Purification of Tengkawang

The physical refining process relies heavily on the deodorization process using steam refining to remove FFA and volatile components, such as secondary oxidation products of fats, aldehydes, and ketones [40]. In the physical process, organic chemicals, namely citric acid, replace phosphoric acid in the degumming process. Because it does not raise the oil's phosphoric content, citric acid is often preferred [41]. In some super degumming processes, citric acid and lecithin remove nonhydratable phosphatides [42]. The oil is heated to a temperature of 70 °C with a mixture of citric acid and lecithin for 5–15 minutes. After that, it is cooled to a temperature of 25 °C. Water is added for three hours to form phosphatide crystals separated by centrifugation. In the physical process, neutralization with base is not carried out so that there is no residual oil or fat in the soap stock. Degummed oil or fat is then bleached using an adsorbent to remove the dye and residual impurities. The process after bleaching is deodorization to remove the remaining FFA and volatile impurities. Deodorization is a steam distillation procedure at temperatures between 210 and 270 °C and a vacuum of 1–5 mm [43]. The bland product is produced due to the steam's role as a carrier to remove odoriferous chemicals. The oil is following steam distilled using dry saturated sanitary stripping steam injected at the base of the oil bed in the deodorizer at extremely low pressure. Usually, the temperature is kept below 500 °F (260 °C) [44]. After deodorization, the oil also has greater trans fatty acid content. However, a higher temperature is required for deodorization due to the increased FFA concentration in physically refined oils.

12.5 Economic Feasibility Based on Process Simulation

The simulation process of processing tengkawang fat with the demand and production capacity of raw TB and RBD of TB is 206 and 169.056 tons/year, respectively. The process flow diagram (PFD) of the processing of tengkawang fat can be seen in Figure 12.6. The pure tengkawang fat produced has a yield of 78.81 kg/batch from raw TB of 100 kg/batch by removing the gum (3 kg/batch), carotene (2.63 kg/batch), and FFA (14.37 kg/batch) [45]. The purity of tengkawang fat produced is greater than that of raw TB, which is 99.25% (w/w) with a FFA composition of 0.5% (w/w).

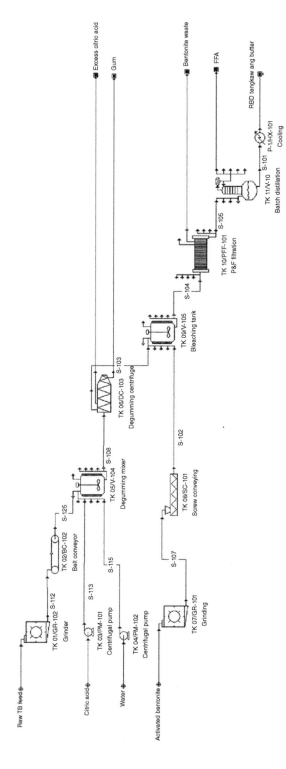

Figure 12.6 Process flow diagram (PFD) of tengkawang butter refining. Source: Muhammad Arif Darmawan (author).

Reducing gum, carotene, and FFA improves the quality of TB, namely reduced acidity and PV. FFA contained in tengkawang fat have a maximum standard of 3.5% (w/w) based on SNI 2903 – 2016.

CAPEX comprises components: equipment, installation, piping, instrumentation, insulation, electricity, buildings, land development, supporting facilities, engineering costs, construction costs, contractor fees, emergency costs, working capital, and start-up costs [45]. Table 12.6 is a simulation of the CAPEX value

Table 12.6 Simulation of the investment value of the tengkawang fat processing plant.

Category	Amounts (US$)	Description
Total plant direct cost (**TPDC**)		
Total bare module cost (**TBMC**)	130,000	Total equipment price
Installation	52,000	39.09% of TBMC
Piping process	45,000	34.58% of TBMC
Instrumentation	52,000	39.84% of TBMC
Insulation	4,000	3.00% of TBMC
Electrical	13,000	9.77% of TBMC
Buildings	59,000	45.11% of TBMC
Yard Improvement	19,000	15.03% of TBMC
Auxiliary facilities	52,000	39.84% of TBMC
TOTAL TPDC	**425,000**	
Total plant indirect cost (**TPIC**)		
Engineering	106,000	24.88% of TPDC
Construction	149,000	35.02% of TPDC
TOTAL TPIC	**255,000**	
Total plant cost/fixed capital investment (**TPC = TPDC + TPIC**)	681,000	
Contractor's Fee and Contingency (CFC)		
Contractor's fee	34,000	5.04% of TPC
Contingency	68,000	9.94% of TPC
TOTAL CFC	**102,000**	
Direct fixed capital cost (**DFC = TPC + CFC**)		
Direct fixed capital (**DFC**)	783,000	100% of DFC
Working capital (**WC**)	153,000	19.79% of DFC
Start-up Cost (**SC**)	39,000	
Total capital investment (**TCI**) (**TCI = DFC + WC + SC**)	**975,000**	

Table 12.7 Simulation of annual operating cost (OPEX) from the tengkawang fat processing plant.

Cost item	Amounts (US$)	Percentage (%)
Raw materials	1,496,000	80.65
Labor-dependent	168,000	9.07
Facility-dependent	147,000	7.95
Laboratory/QC/QA	25,000	1.36
Waste treatment/Disposal	13,000	0.69
Utilities	5,000	0.28
Total	1,854,000	100

at the tengkawang fat processing plant. The investment cost required to build a tengkawang fat processing plant is US$975,000, which is dominated by the total equipment price and construction.

OPEX comprises components: raw materials, workers, facilities, laboratories, waste treatment, and utilities. Table 12.7. shows the OPEX values of the tengkawang fat processing plant. The largest OPEX value is raw materials, with a value of US$1,496,000 and 80.65% of the total OPEX of US$1,854,000.

The parameters for economic analysis include gross margin, return on investment (ROI), payback period (PBP), internal rate of return (IRR), and net present value (NPV) are presented in Table 12.8.

The parameters for economic analysis include gross margin, ROI, PBP, IRR, and NPV. An NPV value greater than 0 indicates that this investment will generate a profit greater than the minimum value [46]. In addition, this value also shows the money or profits obtained by investors when the investment is made. The IRR value is obtained based on the calculation of the entire cumulative cash flow of the tengkawang fat processing plant, which is then compared with the calculated MARR value (MARR = 17.70%) [47]. This condition can be concluded that investing money in this factory will provide greater profits than investing the same amount in the bank. The last two parameters reviewed are ROI and PBP as

Table 12.8 Profitability analysis of tengkawang fat processing plant.

Economic parameter	Value (calculated)	Value [46]
Net Present Value (NPV)	US$2,458,000	US$174,000
Internal Rate of Return (IRR)	35.70%	11.70%
Return on Investment (ROI)	49.57%	17.97%
Gross Margin	26.87%	11.18%
Payback Period (PBP)	2.02 yr	5.56 yr
MARR	17.70%	—

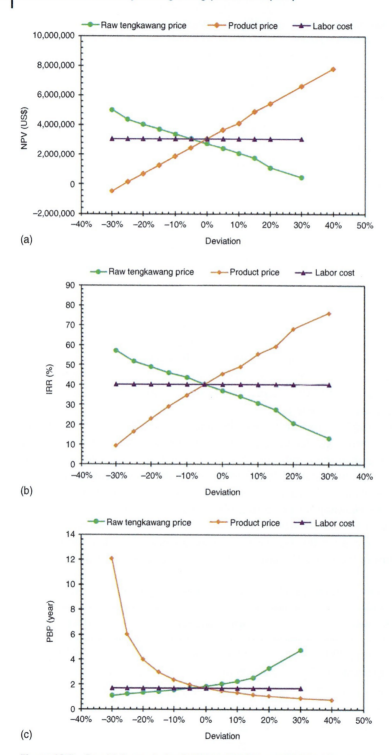

Figure 12.7 Sensitivity analysis of NPV (a), IRR (b), and PBP (c) of tengkawang fat processing plant. Source: Illustration and drawn by Muhammad Arif Darmawan.

shown in. Table 12.8. ROI is the amount of profit after tax obtained by the project each year compared to the investment value at the beginning of the project [48]. PBP is the time required to recover the capital invested at the beginning of the project period [48]. PBP gives an illustration of how quickly the investment capital returns. Projects with high risk will be expected to be able to return capital in a shorter time than projects with low-risk levels. The longer the PBP, the greater the level of uncertainty. These two parameters also support that investing in a tengkawang fat processing plant is feasible and attractive. Capital recovery time is 2.02 years with an ROI of 49.57%. Based on the four parameters calculated in the profitability analysis, it can be concluded that the tengkawang fat processing plant is economically feasible to establish. The data generated in the simulation process is better than previous research conducted by Ramadhani et al. [46] with better IRR, gross margin, ROI, and NPV values and lower PBP times. This is because the simulation process is optimized by adjusting the price of equipment, material prices, labor costs, and operating conditions, which are better than in the research of Ramadhani et al. [46].

Sensitivity analysis is carried out to see the effect of the project's cost and revenue components on the project's economic value. This analysis will show the effect of price changes on the project's profitability. The sensitivity analysis is shown in Figure 12.7a–c.

A profitable but susceptible project indicates a high risk of investment. Figure 12.7 shows the influence of several factors in the tengkawang fat processing plant on the NPV, IRR, and PBP values, which are common indications used to show the economic value being studied. This analysis also plays a role in seeing whether the estimation error used during the calculation process can still be tolerated. The analysis observed the effect of three parameters: the selling price of the product, the cost of raw materials, and the cost of labor on the NPV, IRR, and PBP. The selling price was chosen to test the effect of market price fluctuations on the sustainability of purified tengkawang fat production. Things, such as competitors selling products at lower prices or emerging technologies other than cheaper extraction can affect the selling price of purified tengkawang fat products. In addition to the selling price of the product, price fluctuations can also occur in the price of raw tengkawang fat as a raw material, where the cost of raw tengkawang fat dominates 80.65% of the total factory operating costs. The third parameter that needs to be tested is the labor cost. This is due to changes in regional/district minimum wages that occur periodically, so changes in labor wages are very likely to occur during the project.

12.6 Benefits and Future Outlook of Tengkawang Butter

TB is widely used as a food ingredient and an essential ingredient for cosmetics. Several studies report using TB as a mixture in manufacturing CBE and substitute CBS. The mixture of TB with palm mid-fraction (PMF) produces a product with physicochemical characteristics similar to CBE. The mixture of TB and PMF components has a melting point, TAG composition, and crystalline form similar to CB. The SFC value of a mixture of TB and PMF also has a similarity to CB, namely the composition of 75 : 25 (TB:PMF) [49]. TB can make wet noodles and black butter rice [11]. Waste from traditional tengkawang processing can be used as animal and fish feed.

Apart from being a food source, TB can be used as a cosmetic ingredient. The mixture of TB and lignin can improve the quality of TB, such as the acid number' and PV [50]. Reducing the acid number and PV can improve the quality of TB. The mixture of TB and lignin can also increase the sun protection factor (SPF) value of TB from 4 to 5 to 13–16, depending on the composition and type of lignin added [6].

The challenge in the future faced is that there has not been much increase in the economic value of tengkawang. So, there is a decrease in farmer's interest in cultivating tengkawang. The low economic value makes some farmers sell processed wood from the tengkawang tree instead of collecting tengkawang fruit, a perennial plant harvested once a year. The result is a decrease in the number of tengkawang trees. Amid the challenges and economic threats of this tengkawang commodity, there is still local wisdom consistent in preserving the customary forest area where tengkawang is dominant. At the same time, tengkawang oil can be used as culinary and beauty processed ingredients and traditional herbs. The role and cooperation of the government, research institutes, industry, and the community are very much needed for the conservation and enhancement of the potency of the tengkawang tree sustainably. The government issued a law to maintain the sustainability of the tengkawang tree and regulations regarding the utilization and processing of tengkawang products. Research institutes and universities conduct research and process engineering in tengkawang products to increase the economic value of tengkawang. Industries can use tengkawang as one of the raw or semifinished materials used in food, cosmetics, and pharmaceutical production. The community is the front line of planting, maintaining, and protecting the customary forest, which is the place where tengkawang grows.

12.7 Conclusions

Tengkawang is a natural ingredient endemic to Kalimantan with potential and economic benefits for humans. Tengkawang seeds can be used as a source of lipid biomass to produce TB. The processing of tengkawang fat can be divided into chemical and physical processing. Chemical processing uses an alkaline neutralization process, while physical processing uses a distillation process to remove FFA. The simple simulation calculations show that the investment required to build a tengkawang fat processing plant is feasible. Financing equipment prices and construction dominate investment costs. Tengkawang fat can be used as CBE and cosmetic raw materials, such as sunscreen and lotion. Reducing the acid number and PV can improve the quality of TB, which then increases the economic value of tengkawang.

References

1 Purwaningsih, P. (2004). Review: Sebaran ekologi jenis-jenis dipterocarpaceae di Indonesia. *Biodivers. J. Biol. Diversity* 5 (2): 89–95.
2 Fambayun, R.A. (2014). *Budidaya tengkawang untuk kayu pertukangan, bahan makanan dan kerajinan*. IPB Press.
3 Rizki Maharani, P.H. and Hardjana, A.K. (2013). *Panduan identifikasi jenis pohon tengkawang* (ed. B.B.P. Dipterokarpa), 65. Samarinda: Balai besar penelitian Dipterokarpa https://www.worldcat.org/title/panduan-identifikasi-jenis-pohon-tengkawang/oclc/959964780.
4 Muhammad, B.Z., Darmawan, M.A., and Gozan, M. (2019). Reduction of beta-carotene with thermal activated bentonite in Illipe butter from Nanga Yen, Kalimantan Barat. *AIP Conference Proceedings*. AIP Publishing LLC, Serpong, Banten, Indonesia (23 – 24 October 2019). https://aip.scitation.org/doi/pdf/10.1063/1.5134610.
5 Kusumaningtyas, V. and Sulaeman, A. (2012). Potensi lemak biji tengkawang terhadap kandungan mikroba pangan pada pembuatan mie basah. *Bionatura* 14 (2): https://www.neliti.com/id/publications/217894/potensi-lemak-biji-tengkawang-terhadap-kandungan-mikroba-pangan-pada-pembuatan-m.
6 Darmawan, M.A., Ramadhani, N.H., Hubeis, N.A. et al. (2022). Natural sunscreen formulation with a high sun protection factor (SPF) from tengkawang butter and lignin. *Ind. Crops Prod.* 177: 114466.
7 Athar, M. and Nasir, S.M. (2005). Taxonomic perspective of plant species yielding vegetable oils used in cosmetics and skin care products. *Afr. J. Biotechnol.* 4 (1): 36–44.
8 Kochhar, S.P. (2011). Minor and speciality oils. In: *Vegetable Oils in Food Technology* (ed. F.D. Gunstone), 291–341. Wiley.
9 Darmawan, M.A., Ramadhan, M.Y.A., Curie, C.A. et al. (2022). Physicochemical and oxidative stability of indigenous traditional tengkawang butter as potential cocoa butter equivalent (CBE). *Int. J. Food Prop.* 25 (1): 780–791.
10 Ketaren, S. (2008). *Pengantar teknologi minyak dan lemak pangan*. Jakarta: UI Press https://lib.ui.ac.id/detail?id=13068.
11 Darmawan, M.A., Muhammad, B.Z., Harahap, A.F.P. et al. (2020). Reduction of the acidity and peroxide numbers of tengkawang butter (*Shorea stenoptera*) using thermal and acid activated bentonites. *Heliyon* 6 (12): e05742.
12 Bresson, S., Lecuelle, A., Bougrioua, F. et al. (2021). Comparative structural and vibrational investigations between cocoa butter (CB) and cocoa butter equivalent (CBE) by ESI/MALDI-HRMS, XRD, DSC, MIR and Raman spectroscopy. *Food Chem.* 363: 130319.
13 Jia, C.-H., Shin, J.-A., and Lee, K.-T. (2019). Evaluation model for cocoa butter equivalents based on fatty acid compositions and triacylglycerol patterns. *Food Sci. Biotechnol.* 28 (6): 1649–1658.
14 Gusti, R.E.P. and Zulnely, Z. (2015). Pemurnian beberapa jenis lemak tengkawang dan sifat fisiko kimia. *J. Penelitian Hasil Hutan* 33 (1): 61–68.

15 Wang, H. and Maleky, F. (2018). Effects of cocoa butter triacylglycerides and minor compounds on oil migration. *Food Res. Int.* 106: 213–224.

16 Zhang, Z., Ma, X., Huang, H., and Wang, Y. (2017). Shea olein based specialty fats: preparation, characterization and potential application. *LWT* 86: 492–500.

17 FAO. (2016). Codex Standard for Cocoa butters. https://www.fao.org/input/download/standards/66/CXS_086e.pdf

18 BSN. SNI 2903:2016. (2016). Lemak tengkawang sebagai bahan baku (ed. BSN). https://bsilhk.menlhk.go.id/standarlhk/2022/07/25/sni-2903-2016-lemak-tengkawang-sebagai-bahan-baku/#dearflip-df_1343/1

19 Maharani, R., Fernandes, A., and Pujiarti, R. (2016). Comparison of Tengkawang fat processing and its effect on Tengkawang fat quality from Sahan and Nanga Yen villages, West Kalimantan, Indonesia. *AIP Conference Proceedings*. AIP Publishing LLC, Yogyakarta, Indonesia (18 – 19 September 2015).

20 Galindo-Cuspinera, V., Valenca de Sousa, J., and Knoop, M. (2017). Sensory and analytical characterization of the "cool-melting" perception of commercial spreads. *J. Texture Stud.* 48 (4): 302–312.

21 Kodali, D.R. (2005). Trans fats—chemistry, occurrence, functional need in foods and potential solutions. *Trans fats Altern.* 1–25.

22 Kong, F. and Singh, R.P. (2011). Chapter 12 – Advances in instrumental methods to determine food quality deterioration. In: *Food and Beverage Stability and Shelf Life* (ed. D. Kilcast and P. Subramaniam), 381–404. Woodhead Publishing.

23 Patterson, H.B.W. (2011). Chapter 12 – Quality and control. In: *Hydrogenation of Fats and Oils*, 2e (ed. G.R. List and J.W. King), 329–350. AOCS Press.

24 Bart, J.C.J., Palmeri, N., and Cavallaro, S. (2010). Chapter 6 – Emerging new energy crops for biodiesel production. In: *Biodiesel Science and Technology* (ed. J.C.J. Bart, N. Palmeri, and S. Cavallaro), 226–284. Woodhead Publishing.

25 Torbica, A., Jambrec, D., Tomić, J. et al. (2016). Solid fat content, pre-crystallization conditions, and sensory quality of chocolate with addition of cocoa butter analogues. *Int. J. Food Prop.* 19 (5): 1029–1043.

26 Shin, J.-A., Hong, Y.-J., and Lee, K.-T. (2021). Development and physicochemical properties of low saturation alternative fat for whipping cream. *Molecules* 26 (15): 4586.

27 Naeli, M.H., Farmani, J., and Zargaraan, A. (2018). Prediction of solid fat content curve of chemically interesterified blends of palm stearin and soyabean oil. *J. Oil Palm Res.* 30 (4): 579–590.

28 Fiebig, H.-J. and Lüttke, J. (2003). Solid fat content in fats and oils – determination by pulsed nuclear magnetic resonance spectroscopy [C-IV 3g (2003)]. *Eur. J. Lipid Sci. Technol.* 105 (7): 377–380.

29 Dos Santos, M.T., Gerbaud, V., and Le Roux, G.A.C. (2014). Solid fat content of vegetable oils and simulation of interesterification reaction: predictions from thermodynamic approach. *J. Food Eng.* 126: 198–205.

30 Rao, R., Sankar, K.U., Sambaiah, K., and Lokesh, B.R. (2001). Differential scanning calorimetric studies on structured lipids from coconut oil triglycerides containing stearic acid. *Eur. Food Res. Technol.* 212 (3): 334–343.

31 Jahurul, M., Zaidul, I., Norulaini, N.N. et al. (2014). Hard cocoa butter replacers from mango seed fat and palm stearin. *Food Chem.* 154: 323–329.

32 Ribeiro, A.P.B., Basso, R.C., Grimaldi, R. et al. (2009). Effect of chemical interesterification on physicochemical properties and industrial applications of canola oil and fully hydrogenated cottonseed oil blends. *J. Food Lipids* 16 (3): 362–381.

33 Karabulut, I., Turan, S., and Ergin, G. (2004). Effects of chemical interesterification on solid fat content and slip melting point of fat/oil blends. *Eur. Food Res. Technol.* 218 (3): 224–229.

34 Lida, H.M.D.N. and Ali, A.R.M. (1998). Physico-chemical characteristics of palm-based oil blends for the production of reduced fat spreads. *J. Am. Oil Chem. Soc.* 75 (11): 1625–1631.

35 Afoakwa, E.O., Paterson, A., Fowler, M., and Vieira, J. (2008). Characterization of melting properties in dark chocolates from varying particle size distribution and composition using differential scanning calorimetry. *Food Res. Int.* 41 (7): 751–757.

36 Badu, M. and Awudza, A.J. (2017). Determination of the triacylglycerol content for the identification and assessment of purity of shea butter fat, peanut oil, and palm kernel oil using maldi-tof/tof mass spectroscopic technique. *Int. J. Food Prop.* 20 (2): 271–280.

37 Darmawan, M.A., Siregar, K., and Gozan, M. (2022). Chapter 6 – Coconut oil. In: *Biorefinery of Oil Producing Plants for Value-Added Products*, vol. 1 (ed. S. Abd-Aziz, M. Gozan, M.F. Ibrahim, and P. Lai-Yee). Print ISBN 978-3-527-34876-3, 99–122. Wiley-VCH GmbH.

38 Foletto, E., Colazzo, G., Volzone, C., and Porto, L. (2011). Sunflower oil bleaching by adsorption onto acid-activated bentonite. *Braz. J. Chem. Eng.* 28 (1): 169–174.

39 Hidayat, N., Darmawan, M., Intan, N., and Gozan, M. (eds). (2019). Refining and physicochemical test of tengkawang oil *Shorea stenoptera* origin Sintang District West Kalimantan. *IOP Conference Series: Materials Science and Engineering.* IOP Publishing, Medan, West Sumatera, Indonesia (4 – 6 October 2018).

40 Carvajal, A.K. and Mozuraityte, R. (2016). Fish oils: production and properties. In: *Encyclopedia of Food and Health* (ed. B. Caballero, P.M. Finglas, and F. Toldrá), 693–698. Oxford: Academic Press.

41 O'Brien, R.D. (2008). Chapter 12 – Soybean oil purification. In: *Soybeans* (ed. L.A. Johnson, P.J. White, and R. Galloway), 377–408. AOCS Press.

42 Chakrabarti, P.P. and Jala, R.C.R. (2019). Chapter 3 – Processing technology of rice bran oil. In: *Rice Bran and Rice Bran Oil* (ed. L.-Z. Cheong and X. Xu), 55–95. AOCS Press.

43 Erickson, D.R. (2007). Chapter 1 – Production and composition of frying fats. In: *Deep Frying*, 2e (ed. M.D. Erickson), 3–24. AOCS Press.

44 Gupta, M.K. (2017). Chapter 8 – Deodorization. In: *Practical Guide to Vegetable Oil Processing*, 2e (ed. M.K. Gupta), 217–247. AOCS Press.

45 Darmawan, M. (2022). *Development of Production and Storage Processes of Tengkawang Fat Products (Shorea stenoptera) Fulfilling the Quality of Acid*

Value and Peroxide Numbers in SNI 2903: 2016. Depok, Indonesia: Universitas Indonesia.

46 Ramadhani, N.H., Darmawan, M.A., Harahap, A.F.P. et al. (2021). Simulation of illipe butter purification originated from West Kalimantan by SuperPro Designer. *Journal of Physics: Conference Series*. IOP Publishing, Balikpapan, East Kalimantan, Indonesia (9 – 10 September 2020).

47 Ben-Horin, M. and Kroll, Y. (2017). A simple intuitive NPV-IRR consistent ranking. *Q. Rev. Econ. Finance* 66: 108–114.

48 Max, S.P., Klaus, D.T., and Ronald, E.W. (2003). *Plant Design and Economics for Chemical Engineers*. McGraw-Hill Companies.

49 Aumpai, K., Tan, C.P., Huang, Q., and Sonwai, S. (2022). Production of cocoa butter equivalent from blending of illipé butter and palm mid-fraction. *Food Chem.* 384: 132535.

50 Ramadhani, N.H., Darmawan, M.A., Harahap, A.F.P., et al. (2021). Effect of purification and lignin addition of Illipe butter on beta-carotene level and free fatty acid content. *AIP Conference Proceedings*. AIP Publishing LLC, Depok, West Java, Indonesia (28 – 29 July 2020).

13

Bio-succinic Acid Production from Biomass and their Applications

Abdullah A. I. Luthfi[1,2], Jian P. Tan[3,4], Wen X. Woo[3], Nurul A. Bukhari[5], and Hikmah B. Hariz[1]

[1]*Universiti Kebangsaan Malaysia, UKM, Department of Chemical and Process Engineering, Faculty of Engineering and Built Environment, Bangi 43600, Selangor, Malaysia*
[2]*Universiti Kebangsaan Malaysia UKM, Research Centre for Sustainable Process Technology (CESPRO), Faculty of Engineering and Built Environment, Bangi 43600, Selangor, Malaysia*
[3]*Xiamen University Malaysia, Faculty of Chemical Engineering, School of Energy and Chemical Engineering, Jalan Sunsuria, Bandar Sunsuria, Sepang, 43900, Selangor, Malaysia*
[4]*Xiamen University, College of Chemistry and Chemical Engineering, Xiamen, 361005, China*
[5]*Malaysian Palm Oil Board (MPOB), Energy and Environment Unit, Engineering & Processing Research Division, 6, Persiaran Institusi, Bandar Baru Bangi, Kajang 43000, Selangor, Malaysia*

13.1 Introduction

Concerns about the environment, energy security, and depleting reserves signified the need for research into an alternative pathway for producing valuable dicarboxylic acids, such as succinic acid. Among others, fermentation represents an environmentally clean biocatalytic technology for producing valuable fine chemicals as evidenced by lower emissions and energy intensity, thus making it applicable to a wide range of applications [1]. Various second-generation (2G) biomass was investigated to replace the traditional refined sugars as substrate in the fermentation process due to their sustainability, nonfood source, low cost, ease of accessibility, and richness in beneficial nutrients for microbial cultivation. In contrast to first-generation (1G) edible biomass-to-succinic acid processing, lignocellulosic biomass conversion is more challenging due to the intrinsic recalcitrant structures and complexity of the chemical component, as well as the emergence of a wide range of intermediate inhibitory compounds during processing. This chapter aims to highlight the potential of biomass waste utilization for succinic acid production and its applications in various industries. The review also focuses on biomass pretreatment and hydrolysis, as well as fermentation strategy and downstream processing techniques currently in place for succinic acid production.

13.1.1 Background of Succinic Acid Production

Succinic acid is a four-carbon dicarboxylic acid (molecular formula: $C_4H_6O_4$) with a linear saturated structure that exists in nature [2]. Pure succinic acid is in solid

Chemical Substitutes from Agricultural and Industrial By-Products: Bioconversion, Bioprocessing, and Biorefining,
First Edition. Edited by Suraini Abd-Aziz, Misri Gozan, Mohamad Faizal Ibrahim, and Lai-Yee Phang.
© 2024 WILEY-VCH GmbH. Published 2024 by WILEY-VCH GmbH.

form at room temperature, and its low solubility causes it to dissolve into an aqueous solution. Traditionally, succinic acid is a petro-based by-product from chemical catalytic processes, such as the hydrogenation of maleic anhydride/maleic acid. However, there have been numerous limitations to the chemical production of succinic acid, including a high intensity of energy and environmental pollution concerns [1]. Bio-succinic acid, on the other hand, exhibits the highest potential yield (as much as 1.12 kg of bio-succinic acid for every kg glucose) attainable from biomass sugars compared to any other bio-based chemical products. Some industrial actors have managed to commercialize the production of biotechnologically derived succinic acid, particularly anaerobic fermentation using lignocellulosic biomass as the raw material. This simultaneously led to the massive technological shifts by leaps and bounds toward biological processes involving microbial catalysis. Unlike traditional production pathways, one of the obvious advantages of anaerobic fermentation is the much milder operating condition coupled with the fastidious growth of microorganisms as biocatalysts [3].

13.1.2 Global Market and Demands for Bio-succinic Acid

Bio-succinic acid is a valuable chemical, which contributed significantly to the bio-based economy, and it is expected to be one of the future platform chemicals derived from renewable resources [1]. The global market for bio-succinic acid is valued at US$ 158.96 million in 2021 and is predicted to reach US$ 315.89 million by 2030 [4]. The expanding applications of bio-succinic acid are driven by the chemical industry's shift toward environmentally friendly chemicals obtained through biological means. BioAmber (Canada), Myriant Technologies (United States), Succinity (Spain), Reverdia (Italy), Michigan Institute of Biotechnology (United States), Mitsubishi–Ajinomoto (Japan), and MBEL-KAIST (South Korea) are among the big players, which accommodate a large production scale of succinic acid and its derivatives. At the moment, the main bottleneck in the mass production of bio-succinic acid is suboptimal productivity and relatively costly. The cost of bio-succinic acid production was reported at US$ 1.66–2.2 per kg. On the other hand, petrochemically-derived succinic acid was valued around US$ 1.05–1.29 per kg [5]. This could be attributed to the requirement for microbial growth and downstream processing, wherein the latter is said to account for 60% of the total processing cost to achieve high purity (>98% purity) that can be accomplished via multiple stages of purification and recovery processes [6, 7].

Nowadays, the international demand for succinic acid is estimated at ca. 20 000–30 000 tons per year worldwide, which is dominated by the chemical processing pathways [2]. However, since succinate is an endogenous compound of microbial fermentation and an intermediary/precursor component of the tricarboxylic acid (TCA) cycle, its production could be successfully realized through a wide range of microorganisms. The biological succinic acid production at a commercial scale, however, still requires a lot of work and calls for developing microorganism-based technologies and improving feedstock utilization to justify economically feasible recovery from a biorefinery processing perspective.

13.2 Valorization of Biomass to Bio-succinic Acid

Biorefineries were one of the top priorities in the history of biomass conversion for utilizing organic wastes as sustainable feedstocks. In practice, recovering fermentable sugars from lignocellulose is more difficult than the 1G starch conversion. Figure 13.1 shows the biorefinery concept employing lignocelluloses. Existing biomass conversion technologies emphasize four basic steps: (i) analyzing the chemical composition of biomass, (ii) fractionating beneficial components through pretreatment and/or hydrolysis, (iii) fermenting the fermentable constituents to generate a product using microorganisms, and (iv) purifying the resultant products for commercialization. There is evidence in the literature that lignocellulosic biomass, such as, but not limited to sweet sorghum bagasse [8], sugarcane bagasse

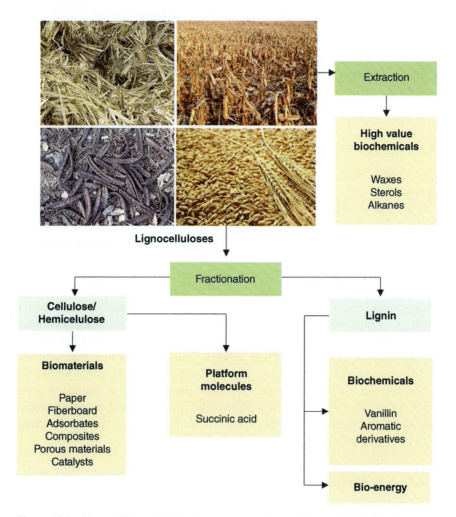

Figure 13.1 Lignocelluloses in biorefinery concept. Source: Drawn by Luthfi et al.

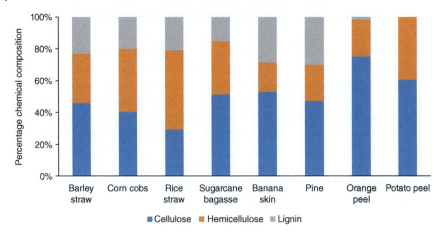

Figure 13.2 Compositional attributes of biomass. Source: Drawn by Luthfi et al. (author).

[9, 10], oakwood [11], giant reed [12], and wheat hydrolysate [13] could be used as potential feedstocks for platform molecule succinic acid production. It is also possible to use industrial by-products, such as spent sulfite liquor, sake lees, rapeseed meal, cheese whey, and crude glycerol as feedstock [14].

13.2.1 Compositional Attributes of the Biomass

Lignocellulose materials are made up of three structural components: cellulose, hemicellulose, and lignin. Figure 13.2 shows the typical compositional attributes of lignocellulosic biomass, with cellulose being the main constituent, making up about 35–50% of the plant cell wall of biomass. Cellulose molecules are high molecular polymers composed of glucose. Whereas, hemicellulose makes up around 20–35% of the cell wall. Hemicellulose is composed of a wide range of sugars including pentoses (such as xylose and arabinose) and hexoses (such as glucose, galactose, mannose, glucuronic acid, and galacturonic acid) [15]. Lastly, lignin is a non-saccharide compound and a highly cross-linked aromatic heteropolymer, which is distinctively different from other macromolecules of lignocellulosic biomass. Lignin acts as a natural binding agent for both cellulose and hemicellulose that are arranged in a 3-dimensional network, thus offering a firm structural and mechanical support [15]. As a result, high lignin content is unfavorable for bioconversion because it reduces substrate bioavailability due to its resistance to water and enzymes. Nonetheless, the richness in carbohydrates of lignocelluloses (>60%) indicates its potential in bio-based chemical manufacturing, e.g. bio-succinic acid, as part of the successful development of a biorefinery.

13.2.2 Size Reduction

Biomass mechanical preprocessing methods like size reduction are important for feedstocks to meet the quality requirements for subsequent biochemical conversion to bio-succinic acid. Size reduction will impact the particle size and distribution. It is usually performed via chipping, grinding, and milling to break down the materials to a much-reduced size, thus minimizing the crystallinity of the lignocellulosic materials. The size of materials is usually between 1.0–3.0 cm, 0.2–2.0 mm, and <2.0 mm after chipping, grinding, and milling, respectively [16]. Sieving is used to divide the materials into parts with different particle sizes. The final particle size and the variety of feedstock have a direct impact on the amount of energy used during the mechanical preprocessing of lignocelluloses. Energy requirement differs from 130 kWh/t for hardwood (i.e. diminution in particle size to ~1.60 mm) to 3.2 kWh/t for corn stover (i.e. diminution in particle size to ~9.5 mm). From an economic standpoint, it is desirable to reduce or eliminate the steps for comminution or milling during the preprocessing step prior to bio-succinic acid production.

13.2.3 Pretreatment

The barrier to unlocking valuable materials from lignocellulosic materials is the degradation-resistant structure of biomass due to the presence of ester- and ether-based cross-linkages between the polysaccharides (cellulose and hemicellulose) and lignin. Therefore, pretreatment is necessary to alter such structural and compositional impediments to hydrolysis in order to improve enzymatic digestibility and increase recovery yields of fermentable sugars [17]. Pretreatment overcomes biomass recalcitrance, which is driven by factors, such as high lignin content, hemicellulose sheathing, high cellulose crystallinity and degree of polymerization, the low accessible surface area of cellulose, and strong fiber strength [16].

Pretreatment can be classified into four categories: physical, physicochemical, chemical, and biological. Pretreatment aims to remove lignin, decrease cellulose crystallinity, increase accessible surface areas, and enhance the porosity of the materials, in order to facilitate subsequent enzymatic hydrolysis [16]. Inefficient pretreatment leads to not only difficulties in hydrolyzing the resultant residues by hydrolytic enzymes, but also the production of a considerable amount of toxic compounds that could inhibit subsequent microbial fermentation to produce bio-succinic acid [18]. Therefore, it is critical to optimize the pretreatment conditions to match the chemical composition and internal structure of the corresponding biomass. In addition, the effectiveness of downstream methods for recovering bio-succinic acid is dependent on the pretreatment stage.

13.2.4 Hydrolysis

Enzymatic hydrolysis is the subsequent step after the pretreatment of lignocellulosic biomass to yield fermentable sugars. The use of enzymes is advantageous over concentrated acid because they require relatively mild process conditions with high substrate specificity, and are utterly green [19]. Enzymatic hydrolysis can also be performed to gauge improvement in sugar conversion, which relates to pretreatment efficiency.

In enzymatic hydrolysis of lignocellulosic biomass, three primary types of cellulases are responsible for the conversion of carbohydrates into sugars, i.e. endoglucanases, exoglucanases, and β-glucosidase.

Endoglucanases attack regions of the internal amorphous sites amidst cellulose chains, generating oligosaccharides of various chain lengths. Exoglucanases act on the reducing and nonreducing ends of cellulose chains, releasing either glucose (via glucanohydrolases) or cellobiose (via cellobiohydrolase) as the major products, while β-glucosidase hydrolyses cellodextrin and cellobiose to glucose [19]. The cost of cellulases accounts for a substantial proportion of the total processing costs in bioconversion of lignocellulosic materials. Cost-saving considerations must be put in place to provide the impetus to reducing enzymatic loadings and recovering used enzymes. Potential enhancement and detailed mechanism of enzymatic hydrolysis remain a research focus in the bio-succinic acid production technology [19–21].

13.3 Bio-succinic Acid as Fermentative Metabolite

Fermentation of substrates to succinic acid by microorganisms follows either one of the following: (i) the TCA or Krebs cycle, (ii) the glyoxalate cycle, and (iii) the reductive TCA cycle. However, during the Krebs and glyoxalate cycles, cells do not accumulate succinate as it will further convert into other metabolites. Accordingly, a successful metabolic flux shunt toward succinic acid production can only be realized by the reductive TCA pathway, during which oxaloacetate is formed from the breakdown of PEP under anaerobic conditions. As demonstrated in the following chemical reaction (Eq. (13.1)), one mol of glucose requires four electron donors and two mols of CO_2, resulting in succinic acid accumulation [22].

$$C_6H_{12}O_6 + 2CO_2 + 4H^+ \rightarrow 2C_4H_6O_4 + 2H_2O \tag{13.1}$$

13.3.1 Microbial Workhorse as Potential Biocatalyst

Bio-succinic acid is found in microbes, plants, and animal cells as a terminal point of metabolic pathways and as a precursor to the TCA pathway. The best hosts for succinic acid accumulation have been identified as bacteria and fungi, wherein both wild-type and bioengineered strains have been extensively researched. Succinic acid is produced alongside other metabolites, such as acetic and formic acids, and, in some cases, ethanol, pyruvic acid, and lactic acid in small quantities.

13.3.1.1 Bacteria

In recent investigation, rumen-type bacteria *Actinobacillus succinogenes*, *Enterobacter aerogenes*, *Mannheimia succiniciproducens*, and *Basfia succiniciproducens* have been a widely researched bacteria that are able to accumulate succinic acid at high levels. It has been reported that they mainly produce bio-succinic acid from carbohydrates during anaerobic fermentation and are capable to achieve a yield of 0.61–0.84 g/g glucose. They are facultative anaerobes that are able to consume numerous reducing sugars in the presence of excess CO_2 supply. Particularly, *A. succinogenes* is one of the most appealing wild-type succinic acid-producing strains. Recent studies have shown that *A. succinogenes* has a wide acceptance of carbon sources, including glucose, xylose, and arabinose, thereby elevating the convertibility of substrate and thus, boosting the yield and productivity of bio-succinic acid when lignocellulosic biomass is used as the raw material [23]. However, industrial application of these strains is limited by poor cell growth, low tolerance toward harsh pH conditions and by-product accumulation, and lack of effective genetic engineering tools [24]. A better understanding of succinic acid synthesis still necessitates a thorough understanding of this species' characteristics as well as genetic techniques.

13.3.1.2 Fungi

Fungi and yeasts are potential candidates as a viable alternative to bacteria for fermentative succinic acid production due to advantages, such as the accessibility of genetic engineering tools and resistance to low pH environments [24]. Generally, fungi prefer acidic pH for their optimum growth. This has led to the avoidance of neutralization agents, risk of contamination, and the overall processing costs, thereby making the downstream processing more favorable. Much effort has been devoted to developing bio-succinic acid production employing several fungi/yeasts, including *B. nivea*, *L. degener*, *P. viniferum*, *Ammophilus fumigatus*, *A. niger*, and *Saccharomyces cerevisiae* [24, 25]. Under aerobic and/or anaerobic conditions, these fungi produce bio-succinic acid as a metabolic by-product. A recent study has shown that the engineered *Yarrowia lipolytica* could metabolize a large variety of substrates, including xylose, which normally appears as the second major component in lignocellulosic biomass hydrolysate. Without any pH control strategy, engineered *Yarrowia lipolytica* attain 0.19 g/g with productivity amounting to 0.13 g/l.h, proving the advantages of fungi in accumulating succinate even at low pH [9].

13.3.2 Fermentation Conditions

Unlike conventional process, biotechnologically derived succinic acid production using microbes, usually operate under milder operating condition. The most obvious difference is the operating temperature since the succinic acid-producing strains are mostly mesophilic microbes. As demonstrated in Table 13.1, the optimum temperature for microbial fermentation ranges from 30 to 40 °C.

Theoretically, the succinate is formed within the cell's metabolic pathways, which occur through the reductive TCA via CO_2 assimilation. Thus, the fermentation process is usually supplied with either CO_2 or air, depending on the type of microbes and their ideal conditions (i.e. aerobic or anaerobic). The extent of CO_2 in the broth culture directly controls metabolism as well as fermentation product selectivity [22].

Apart from that, most succinic acid-producing bacteria are pH sensitive, with an optimum range of 6–7. Lowering or raising the pH value within such range would pose a negative effect on gas solubility. The availability of dissolved CO_2 is

Table 13.1 Fermentation conditions of succinic acid-producing strains and their performance.

Strain type	Substrate	C/N ratio	Temperature (°C)	Gas supply	pH	Yield (g/g glucose)	References
A. succinogenes 130Z	Sweet sorghum bagasse	27.4	37	CO_2 (0.5 vvm)	6.8–7.2	0.61	[8]
M. succiniciproducens MBEL55E	Oakwood	43.6	39	CO_2	6.5	0.56	[11]
B. succiniciproducens BPP7	Energy crops (giant reed)	23.9	37	CO_2 (0.5 vvm)	6.5	0.84	[12]
E. aerogenes LU2	Lactose	98.9	34	—	7	0.62	[26]
E. coli K3OS	Strophanthus preussii	25.8	37	Air (5 l/min)	—	1.1 (mol/mol sugar)	[27]
C. glutamicum BL-1/pVWEx1-glpFKD	Crude glycerol	—	30	Air (0.9 vvm)	7	0.21 (mol/mol)	[14]
Y. lipolytica PSA02004PP	Xylose (sugarcane bagasse)	19.6	30	Air (2 l/min)	6.8	0.13	[9]

directly related to the pH of the culture medium, indicating that the pH was the most important feature influencing bio-succinic acid accumulation. Due to the inhibition brought on by fermentation by-products like acetic acid and formic acid, the synthesis of fermentative succinic acid is typically linked to pH regulation. Moreover, the application of neutralizers and the neutral pH of the fermentation medium created by bacterial cultures influence both fermentation and downstream separation expenses [28]. As with furfural and 5-hydroxymethylfurfural, the by-products of lignocellulosic hydrolysate were also discovered to be harmful to the microorganisms by preventing their proliferation and so lowering product yield [18]. The main impediment to fermentative succinate production is by-product inhibition. By contrast, application of fungi (*S. cerevisiae* and *Y. lipolytica*) as the biocatalyst in producing succinic acid can be considered an attractive option for commercialization because of the feasibility of low pH fermentation (pH 3–5); however, major bottlenecks, such as low yield and productivity seemed to limit its application [29].

13.3.3 Media Composition

A microbial culture medium is a mixture of substances that promotes and supports the growth and differentiation of microorganisms. Culture media should contain nutrients, energy sources, growth-promoting factors, minerals, metals, and buffer salts. As seen in Table 13.1, the microorganism and type of feedstock involved exhibit a direct effect on the yields. Carbon and nitrogen (C/N) sources, in this regard, are the most essential media components governing the yield for each type of microorganism and feedstock. Nevertheless, both organic and inorganic sources have a significant effect on the growth of succinic acid-producing strains and the subsequent succinic acid production. The culture medium must be adequately designed and optimized for the successful cultivation of bacteria and/or fungi.

13.3.3.1 Carbon Sources

Succinic acid-producing bacteria and fungi utilize carbohydrates to produce succinic acid, signifying that the quality of the carbon source is the most crucial factor for cell growth. The valorization of lignocellulosic biomass has received huge attention for bio-succinic acid production as it contains an attractive amount of cellulose and hemicellulose, which can be hydrolyzed into its monomer units as carbon sources for consecutive bioprocessing stages. It has been reported that *A. succinogenes* 130Z can metabolize both glucose and xylose due to its wide acceptability of carbon sources [30]. Similarly, Andersson et al. [31] have reported that the *E. coli* strain AFP184 was able to produce succinic acid in a low-cost medium from a variety of sugars (sucrose, glucose, fructose, and xylose) with only small amounts of by-products formed. In general, the higher sugar concentration is favored for high succinic acid production until it reached a saturation point. For instance, it was reported that the productivity and yield of succinic acid reached a maximum at a glucose concentration of 60 g/l. Further increasing the glucose concentration to 85 g/l might lead to detrimental effect on the growth of *A. succinogenes* [3].

13.3.3.2 Nitrogen Sources

Nitrogen source is considered another key factor due to its importance for cell nutrition and growth. Thus, an appropriate level of nitrogen addition is beneficial to the growth of bacteria and fungi as well as the yield of subsequent bio-succinic acid production. Ammonia–nitrogen is the most widely investigated nitrogen source for fermentative succinic acid production. There are two types of nitrogen sources: organic (yeast extract [YE], corn steep liquor [CSL]) and inorganic (ammonium salt, urea). Typically, fermentations of *A. succinogenes* utilize YE or CSL as the nitrogen source. When YE was compared to a variety of nitrogen sources, including CSL, peptone, tryptone, and ammonium chloride, it produced the highest biomass viability and succinic acid accumulation. Increasing the YE composition between 2.5 and 25 g/l had a positive effect on the growth of *A. succinogenes*. This is because YE encompasses traces of vital microelements, such as vitamins B1, B2, B6, and B12, pantothenic acid, biotin, as well as folic acid, thereby allowing the addition of several nutrients to the medium dispensable [32]. However, the application of YE is limited by the high cost of fermentative media on industrial scale. Therefore, exploring and developing other cost-effective nitrogen sources derived from biomass is required for commercialization.

13.3.3.3 Carbon Dioxide

To achieve the highest possible yield of succinic acid, the central metabolism must flux toward succinate via the reductive part of the TCA cycle, resulting in the net fixation of CO_2. For industrial applications, a carbonate salt is a good option, which can supply CO_2 and regulate pH simultaneously. Na_2CO_3, $NaHCO_3$, $MgCO_3$, and $CaCO_3$ are the most commonly used CO_2 sources. For instance, magnesium carbonate ($MgCO_3$) is usually supplemented in excess as a pH buffer and serves as the CO_2 supplier once it reacts with the fermentation metabolites, such as acetic and formic acids [22].

13.3.3.4 Additives

Other additives, such as mineral salts and metal ions are relatively less significant than the aforementioned nutrients, but they still contribute to cell growth as supplements. The formulation includes, but is not limited to, KH_2PO_4, K_2HPO_4, $(NH_4)_2HPO_4$, $MgSO_4 \cdot 7H_2O$, $CaCl_2:2H_2O$, $MgCl_2 \cdot 6H_2O$, NaCl, and NaH_2PO_4 [27, 33]. Other nutrients, such as vitamins, biotin, and heme as growth factors were also added to the medium [34]. In some cases, an anti-foaming agent is added when necessary to prevent foaming spillovers and contamination [3].

13.3.4 Fermentation Configuration and Strategies

Many different fermentation configurations and strategies for bio-succinic acid accumulation were developed. There are 4 primary process configurations and 3 primary operational modes for the fermentation of pretreated and hydrolyzed biomass. The configurations can be divided into SHF (separate hydrolysis and fermentation), SSF (simultaneous saccharification and fermentation), SSCF (simultaneous saccharification and co-fermentation), and CBP (consolidated bioprocessing). The three operational modes include batch, fed-batch, and continuous process.

13.3.4.1 Separate Hydrolysis and Fermentation

A key distinguishing feature of SHF is that multiple steps could be carried out under their respective optimal conditions to produce the highest possible yield of reducing sugars and succinic acid, as enzymatic hydrolysis and microbial fermentation usually have a different operating conditions (45–55 °C, pH 4.5–5.5 and 30–37 °C, pH 7.0 for enzymatic saccharification and fermentation, respectively) [35]. A succinic acid yield of 0.73 g/g was recently discovered via SHF method. The research discovered that the main by-products were 0.7 g/l formic acid, 0.6 g/l acetic acid, and 0.4 g/l ethanol [22]. Although SHF involves higher equipment investment costs and a longer operating duration, it is an ideal strategy due to its simplicity for improving one's knowledge of the process mechanism, such as finding the suitable microorganism that is able to produce succinate efficiently [12]. Therefore, SHF is the primary focus of research on lignocellulose bioconversion into succinate.

13.3.4.2 Simultaneous Saccharification and Fermentation/Co-fermentation

In SSF, hydrolysis and fermentation occur within the same bioreactor using single or multiple enzymes and microorganisms. SSF outperforms SHF in terms of reduced hydrolytic enzyme supplementation and increased product yield given the absence of product feedback inhibition to enzymes. In batch fermentation using corn straw hydrolysate as the carbon source, succinic acid productivity (0.99 g/l/h) by SSF was slightly higher than that by SHF (0.95 g/l/h) [36]. Apart from that, semi-SSF was performed on duckweed biomass by combining 5.90 Amyloglucosidase Units (AGU)/g dry matter glucosidase with 31.86 μ/g dry matter pullulanase at pH 4.5 at 2, 4, and 6 hours at 60 °C. It was found that the titer of succinic acid obtain (65.31 g/l) was relatively higher than that using SHF (62.12 g/l). The yield and productivity attainable through this method were 82.87% (g/g) and 1.35 g/l/h, respectively, after 56 hours fermentation. However, it should be noticed that due to the different optimum temperatures, optimizing the SSF to compromise the two optimal temperatures is complicated, thus limiting its potentiality in industrial bioprocessing. Hence, a robust biocatalyst can be isolated or genetically modified to withstand severe temperature conditions to elevate succinic acid yields [35].

13.3.4.3 Consolidated Bioprocessing

CBP is another alternative to SHF and SSF, which employs organisms to produce hydrolases, perform lignocellulose hydrolysis, and microbial fermentation all in a single pot that can improve product yield and productivity while also mitigating separate costs of operation and equipment [35]. A recent study successfully designed a mixed fungi consortium to produce succinic acid derived from minimally pretreated lignocellulose liquor via CBP [37]. This process included three different microbial cultures, *A. niger*, *T. reesei*, and *P. chrysosporium*. *A. niger* and *T. reesei* secreted cellulase and hemicellulase, which were primarily accountable for saccharifying lignocellulose. *P. chrysosporium*, on the other hand, could evoke ligninolytic enzymes for delignification, allowing the enzyme easy access to the substrate. After two-step fermentation, 32.58 g succinic acid/kg dry substrate was generated [37]. This result demonstrated that the microbial cocultivation system in an intensified process was easily adaptable for the lignocellulosic biomass valorization, representing a viable strategy for succinic acid generation from lignocellulosic materials.

13.3.4.4 Operational Modes

The most widely employed technique is batch fermentation. In batch mode, all the components in fermentative medium are initially charged in the reactor. This mode is ideal for instant processes like nutrient medium optimization and strain characterization.

In addition to the common batch approaches, more dynamic systems, such as fed-batch, continuous, and immobilization of cells, which allow for recycling and reusing the viable cells, have been proposed for bio-succinic acid fermentation [38]. With controlled sequential additions of nutrients, a fed-batch culture is generally more productive and yields higher cell densities. This is becoming more important when designing a large-scale production system aiming to process as much biomass loading as possible, which is not feasible in a batch process due to limited space. In this regard, the fed-batch strategy has been proven to overcome this bottleneck and process biomass with solid concentrations of up to 17% [39]. For example, Wang et al. [40] reported that the fed-batch fermentation of pretreated residues of *Glycyrrhiza uralensis Fisch* (GUR) achieved a succinic acid yield of 0.89 g/g total sugar, while the batch fermentation only account for 0.69 g/g total sugar. However, the fed-batch process might indeed lengthen the fermentation time and result in product inhibition due to the buildup of toxic metabolic ends.

Continuous mode is a process in which new media and nutrients are constantly charged to the fermenter while the spent culture medium is removed, keeping the reaction volume constant at steady state. When compared to batch and fed-batch operations, continuous operations generally have suboptimal yields but greater productivity. For instance, Kim et al. [11] asserted that in a batch process, the pretreated wood hydrolysate yielded a final succinic acid concentration of 11.73 g/l, resulting in a succinic acid yield and productivity of 56% and 1.17 g/l h, respectively. The continuous process, on the other hand, achieved a succinic acid yield of 55% and productivity of 3.19 g/l h [11]. Table 13.2 summarizes the succinic acid production efficiency under different operational modes.

13.3.4.5 Immobilization

Immobilization of cells uses physical or chemical methods to bind or entrapped cells to a macroscopic support matrix so that they can be used repeatedly [38]. Unlike batch and fed-batch, the continuous process suffers from cell washout due to the high dilution rate beyond the specific growth rate in high throughput operation. In this regard, an immobilization system associated with a continuous mode is a potential strategy to exploit economies of scale for succinic acid production. Cell immobilization on support carriers is mainly employed by physical adsorption, entrapment, biofilm formation, covalent attachment, and cross-linking [38, 43, 44].

Table 13.2 Succinic acid production efficiency under different operational modes.

Substrate	Operational modes	Titer (g/l)	Yield (g/g sugar)	Productivity (g/l/h)	References
Glycyrrhiza uralensis Fisch	Batch	35	0.69	—	[40]
	Fed-batch	65	0.89	—	
Glucose	Batch	40.5	0.81	1.3 (32 h)	[41]
	Fed-batch	52.1	0.82	1.1 (48 h)	
Oakwood	Batch	11.73	56	1.17	[11]
	Continuous	—	55	3.19	
Organic fraction of municipal solid waste	Batch	29.4	56	0.89	[42]
	Continuous	21.2	47	1.27	
Xylose	Batch	38.4	0.70	0.94	[30]
	Continuous	32.5	0.77	2.64	

Ercole et al. [43] investigated bio-succinic acid production by entrapping *A. succinogenes* cells in alginate beads in a fluidized bed reactor (FBR). The performances of the FBR were particularly attractive because it was able to achieve high succinic acid productivity (35.6 g/l h) with high yield (0.86 g/g) and substrate conversion (76.4%). Alexandri et al. [44] employed fed-batch fermentation with a cell immobilization system, where *A. succinogenes* and *B. succiniciproducens* were immobilized on two different supports, namely delignified cellulosic material (DCM) and alginate beads. The final succinic acid concentration and yield of 45 g/l and 0.66 g/g, respectively, were achieved by *B. succiniciproducens* in alginates, which were higher than *A. succinogenes* immobilized cultures (35.4 g/l and 0.61 g/g). It was demonstrated that the immobilized cultures improved the efficiency of succinic acid production as compared to free cell cultures [38].

13.4 Purification and Recovery of Succinic Acid

13.4.1 Fermentation Broth Constituents

The final checkpoint for succinic acid production is its downstream processing to generate high recovery yield and purity of succinic acid crystals as the final, commercializable product. As demonstrated in Table 13.3, the fermentation broths generated from the studies contain considerable amounts of contaminants and by-products such as acetic acid, formic acid, residual sugars, lignosulfonates, and protein. Apart from that, the fermentation broth consists of a copious amount of pigments, colloids, and multivalent ionic compounds [48].

As could be seen in Figure 13.3, several purification processes are necessary to generate succinic acid in its crystal form. It starts with the primary treatment, which consists of pH adjustment, contaminant separation through centrifugation or filtration,

Table 13.3 The studies conducted on downstream processing of succinic acid from various agricultural biomass source.

Carbon source	Fermentation broth constituents (g/l)	Purification and recovery method	Yield (%)	Purity (%)	References
Wheat-derived hydrolysate	Succinic acid: 46.0 Acetic acid: 11.0 Formic acid: 9.2 Pyruvic acid: 2.6	Direct crystallization combined with ion-exchange resins	89.5	99	[45]
Wheat-derived hydrolysate	Succinic acid: 46.0 Acetic acid: 11.0 Formic acid: 9.2 Pyruvic acid: 2.6	Direct acid method	35	46	[45]
Wheat-derived hydrolysate	Succinic acid: 35.7 Acetic acid: 16.9 Formic acid: 9.4 Pyruvic acid: 4.1	Direct vacuum distillation combined with crystallization	28	45	[13]
Oil palm frond hydrolysate	Succinic acid: 29.16 Acetic acid: 3.74 Formic acid: 0.25 Glucose: 3.35	Forward osmosis combined with crystallization	67.09	90.52	[46]
Olive pits	Succinic acid: 33.6	Direct crystallisation	76.5	—	[10]
Sugarcane bagasse	Succinic acid: 28.7	Direct crystallisation	75.2	—	[10]
Spent sulfite liquor	Succinic acid: 41.2 Acetic acid: 9.9 Formic acid: 2.0 Lactic acid: 12.3 Xylose: 9.3 Lignosulfonates: 10.0	Reactive extraction with back extraction	73	97.2	[47]
Spent sulfite liquor	Succinic acid: 41.2 Acetic acid: 9.9 Formic acid: 2.0 Lactic acid: 12.3 Xylose: 9.3 Lignosulfonates: 10.0	Direct crystallization combined with cation-exchange resins	74.6	99.3	

followed by decolorization for the removal of pigments and the purification process. Since succinic acid is an extracellular product, cell disruption is not required. Meanwhile, the secondary treatment involves the recovery of succinic acid through chromatography and crystallization.

Achieving efficient purification and recovery strategies that are cost-effective and environmentally benign remains a challenge for downstream processing of succinic acid, given the fact that it accounts for 60% of succinic acid's total production

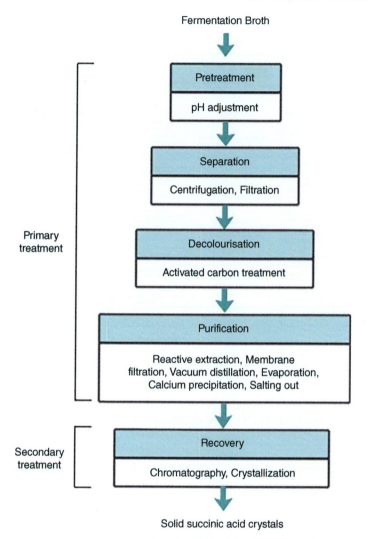

Figure 13.3 Process flow of the downstream processing of succinic acid from fermentation broth. Source: Oreoluwa Jokodola et al. [10]; Luque et al. [13]; Alexandri et al., [44]; Lin et al. [45]; Law et al. [46].

costs [6]. Furthermore, the purification steps of succinic acid from the by-products are riddled with the presence of organic acids by-products of similar characteristics like acetic acid and formic acid, as well as salt that was generated during pH adjustment. These impurities must be removed as they tend to influence the final attributes of succinic acids, such as size, shape, and crystal structure [48].

13.4.2 Downstream Processing Method

The studies listed in Table 13.3 have investigated the production of succinic acid from agricultural biomass such as wheat, oil palm frond, olive, and sugarcane along

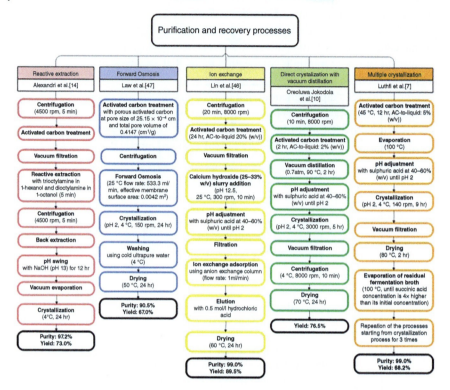

Figure 13.4 The overall processes involved in the downstream processing of succinic acid. Source: Luthfi et al. [7]; Alexandri et al. [44]; Lin et al. [45]; Law et al. [46].

with its downstream processing strategies. In all fermentation broths, succinic acid serves as the main product with a concentration range from 28 to 46 g/l. Among the included studies, the highest purity of 99% was achieved by Lin et al. [45] through direct crystallization combined with ion-exchange resins. The overall processes of some of the studies listed in Table 13.3 are demonstrated in Figure 13.4.

13.4.2.1 Reactive Extraction

Alexandri et al. [47] have conducted reactive extraction in purifying the succinic acid generated from spent sulfite liquor, which is the waste stream from pulp processing that contains a high amount of xylose from the dissolved cellulose [49]. The study has reported the ability of the amine used (trioctylamine in 1-hexanol and dioctylamine in 1-octanol) that can be recycled back into the system. This would serve as a great advantage in maintaining an environmentally benign purification process while contributing to high succinic acid purity. However, the recycling process is energy consuming as it requires an additional step in recovering the succinic acid–amine complex. It is also plagued with the high cost of amine and major chemical usage from the acidification process conducted to recover succinic acid from its salt form after the pH swing method.

13.4.2.2 Forward Osmosis

The capability of the forward osmosis membrane system to purify succinic acid broth was tested by Law et al. [46], prior to the crystallization process. Its usage is proven as an efficient method in concentrating succinic acid to a final concentration of 111.26 g/l from an initial concentration of 28.88 g/l. The method maybe considered economically advantageous due to its low chemical and energy consumption because it can be employed at atmospheric pressures and temperatures. Although membrane technology often faces complications, such as flux reduction and cake formation due to membrane fouling, the utilization of forward osmosis had low membrane fouling. This is contributed by the pretreatment process of the fermentation broth using porous activated carbon right at the beginning of the purification process.

13.4.2.3 Ion Exchange

In a study done by Lin et al. [45], the ion-exchange technology is incorporated prior to the crystallization process, as a method of purification. Since calcium precipitation is adopted as a strategy in separating succinic acid in the form of calcium succinate from other contaminants, it requires a follow-up step in liberating the succinic acid back into its free acid form, which can be adsorbed onto the anion-exchange column during the ion- exchange adsorption process. In this case, the positively charged free acids including succinic acid will be attracted to the anion column. The captured succinic acid ions are released with the help of hydrochloric acid as the elution agent. This strategy has granted higher purity and recovery yield of succinic acid compared to the direct acid method adopted in the same study. The only catch is gypsum formation in the study, which demanded additional handling and material's disposal costs.

13.4.2.4 Crystallization

The crystallization process of succinic acid operates on the concept of its solubility at different pH and temperatures [46]. The solubility of succinic acid at 4 °C is around 30 g/l, in between the pH of 2–8, and it increases to 75 g/l as the temperature is increased to 25 °C [7]. Oreoluwa Jokodola et al. [10] have directly crystallized succinate from the fermentation liquor by adopting vacuum distillation at 60 °C as a technique to concentrate succinic acid by evaporating carboxylic acids like formic and acetic acid. Meanwhile, the multiple crystallization strategy has conferred an improvement in succinic acid recovery yield by 54.18%, as well as high succinic acid purity and recovery of 99.7% and 84.8%, respectively [7]. It is achieved by repeating the crystallization process three times, where each process consists of broth evaporation (i.e. until the succinic acid concentration is fourfolds its initial concentration), vacuum filtration, and, finally, drying to obtain succinic acid crystals. Hence, the studies conducted on the downstream processing of succinic acid derived from agricultural biomass have successfully demonstrated the ability of the implemented strategies in generating high purity and recovery yield of succinic acid. Such findings were some of the major contributions that are highly required toward establishing an efficient and economical succinic acid production as its demand are growing by leaps and bounds. The optimization of the cost for downstream processing strategies of the biotechnologically derived succinic acid is an important aspect to consider in order to attain its economic advantage over the petrochemically derived ones.

13.5 Application of Bio-succinic Acid

Bio-succinic acid is one of the most competitive platform chemicals alongside 2,5-furan dicarboxylic acid, 3-hydroxypropanoic acid, aspartic acid, glucaric acid, glutamic acid, itaconic acid, levulinic acid, 3-hydroxybutyrolactone, glycerol, sorbitol, xylitol, and arabinitol, as per the US Department of Energy (DOE)'s list [49]. Succinic acid comes with a great deal of application in industrial solvents and chemicals (57.1%), pharmaceuticals (15.91%), food and beverages (13.07%), and other industries (13.92%) [4]. Specialty chemicals including 1,4-butanediol (BDO), tetrahydrofuran, adipic acid, N-methyl pyrrolidinone, 2-pyrrolidinone, gamma-butyrolactone, succinamide, putrescine, succinonitrile, and succinate salts have tremendous growth potential, which drives the succinic acid market [2]. The solvent dimethyl succinate also possesses high potential to be marketed as an environmentally friendly solvent by esterifying succinic acid [4]. On the other hand, BDO, a product derived from the hydrogenation of succinic acid, accounts for 45–53% of the global market [2]. Realizing this as a golden opportunity for revenue generation, BioAmber and Mitsui & Co. had jointly planned to pave their way into the additional BDO plants next to the bio-based succinic acid plant [50]. With the huge amount of bio-succinic acid generated, the production of BDO can then be sustained in the long run.

Apart from industrial solvents, the market for succinic acid is expected to grow as a result of its expanding use in the pharmaceutical industry. Thus, its application can be extended to the production of antibiotics, amino acids, vitamins, sedatives, antispasmodics, antirheoters, contraceptives, and several curative agents [4]. In the pharmaceutical industry, it functions as an adjuvant to formulations, an insulinotropic agent (succinic acid monoethyl ester), a cross-linking agent, and a precursor for active pharmaceutical ingredients. In addition, succinic acid is used as a potassium ion inhibitor and an antioxidant.

In the food industry, it serves as a flavor enhancer in salt substitutes, as well as an antimicrobial agent. It has also been demonstrated that succinate salts could increase rumen propionate production. It functions both as glycogenic material and a protein synthesis precursor. Succinate salts, thus serve as feed additives for ruminants and monogastric animals. As a result, raw succinate derived from monosaccharides may enter new markets as animal nutrition products, contributing to the reduction of antimicrobial drugs in certain animal feeds (such as monocin and lasalocid) [2].

As for the synthesis of biodegradable plastics and plasticizing agents, bio-based succinic acid is known as an alternative to several chemicals, such as adipic acid and petroleum-derived phthalates. Polybutylene succinate (PBS) and the copolymers thereof are among succinic acid's most widely used applications, which are a type of biodegradable polyester used by Thailand-based PTT MCC Biochem in the production of butterfly cups and other innovative food packaging. Given their thermal stability, flexibility, and biodegradability, PBS and its copolymers are useful for various applications [15]. Bionelle, GS Pla (green and sustainable plastic), and BioPBS are well-known biodegradable plastic brands derived from succinic acid as the monomer. Several PBS producers are actively involved in this area, including Hexing Chemical (Anhui, China), Xinfu Pharmaceutical (Hangzhou, China), and IRe Chemical (South Korea), with Xinfu Pharmaceutical dominating the market, which had the world's largest continuous PBS production line with an annual capacity of 20 000 tonnes. Succinic acid as well as its derivatives have also been employed to improve water-repellent activity and wet strength in the leather industry, as well as in the metallurgical industry to improve the froth hanging of numerous ores. It is also used as an emulsion coalescing agent in paints and antifreeze cooling liquid in automobile water-cooling systems [50].

13.6 Conclusions

Massive accumulation of underutilized biomass is a prime issue faced by many nations worldwide. Its bioconversion into value-added products, however, can turn such a problem into an opportunity by reducing waste and minimizing the dependency on fossil fuels for succinic acid production. Accordingly, many efforts have been devoted for producing succinic acid from biomass on a laboratory scale, which seems promising and technically feasible. Further process development and product quality improvement are necessary for realizing its maximum commercial potential. In conclusion, this chapter identified several important aspects of competitive succinic acid production from biomass feedstock fermentation. In light of the economic and sustainability metrics of bioprocesses, there has been a surge of interest in the biological synthesis of succinic acid. Despite considerable advances in research activities, challenges remain in developing environmentally sustainable succinic production with low production costs. Other systematic studies on the optimization of fermentation conditions and configuration, and developing appropriate media composition are required for commercialization. Moreover, future and ongoing progressions in biotechnological processes should indeed account for the complexity of lignocellulosic processing. This can be achieved by developing a diverse set of tools, some of which should be versatile, while others should be tailored specifically to particular biomass components.

References

1 Pinazo, J.M., Domine, M.E., Parvulescu, V., and Petru, F. (2015). Sustainability metrics for succinic acid production: a comparison between biomass-based and petrochemical routes. *Catal. Today* 239: 17–24.

2 Nghiem, N.P., Kleff, S., and Schwegmann, S. (2017). Succinic acid: technology development and commercialization. *Fermentation* 3 (2): 26.

3 Luthfi, A.A.I., Jahim, J.M., Harun, S. et al. (2018). Kinetics of the bioproduction of succinic acid by *Actinobacillus succinogenes* from oil palm lignocellulosic hydrolysate in a bioreactor. *BioResources* 13 (4): 8279–8294.

4 Saxena, R., Saran, S., Isar, J., and Kaushik, R. (2017). Production and applications of succinic acid. In: *Current Developments in Biotechnology and Bioengineering*, 601–630. Elsevier https://doi.org/10.1016/B978-0-444-63662-1.00027-0.

5 Sadare, O.O., Ejekwu, O., Moshokoa, M.F. et al. (2021). Membrane purification techniques for recovery of succinic acid obtained from fermentation broth during bioconversion of lignocellulosic biomass: current advances and future perspectives. *Sustainability* 13 (12): 6794.

6 Cheng, K.-K., Zhao, X.-B., Zeng, J. et al. (2012). Downstream processing of biotechnological produced succinic acid. *Appl. Microbiol. Biotechnol.* 95 (4): 841–850.

7 Luthfi, A.A.I., Tan, J.P., Isa, N.F.A.M. et al. (2020). Multiple crystallization as a potential strategy for efficient recovery of succinic acid following fermentation with immobilized cells. *Bioprocess. Biosyst. Eng.* 43 (7): 1153–1169.

8 Lo, E., Brabo-Catala, L., Dogaris, I. et al. (2020). Biochemical conversion of sweet sorghum bagasse to succinic acid. *J. Biosci. Bioeng.* 129 (1): 104–109.

9 Prabhu, A.A., Ledesma-Amaro, R., Lin, C.S.K. et al. (2020). Bioproduction of succinic acid from xylose by engineered *Yarrowia lipolytica* without pH control. *Biotechnol. Biofuels* 13 (1): 1–15.

10 Oreoluwa Jokodola, E., Narisetty, V., Castro, E. et al. (2022). Process optimisation for production and recovery of succinic acid using xylose-rich hydrolysates by *Actinobacillus succinogenes*. *Bioresour. Technol.* 344: 126224.

11 Kim, D.Y., Yim, S.C., Lee, P.C. et al. (2004). Batch and continuous fermentation of succinic acid from wood hydrolysate by *Mannheimia succiniciproducens* MBEL55E. *Enzyme Microb. Technol.* 35 (6): 648–653.

12 Ventorino, V., Robertiello, A., Cimini, D. et al. (2017). Bio-based succinate production from Arundo donax hydrolysate with the new natural succinic acid-producing strain *Basfia succiniciproducens* BPP7. *Bioenergy Res.* 10 (2): 488–498.

13 Luque, R., Lin, C.S., Du, C. et al. (2009). Chemical transformations of succinic acid recovered from fermentation broths by a novel direct vacuum distillation-crystallisation method. *Green Chem.* 11 (2): 193–200.

14 Litsanov, B., Brocker, M., and Bott, M. (2012). Toward homosuccinate fermentation: metabolic engineering of *Corynebacterium glutamicum* for anaerobic production of succinate from glucose and formate. *Appl. Environ. Microbiol.* 78 (9): 3325–3337.

15 Harun, S., Luthfi, A.A.I., Abdul, P.M. et al. (2022). Chapter 11 – Oil palm biomass zero-waste conversion to bio-succinic acid. In: *Value-Chain of Biofuels* (ed. S. Yusup and N.A. Rashidi), 249–275. Elsevier.

16 Akhlisah, Z., Yunus, R., Abidin, Z. et al. (2021). Pretreatment methods for an effective conversion of oil palm biomass into sugars and high-value chemicals. *Biomass Bioenergy* 144: 105901.

17 Agrawal, R., Verma, A., Singhania, R.R. et al. (2021). Current understanding of the inhibition factors and their mechanism of action for the lignocellulosic biomass hydrolysis. *Bioresour. Technol.* 332: 125042.

18 Dessie, W., Xin, F., Zhang, W. et al. (2019). Inhibitory effects of lignocellulose pretreatment degradation products (hydroxymethylfurfural and furfural) on succinic acid producing *Actinobacillus succinogenes*. *Biochem. Eng. J.* 150: 107263.

19 Teugjas, H. and Väljamäe, P. (2013). Selecting β-glucosidases to support cellulases in cellulose saccharification. *Biotechnol. Biofuels* 6 (1): 1–13.

20 Sun, Y. and Cheng, J. (2002). Hydrolysis of lignocellulosic materials for ethanol production: a review. *Bioresour. Technol.* 83 (1): 1–11.

21 Zabed, H., Sahu, J., Suely, A. et al. (2017). Bioethanol production from renewable sources: current perspectives and technological progress. *Renewable Sustainable Energy Rev.* 71: 475–501.

22 Tan, J.P., Luthfi, A.A.I., Manaf, S.F.A. et al. (2018). Incorporation of CO_2 during the production of succinic acid from sustainable oil palm frond juice. *J. CO^2 Util.* 26: 595–601.

23 Bukhari, N.A., Jahim, J.M., Loh, S.K. et al. (2020). Organic acid pretreatment of oil palm trunk biomass for succinic acid production. *Waste Biomass Valorization* 11 (10): 5549–5559.

24 Li, C., Ong, K.L., Cui, Z. et al. (2021). Promising advancement in fermentative succinic acid production by yeast hosts. *J. Hazard. Mater.* 401: 123414.

25 Ahn, J.H., Jang, Y.-S., and Lee, S.Y. (2016). Production of succinic acid by metabolically engineered microorganisms. *Curr. Opin. Biotechnol.* 42: 54–66.

26 Szczerba, H., Komoń-Janczara, E., Dudziak, K. et al. (2020). A novel biocatalyst, *Enterobacter aerogenes* LU2, for efficient production of succinic acid using whey permeate as a cost-effective carbon source. *Biotechnol. Biofuels* 13 (1): 1–12.

27 Olajuyin, A.M., Yang, M., Mu, T. et al. (2018). Enhanced production of succinic acid from methanol–organosolv pretreated *Strophanthus preussii* by recombinant *Escherichia coli*. *Bioprocess. Biosyst. Eng.* 41 (10): 1497–1508.

28 Pateraki, C., Patsalou, M., Vlysidis, A. et al. (2016). *Actinobacillus succinogenes*: Advances on succinic acid production and prospects for development of integrated biorefineries. *Biochem. Eng. J.* 112: 285–303.

29 Otero, J.M., Cimini, D., Patil, K.R. et al. (2013). Industrial systems biology of *Saccharomyces cerevisiae* enables novel succinic acid cell factory. *PLoS One* 8 (1): e54144.

30 Bradfield, M.F.A., Mohagheghi, A., Salvachúa, D. et al. (2015). Continuous succinic acid production by *Actinobacillus succinogenes* on xylose-enriched hydrolysate. *Biotechnol. Biofuels* 8.

31 Andersson, C., Hodge, D., Berglund, K.A., and Rova, U. (2007). Effect of different carbon sources on the production of succinic acid using metabolically engineered *Escherichia coli*. *Biotechnol. Progr.* 23 (2): 381–388.

32 Gonzales, T.A., de Carvalho Silvello, M.A., Duarte, E.R. et al. (2020). Optimization of anaerobic fermentation of *Actinobacillus succinogenes* for increase the succinic acid production. *Biocatal. Agri. Biotechnol.* 27: 101718.

33 Jiang, M., Chen, K., Liu, Z. et al. (2010). Succinic acid production by *Actinobacillus succinogenes* using spent brewer's yeast hydrolysate as a nitrogen source. *Appl. Biochem. Biotechnol.* 160 (1): 244–254.

34 Shen, N., Qin, Y., Wang, Q. et al. (2015). Production of succinic acid from sugarcane molasses supplemented with a mixture of corn steep liquor powder and peanut meal as nitrogen sources by *Actinobacillus succinogenes*. *Lett. Appl. Microbiol.* 60 (6): 544–551.

35 Lu, J., Li, J., Gao, H. et al. (2021). Recent progress on bio-succinic acid production from lignocellulosic biomass. *World J. Microbiol. Biotechnol.* 37 (1): 1–8.

36 Zheng, P., Fang, L., Xu, Y. et al. (2010). Succinic acid production from corn stover by simultaneous saccharification and fermentation using *Actinobacillus succinogenes*. *Bioresour. Technol.* 101 (20): 7889–7894.

37 Alcantara, J. and Mondala, A. (2021). Optimization of slurry fermentation for succinic acid production by fungal co-culture. *Chem. Eng. Trans.* 86: 1525–1530.

38 Lu, J., Peng, W., Lv, Y. et al. (2020). Application of cell immobilization technology in microbial cocultivation systems for biochemicals production. *Ind. Eng. Chem. Res.* 59 (39): 17026–17034.

39 Wu, D., Li, Q., Wang, D., and Dong, Y. (2013). Enzymatic hydrolysis and succinic acid fermentation from steam-exploded corn stalk at high solid concentration by recombinant *Escherichia coli*. *Appl. Biochem. Biotechnol.* 170 (8): 1942–1949.

40 Wang, C., Su, X., Sun, W. et al. (2018). Efficient production of succinic acid from herbal extraction residue hydrolysate. *Bioresour. Technol.* 265: 443–449.

41 Liu, Y.P., Zheng, P., Sun, Z.H. et al. (2008). Strategies of pH control and glucose-fed batch fermentation for production of succinic acid by *Actinobacillus succinogenes* CGMCC1593. *J. Chem. Technol. Biotechnol.* 83.

42 Stylianou, E., Pateraki, C., Ladakis, D. et al. (2020). Evaluation of organic fractions of municipal solid waste as renewable feedstock for succinic acid production. *Biotechnol. Biofuels* 13 (1): 1–16.

43 Ercole, A., Raganati, F., Salatino, P., and Marzocchella, A. (2021). Continuous succinic acid production by immobilized cells of *Actinobacillus succinogenes* in a fluidized bed reactor: entrapment in alginate beads. *Biochem. Eng. J.* 169: 107968.

44 Alexandri, M., Papapostolou, H., Stragier, L. et al. (2017). Succinic acid production by immobilized cultures using spent sulphite liquor as fermentation medium. *Bioresour. Technol.* 238: 214–222.

45 Lin, S.K.C., Du, C., Blaga, A.C. et al. (2010). Novel resin-based vacuum distillation-crystallisation method for recovery of succinic acid crystals from fermentation broths. *Green Chem.* 12 (4): 666–671.

46 Law, J.Y., Mohammad, A.W., Tee, Z.K. et al. (2019). Recovery of succinic acid from fermentation broth by forward osmosis-assisted crystallization process. *J. Membr. Sci.* 583: 139–151.

47 Alexandri, M., Vlysidis, A., Papapostolou, H. et al. (2019). Downstream separation and purification of succinic acid from fermentation broths using spent sulphite liquor as feedstock. *Sep. Purif. Technol.* 209: 666–675.

48 Thuy, N.T.H. and Boontawan, A. (2017). Production of very-high purity succinic acid from fermentation broth using microfiltration and nanofiltration-assisted crystallization. *J. Membr. Sci.* 524: 470–481.

49 Llano, T., Rueda, C., Dosal, E. et al. (2021). Multi-criteria analysis of detoxification alternatives: techno-economic and socio-environmental assessment. *Biomass Bioenergy* 154: 106274.

50 de Jong, E., Higson, A., Walsh, P., and Wellisch, M. (2012). Bio-based chemicals value added products from biorefineries. *IEA Bioenergy*, Task42 Biorefinery 34.

14

Furfural and Derivatives from Bagasse and Corncob

Muryanto Muryanto[1,2], Yanni Sudiyani[1], Andre F. P. Harahap[3], and Misri Gozan[2]

[1] *Research Center for Chemistry, National Research and Innovation Agency, Kawasan Puspiptek, Serpong, Tangerang Selatan, 15314, Indonesia*
[2] *Universitas Indonesia, Department of Chemical Engineering, Faculty of Engineering, Kampus Baru UI, Depok, 16424, Jawa Barat, Indonesia*
[3] *University of Hohenheim, Institute of Food Science and Biotechnology, Bioprocess Engineering (150k), Building 03.26, Room 228, Fruwirthstr. 12, 70599, Stuttgart, Germany*

14.1 Introduction

Furfural ($C_5H_4O_2$), often referred to as furanaldehyde, 2-furfuraldehyde, furaldehyde, and 2-furancarboxaldehyde, is an organic compound with a $C_5H_4O_2$ chemical structure. It is a yellow to brownish liquid with a boiling point of 161.5 °C, with a molecular weight of 96.086 g/mol, and a density at 20 °C of 1.16 g/cm^3. Furfural is a compound with a water solubility of 8.3 g of furfural in 100 g of water solvent (temperature 20 °C) and is easily soluble in alcohol, ether, and benzene [1]. In the structure of furfural, hydrogen at position two is substituted by a formyl group. Furfural is formed by dehydration of five-carbon sugars, such as xylose and arabinose.

Furfural is a selective solvent for petroleum, which can used to separate saturated and unsaturated compounds in the petroleum industry. Furfural has been identified as one of 30 high-value biochemical substances [2]. For example, furfural is generally used in the chemical industry as a chemical intermediate, raw material for adiponitrile, furfuryl alcohol, methyl furan, pyrrole, furoic acid, hydro furamide, and tetrahydrofurfuryl alcohol (THFA). Another example is as a selective solvent in refining petroleum and vegetable oils, manufacture of resins, such as phenol–aldehyde (phenol–furfural), and color removers for wood resins in the soap, varnish, and paper industries [1].

In general, the furfural production process consists of two stages: the hydrolysis of hemicellulose to produce xylose and then conversion to furfural by dehydration process. Until now, the raw material, which is commonly used is corncob. However, hemicellulose is a part of lignocellulosic biomass (LCB). Hemicellulose can be obtained from corncobs and other LCB, such as sugarcane bagasse, rice straw, and

Chemical Substitutes from Agricultural and Industrial By-Products: Bioconversion, Bioprocessing, and Biorefining,
First Edition. Edited by Suraini Abd-Aziz, Misri Gozan, Mohamad Faizal Ibrahim, and Lai-Yee Phang.
© 2024 WILEY-VCH GmbH. Published 2024 by WILEY-VCH GmbH.

oil palm biomass. This study describes furfural and its derivatives, raw materials for furfural production, production processes, and techno-economics.

14.2 Furfural as a Building Block Material

Furfural can be used as a building block material and basic chemical that form more complex products. The building block of furfural is 5-hydroxymethylfurfural (HMF). HMF has been termed the "Sleeping giant of sustainable chemistry" because of its tremendous synthetic and economic potential [3]. However, on the other hand, HMF also has high reactivity and can cause instability of the building blocks. HMF is subject to incidental reactions, such as humification and hydrolysis of levulinic acid [3]. The approach that can be done to solve this problem is to diversify HMF into other building blocks that are more multifunctional as analogs derived from basic chemicals. Figure 14.1 shows the biorefinery biomass to HMF and derivatization.

14.2.1 5-Chloromethylfurfural

The most researched HMF derivative is 5-chloromethylfurfural (CMF), which is a halide and sulfonate derivative. Because of the durability of furan carbocations and the strong reactivity of its hydroxyl groups, HMF is a powerful alkylating agent [5]. Because of its reduced polarity and greater stability with high retention of synthesis potential compared to HMF, CMF is regarded as a novel chemical alternative [6].

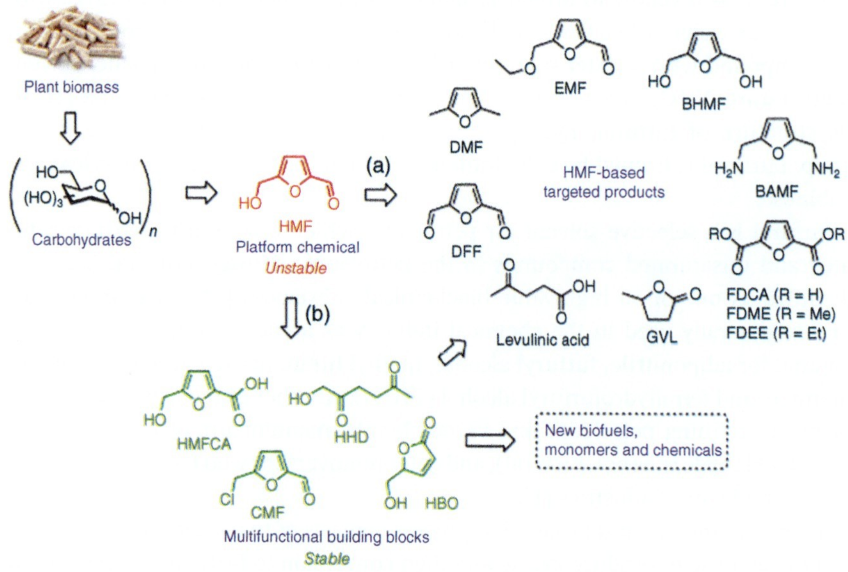

Figure 14.1 (a) The classic method of HMF-based biorefining by direct HMF derivatization. (b) Biorefining is based on the diversification of HMFs to create multifunctional building blocks. Source: Galkin and Ananikov [4]/ from John Wiley & Sons.

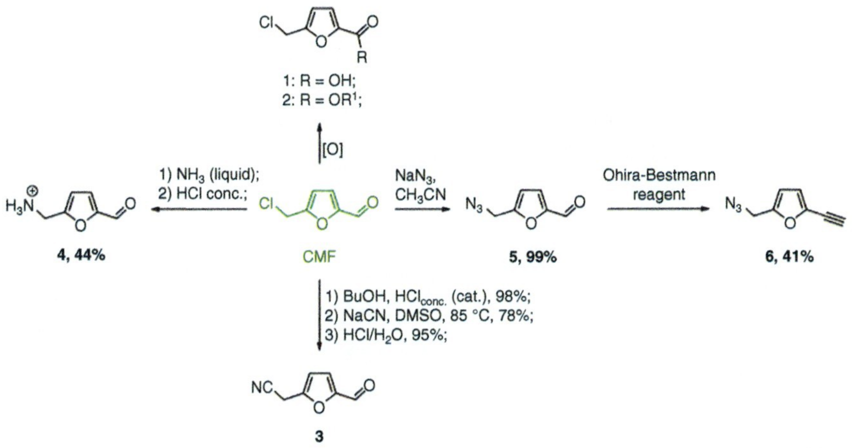

Figure 14.2 CMF multifunctional building blocks synthesis. Source: Galkin and Ananikov [4] Reproduced with permission of John Wiley & Sons Inc.

The synthesis of multipurpose polyfunctional building blocks with CMF modification is shown in Figure 14.2.

14.2.2 HMFCA and Esther

The high synthetic potential of HMF can be attributed in large part to the presence of an aldehyde group within its structure. Metal catalysts can be used to selectively oxidize aldehyde groups by producing 5-Hydroxymethyl-2-furan carboxylic acid (HMFCA) and its esters in the presence of molecular oxygen or other oxidants (Figure 14.3). Because HMFCAs and their esters have reactive groups, which assure their great synthetic potential, they are in high demand as monomers and multifunctional building blocks [7].

14.3 Furfural Derivatives

Furfural as building block material can be converted into several derivative products. Several chemical reactions, including hydrogenation, decarboxylation, decarbonylation, and oxidation, as well as others, can be used to transform furfural into its derivatives. Figure 14.4 shows some reactions to produce furfural derivatives.

14.3.1 Furfuryl Alcohol

Furfuryl alcohol is one of the most developed furfural derivative products. Due to its high reactivity, furfuryl alcohol is important in producing foundry sand binders. Furfuryl alcohol has been widely used to manufacture molds for metals for decades. THFA, which is frequently utilized in the pharmaceutical industry, is made from furfuryl alcohol. The World Foundry Organization (WFO) predicts that India's foundry

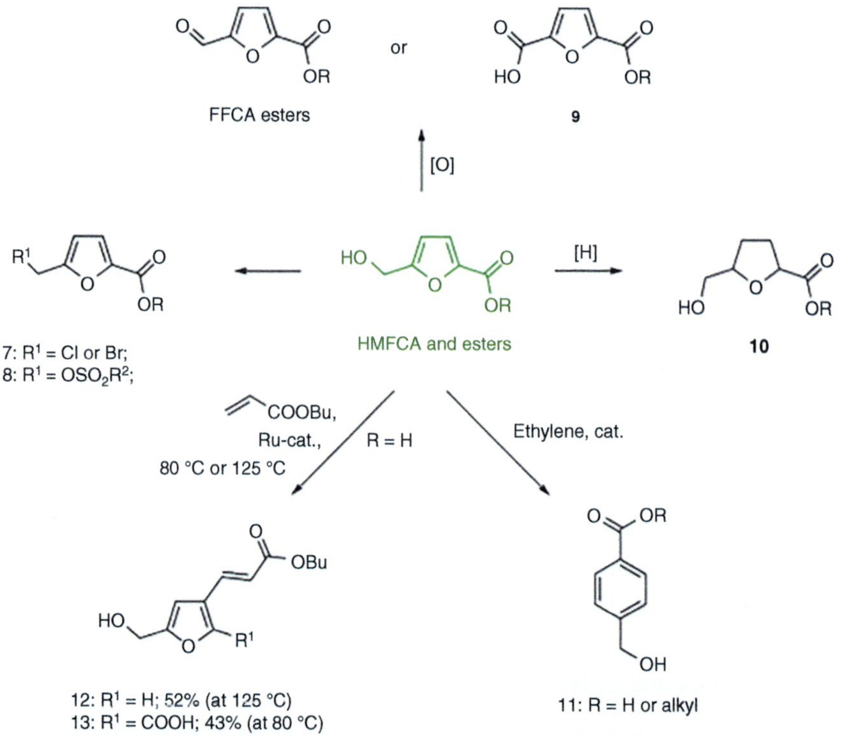

Figure 14.3 HMFCA as a multifunctional building block. Source: Galkin and Ananikov [4] Reproduced with permission of John Wiley & Sons Inc.

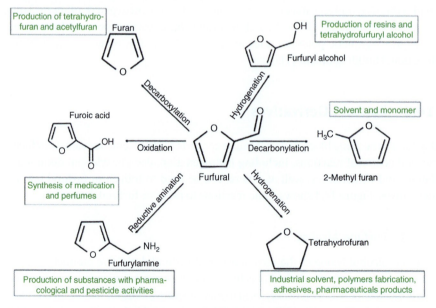

Figure 14.4 Furfural derivatives and their application. Source: Mathew et al. [8] Reproduced with permission of Elsevier.

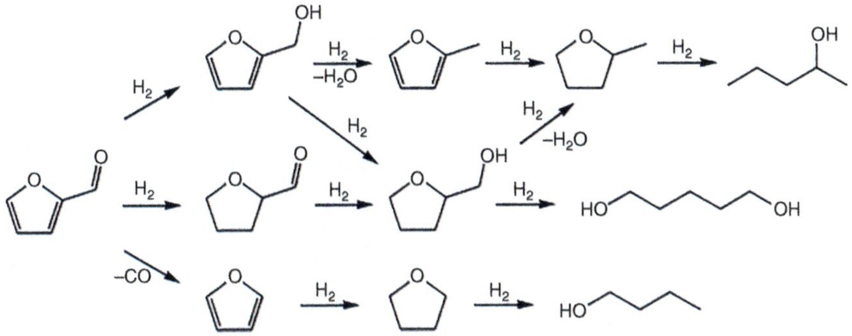

Figure 14.5 Furfural hydrogenation. Source: [11]/MDPI/Licensed under CC BY 4.0.

market will expand by 13.0–14.0% by 2025 and surpass China as the second-largest foundry market in the world. Therefore, the foundry industry's growth leads to the demand for furfuryl alcohol, thereby increasing the furfural market.

About 85% of furfural production is used to synthesize furfural into furfuryl alcohol. This is due to the increasing demand for furfuryl alcohol, which can be used as furan resins, coatings, and adhesives [9]. Furfuryl alcohol can also be used for resins, in components of motor fuels (alkyl levulinates), in the pharmaceutical industry, such as ranitidine, and in biochemistry. In 2021, the global furfuryl alcohol market was more than US$ 500 million and is expected to grow to US$ 821.6 million by 2028 [10]. In the hydrogenation of furfural to furfuryl alcohol, there are several other reactions, such as THFA,2-methoxyflurane, and others, which can be seen in Figure 14.5.

14.3.2 Furan

The furan molecule is a type of heterocyclic organic compound. It is made up of a five-membered aromatic ring that has four carbon atoms and one oxygen atom. Furans are frequently used to describe other chemical compounds that include similar rings. Furan is a colorless, flammable, extremely volatile liquid having a near-room temperature boiling point. Furan is derived from the Latin furfur, which means bran. This is because furan is derived from the bran. Carl Wilhelm Scheele discovered and characterized the first furan derivative, 2-furoic acid, in 1780. While furan was discovered by Heinrich Limpricht in 1870, at that time, Heinrich called it "tetraphenol." Furans are industrially produced by palladium-catalyzed furfural decarbonylation or by copper-catalyzed oxidation of 1,3-butadiene [12].

Furan is classified as an aromatic compound because one of the oxygen atom's lone pairs is delocalized into the ring, resulting in a $4n+2$ aromatic system similar to benzene. The furan molecule is flat and lacks discrete double bonds since it is an aromatic compound. Another lone pair of oxygen atom electrons extends in the plane of the flat ring structure. The purpose of the sp^2 hybridization is to place one of the lone pairs of electrons in the p orbital, allowing it to interact in the conjugated system. Because of the electron donor effect of the oxygen heteroatom, furan

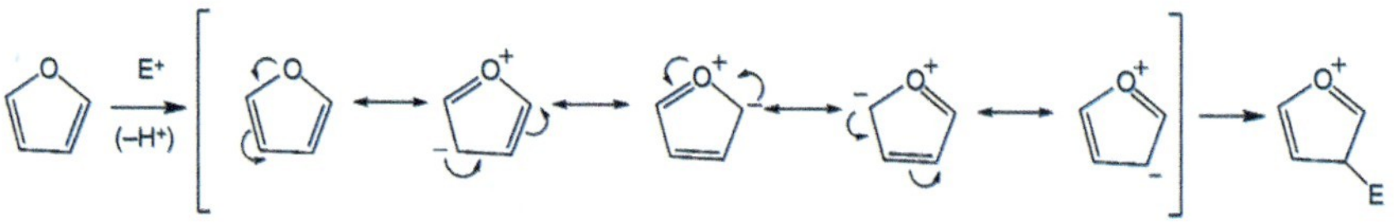

Figure 14.6 Electrophilic substitution reaction on furan. Source: Modified from Bruice [13].

is substantially more reactive than benzene in electrophilic substitution processes. Figure 14.6 shows an example of an electrophilic substitution reaction on furan, with the reaction mechanism in brackets.

14.3.3 Tetrahydrofuran

THF, also known as oxolane, is an organic chemical having the formula $(CH_2)_4O$. These compounds, particularly cyclic ethers, are categorized as heterocyclic. It is a colorless organic liquid that is water soluble and has a low viscosity. The majority of its applications are in the production of polymers. Because of its polar nature and wide liquid range, THF is a very useful and versatile solvent. The dielectric constant of THF is 7.6, making it an aprotic solvent. THF can dissolve a variety of polar and nonpolar chemical substances since it is a moderately polar solvent. Water and THF can combine to produce solid clathrate hydrate structures at low temperatures. THF is a commercial solvent that may be used with varnishes and polyvinyl chloride (PVC).

THF is manufactured in around 200 000 tons annually [14]. The most common commercial method for producing THF uses an acid catalyst to dehydrate 1,4-butanediol. One of the biggest manufacturers of this chemical class is Ashland/ISP. This process is comparable to making diethyl ether from ethanol. By first oxidizing n-butane to crude maleic anhydride and then catalytic hydrogenation, DuPont developed a method for creating THF. The catalytic hydrogenation of furans is another method for making THF. This allows for the conversion of certain sugars to THF via an acid catalyst to furfural and decarbonylation to furan, although it is not a commonly used technique.

14.3.4 Furoic Acid

Furoic acid is a carboxylic acid side group and a furan ring organic compound. Furoates are the names of furoic acid's salts and esters. Food items frequently contain furoic acid as a preservative and flavoring ingredient. Furoic acid is being utilized extensively in the chemical, medicinal, and agricultural fields. Furyl chloride, which is used to make medications, is widely converted from furoic acid.

The oxidation of furfuryl alcohol or furfural can both be used as starting materials for the synthesis of furoic acid. This synthesis can be carried out chemically or enzymatically. In today's industry, furoic acid manufacture involves the Cannizzaro furfural reaction in a liquid NaOH solution. This disproportionation reaction yields furoic acid and furfuryl alcohol in a 1 : 1 ratio (yield 50% [15]. Figure 14.7 shows the synthesis of furoic acid from furfuryl alcohol.

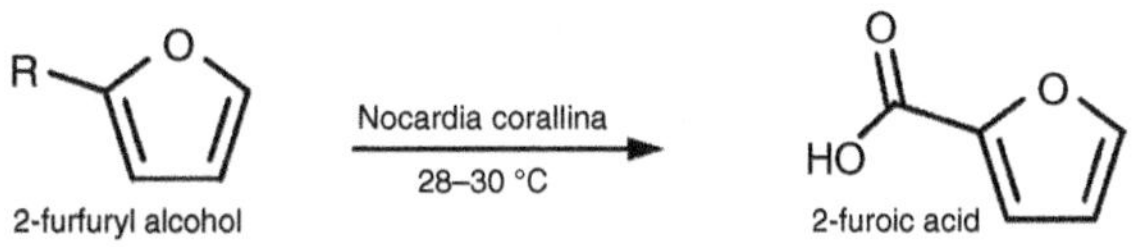

Figure 14.7 Transformation of furfuryl alcohol into 2-furoic acid. Source: Modified from Pérez et al. [16].

14.4 Lignocellulosic Biomass as Raw Material for Furfural Production

LCB can be defined as a natural resource due to its abundance, availability, and renewability, is a promising feedstock for producing bio-based chemicals and material [17]. The use of LCB for biofuels and chemicals raises numerous concerns about major environmental issues, such as food shortages and depletion of nonrenewable natural resources. Furfural, a chemical known as a building block material, is one of several products obtained from LCB. LCB is made up of three major compounds cellulose, hemicellulose, and lignin. The composition of LCB depends on several variables, the most significant of which are the species of plant, its age, the stage of growth it is in, and the conditions in its environment. On a dry basis, the substrate contains about 40–60% cellulose, 20–40% hemicelluloses, and 10–24% lignin [18]. Pectin, proteins, lipids, and nonstructural carbohydrates (glucose, fructose, and sucrose), are also present in lower concentrations.

Cellulose, which may be found in both crystalline and noncrystalline formations and is resistant to hydrolysis, is an important structural element of plant cell walls that gives them mechanical strength and chemical stability. Hemicellulose is a polysaccharide that is abundant in plant cell walls, accounting for 15–30% of LCB by weight. Unlike cellulose, hemicellulose consists of a short, highly branched polymer of different C5 and C6 sugars, such as xylan, mannan, β-glucans, and xyloglucans, with the dominant sugar being xylose. The lack of crystalline structure, owing to the highly branched structure, and the presence of acetyl groups connected to the polymer chain are important aspects of hemicellulose's structure and composition. Hemicellulose is found between the micro and macro cellulose fibrils.

A complex polymer called lignin binds the cellulose and hemicellulose-containing matrix. Given that cellulose makes up the bulk of plant cell walls, lignin is mostly located in the interfibrous region, with a minor amount on the cell surface. Lignocellulose is a combination of three fundamental chemical substances, plus water and an organic substance. Ash is one of the most abundant organic constituents. Ash typically contains calcium, potassium, magnesium, manganese, and sodium oxides, as well as trace amounts of iron and aluminum oxides [17].

Whole plants (dedicated energy crops, perennial grasses, and aquatic plants), agriculture waste (cereal, stovers, and bagasse), forest biomass waste (hardwood, softwood, sawdust residues, pruning, and thinning of bark), and municipal solid waste are the different types of LCB. Agriculture waste, among other LCB, is abundant and inexpensive biomass that can facilitate biofuel production, and biochemical production is the current main research of interest worldwide. The use of

LCB waste as a feedstock for fuels and furfural production has sparked international interest. According to Dahmen et al., 181.5 billion tons of LCB are produced globally each year. Currently, approximately 8.2 billion tons of biomass are used; 7 billion tons come from forest, agriculture, and grass, while the remaining 1.2 billion tons come from agricultural residues [19]. These wastes are simply dumped, left on the field, or incinerated, resulting in additional environmental pollution. In the United States, South America, Europe, and Asia, the most promising and abundant cellulosic feedstock derived from agricultural residues are corncob and corn stover, sugarcane bagasse, and wheat straw, respectively, and today empty fruit bunch of oil palm, it is counting abundant in Asia as feedstock for lignocellulosic biorefinery [20]. Sugarcane bagasse and corncob are the most commonly used materials for furfural production, accounting for more than 98% of all furfural produced [21].

14.4.1 Sugarcane Bagasse

Sugarcane is widely cultivated in tropical and subtropical countries around the world, with annual production increasing. Sugarcane bagasse, a by-product, is now valued by sugar–alcohol producers as the primary feedstock source for bioenergy and biofuel production. Because bagasse is a by-product of the cane sugar industry, the amount produced in each country corresponds to the amount of sugarcane produced. Sugarcane bagasse is available after sucrose extraction from sugarcane milling at approximately 125 kg of dried bagasse per ton of processed sugarcane [18].

The primary sugarcane producer in Asia, followed by South America, which is second. Sugarcane productivity in one year is 14–22 tons/ha, with an average of 17 tons/ha. Food manufacturers consume about 92% of sugarcane (400 kg of sugar per dry tonne). Besides bagasse, solid biomass can be obtained from cane tops and leaves. But these two materials are generally used to feed cattle. Indonesia is now developing integrated farming (cattle and sugar) in several areas.

14.4.2 Corncob and Corn Stover

Agricultural residues left over from corn production are a possible source of lignocellulosic raw materials for furfural production. In the field, the corn grains are separated from the cobs, stalks, and leaves. The corn grain was utilized not only for human food but also used for animal feed. While the corn stover that consists of stalks, leaves, and corncobs is not yet utilized. Corn stover as lignocellulose biomass is a significant amount that is harvested in agricultural countries, such as the United States, China, Brazil, India, Pakistan, and Indonesia. The United States and China are both large agricultural countries, with annual corn stover production totaling 250 million tons in the United States and 220 million tons in China [22].

In the year 2020/2021, the harvested area of corn will increase to 3.85 million ha, and production will reach 12.6 million metric tons. Approximately, 5% of global production is wasted. The grain-to-stover ratio is approximately 1 : 1, accounting for slightly more weight than the stover. The corncob accounts for about 20% of the weight of corn stover [18].

14.4.3 Other Agriculture and Industrial Biomass

14.4.3.1 Rice Straw

Asia is the largest rice contributor in the world, with almost 90% of global production. Roughly 526 million tonnes of dry rice are produced yearly as global production. The rice harvest area in Asia reaches 140 million ha. Rice productivity in Asia is 3.5 tonnes/ha dry based. The vast majority of rice (roughly 88% of the total production worldwide) is consumed by humans. About 2.6% of global production is used for animal feed and 4.8% of world rice production is wasted [18].

Rice straw is a by-product of rice harvesting. Rice straw is commonly used as fertilizer by traditional farmers. Each hectare of farmland yields 10–15 tons of rice straw. Rice straw contains cellulose and hemicelluloses, which can be converted to furfural.

14.4.3.2 Oil Palm Biomass

Oil palm is one of Indonesia and Malaysia's most important commodities. Each kg of CPO produced will generate three times the amount of LCB. Biomass from oil palm plantations is frond and oil palm trunk (OPT). After 25 years, oil palm plantations must replant so that OPT biomass will be produced. Due to the large area of oil palm plantations, the replanting process will occur every year and sustainably deliver OPT. Besides plantations, CPO mills produce biomass, such as oil palm empty fruit bunches (OPEFB), shells, and fiber. But the last two are used as boiler fuel for sterilizing fresh fruit bunches.

Each ton of CPO produced 1.1–1.2 tons of OPEFB as LCB. Glucose (95.48%) was found to be the dominant neutral sugar in the cellulose (insoluble fraction) of OPEFB after pretreatment with 10% sodium hydroxide solution, while xylose (88.39%) was discovered in the hemicellulose [18].

14.4.3.3 Forest/Wood Biomass

Primary and secondary waste from the forest/wood industry can be classified as biomass. Direct waste is a by-product of industrial activities directly related to forestry. Forest residues generally form non-merchantable trees and logging slashes, which occur during tree felling and land clearing. While the biomass produced by the wood industry is generally in the form of sawdust, bark, shavings, solid trim and clarifier sludge, and spent pulping liquors. Secondary lignocellulosic waste can be generated from domestic, commercial, and industrial activities. Domestic waste generally takes the form of paper, cloth, garden debris, and other municipal solid waste. Waste paper, packaging materials, textiles, and demolition wood are biomass produced from commercial and industrial activities [23]. Every logging activity generated 20% of waste in the wood industry [24].

LCB has been investigated as a low-cost feedstock for the production of bioenergy and fine chemicals. It is essentially free of agricultural waste and forest residues. The use of these wastes could solve the disposal problem while also lowering the cost of waste treatment. LCB is a good renewable energy source that can be used to make a variety of compounds. Some lignocellulosic materials contain cellulose and hemicellulose as shown in Table 14.1.

Table 14.1 Composition of lignocellulosic biomass from agriculture [18].

Feed stock	Lignocellulosic component (%)		
	Cellulose	Hemicellulose	Lignin
Sugarcane bagasse	43.6–45.8	31.3–33.5	18.1–22.9
Sweet sorghum bagasse	34.1–38.3	14.7–26.3	20.7–21.4
Corn stover	31.3–49.4	21.1–26.2	3.1–8.8
Corncob	39–43	29–33	18–26
Rice straw	29.2–36	23–26	17–19
OPEFB	41.3–46.5	25.3–33.8	17.5–23.6

14.5 Furfural Production

14.5.1 Furfural Production Process

14.5.1.1 Pretreatment

Pretreatment is carried out for raw materials from lignocellulosic with high lignin content. The lignin content can inhibit the following process, such as hydrolysis and dehydration. Therefore, the lignin content in the raw material must be reduced before being used. The main objective of the pretreatment process on LCB is the degradation of lignin to separate it from holocellulose. Determination of the pretreatment methods depends on the type of LCB. Therefore, no universal pretreatment can be applied to all LCB. There are several pretreatments for LCB, such as mechanical pretreatment, physical pretreatment, and chemical pretreatment.

Mechanical pretreatment includes chipping, grinding, and milling. This pretreatment method aims to increase the specific surface area of the biomass and reduce the degree of polymerization and crystallinity of cellulose. Irradiation pretreatment can use gamma-ray irradiation, as well as electron beams.

Chemical pretreatment using acid will hydrolyze most of the hemicellulose and leave a lot of cellulose and lignin. This process is generally carried out by immersing the LCB in an acid solution and then heating it in a temperature range of 140–200 °C. In general, the pretreatment uses mineral acids, such as sulfuric acid (H_2SO_4) and hydrochloric acid (HCl) with low concentrations (0.5–6%) for 30–60 minutes. However, acid pretreatment can also be carried out using high-concentration acid (>6%) for several minutes. These methods have attracted much attention because they are considered cheaper and more effective against the stripping of hemicellulose components. According to Lloyd and Wyman, pretreatment using H_2SO_4 at low concentrations (0.5–1%) and moderate temperatures (140–190 °C), almost all hemicelluloses are hydrolyzed to soluble pentose sugars (both monomers and oligomers) [25].

Pretreatment using an alkali, such as NaOH aims to reduce the lignin content of acetyl group compounds, dissolve a little hemicellulose, and breakdown down the lignin structure. The saponification of intermolecular ester bonds that cross-link silane and other components, such as lignin and hemicellulose is the mechanism of alkaline pretreatment [26]. Alkaline solution with a low concentration can cause the lignocellulosic expansion to increase the internal surface area, decreasing the degree of polymerization and crystallinity. Besides NaOH, alkaline pretreatment can also use $Ca(OH)_2$, KOH, or alkaline peroxide, a combination of NaOH and H_2O_2 [27].

Generally, pretreatment can be carried out with physical, chemical, physicochemical, and biological processes, where each technique has advantages and disadvantages. The combination of pretreatment methods, which is defined as the merging of two or more types of pretreatment/sequential pretreatment process can solve each pretreatment method's weakness [28].

14.5.1.2 Hydrolysis Process

The furfural production can be carried out in a one- or two-stage process. At one stage, the hemicellulose in lignocellulose will be hydrolyzed to xylose, and then the dehydration process will directly occur to furfural. Meanwhile, the hydrolysis and dehydration processes are carried out separately in the two-stage process. Figure 14.8 shows the simple reaction of furfural from xylose.

The hydrolysis of holocellulose (hemicellulose and cellulose) aims to produce simple sugars. Hydrolysis of cellulose will produce glucose, while hydrolysis of hemicellulose will produce xylose as in the following Eqs. (14.1) and (14.2):

$$(C_6H_{10}O_5)_n + nH_2O \rightarrow nC_6H_{12}O_6 \tag{14.1}$$

$$(C_5H_8O_4)_n + nH_2O \rightarrow nC_5H_{10}O_5 \tag{14.2}$$

The cellulose hydrolysis reaction in Eq. (14.1) shows that each glucose unit in the long chain combines with a water molecule, where 162 mass units of cellulose and 18 mass units of water will release 180 glucose units. This equation shows that the conversion factor of cellulose to glucose is 1.111. Oligomers consisting of glucose can also be produced as intermediate products in the hydrolysis of cellulose.

Hemicellulose can also be hydrolyzed by adding water to release monomeric sugars. The monomer sugar formed from the hydrolysis of hemicellulose is xylose (pentose) or n-arabinose. The reaction Eq. (14.2) shows that 132 mass units of hemicellulose and 18 mass units of water will release 150 mass units of pentose sugar. The conversion factor of hemicellulose to pentose sugar is 1.136 [30].

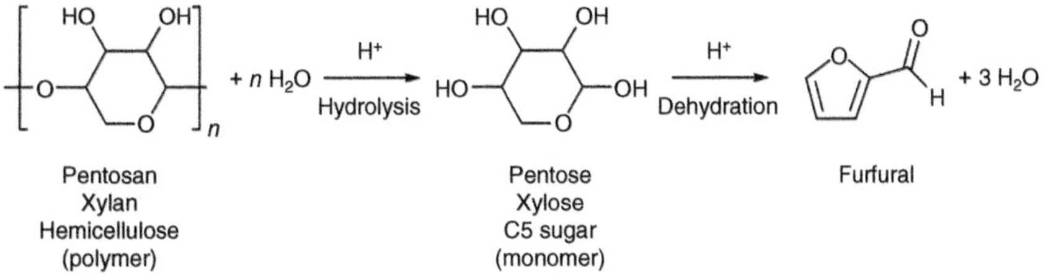

Figure 14.8 Simple reaction mechanism for furfural formation. Source: Metkar et al. [29] Reproduced with permission of Royal Society of Chemistry.

Table 14.2 Advantages and disadvantages of acid hydrolysis [32].

Hydrolysis method	Advantages	Disadvantages	Remarks
Concentrated acid	• Process conditions at low temperatures • High yield of sugar monomers	• High acid consumption • There is corrosion on the equipment • High energy consumption for acid recovery	Acid concentration 10–30%
Dilute acid	Low acid consumption	• Process conditions at high temperatures • Low sugar monomer yield • There is corrosion on the equipment	Acid concentration 1–5%

The hydrolysis of LCB can be carried out using chemicals or enzymes. The chemical hydrolysis process requires chemicals to breakdown cellulose and hemicellulose polymers into sugar monomers (glucose and xylose). Several acids can be used for hydrolysis, such as sulfuric acid, hydrochloric acid, hydrofluoric acid, phosphoric acid, nitric acid, and formic acid [31]. In general, most cellulose is in crystalline form and hard, requiring concentrated acid (high concentration) and high temperatures during hydrolysis. The concentration of concentrated acid used for the hydrolysis process is 10–30% under low temperatures. On the other hand, the dilute acid (low concentration) used for hydrolysis between 2% and 5% requires high temperatures to get high cellulose conversion [32].

Hemicellulose hydrolysis using acids is almost the same as cellulose. However, hemicellulose is generally amorphous, so it does not require high processing conditions as in the hydrolysis of cellulose. For example, yields of up to 85–90% sugar monomers can be obtained from hemicellulose under process conditions of about 160 °C temperature, 10 minutes process, and acid concentration of only 0.7% [30]. Table 14.2 shows the advantages and disadvantages of the hydrolysis method using concentrated and dilute acid.

Enzymatic hydrolysis or saccharification of lignocellulose uses cellulase and hemicellulase enzymes that work specifically. Cellulase can convert cellulose to glucose, while hemicellulase is used to convert hemicellulose to xylose.

Enzymatic hydrolysis has several advantages over acid hydrolysis. It can reduce the equipment's risk of corrosion and energy loss in production. In addition, the advantage of enzymatic hydrolysis is that it will produce uniform sugar and does not contain other products due to glucose and xylose degradation. The disadvantage of enzymatic hydrolysis is the reduced rate of hydrolysis caused by enzyme inhibition. This enzyme inhibition was due to the sugar concentration being too high in the reactor. In addition, the enzymatic hydrolysis process can take several days compared to acid, which only takes a few minutes. Also, the price of the enzyme is relatively high [26].

14.5.1.3 Dehydration Processes

The mechanism of the dehydration process of xylose to furfural is currently divided into two approaches: the open-chain mechanism and the closed-chain mechanism. The mechanism of open-chain dehydration was proposed by Binder et al. in their research to produce furfural by dehydration of xylose with a $CrCl_3$ as a catalyst. According to this open chain mechanism, Cr^{3+} ions can cause a change in the formal charge of the C1 carbon atom, which in turn leads to a 1,2-hydride shift to produce xylulose. After that, further dehydration is continued to convert xylulose into furfural (Figure 14.9) [33]. Meanwhile, the closed-chain mechanism was proposed by Nimlos et al. [34]. According to this mechanism, the reaction begins with the protonates oxygen atom in the hydroxyl group connected to the second carbon in the xylose molecule. The following process occurs intramolecular rearrangement of the molecule (ring contraction) to form dehydrofuranose, then ends with a hydride shift reaction in dehydrofuranose resulting in the formation of furfural, as shown in Figure 14.10.

The mechanism of furfural formation from glucose has not been widely reported. The furfural formation process only comes from xylose. However, one mechanism related to the process of furfural formation from glucose was reported by Asakawa et al. via fructose [35]. Lewis acid catalyst is needed to produce furfural from glucose. Some of the catalysts derived from Lewis acid are aluminum chloride ($AlCl_3$), iron (III) chloride ($FeCl_3$), copper (II) chloride ($CuCl_2$), boron trifluoride (BF_3), and zinc chloride ($ZnCl_2$) and others. The process of furfural formation from glucose is influenced by the polar aprotic solvent used. The formation of furfural is almost the same as the formation of 5-HMF until the formation of a furan ring. Furfural is produced by removing the hydroxymethyl group as formaldehyde (Figure 14.11). Furfural formation is preferred in the presence of a polar aprotic solvent with very low basicity [35].

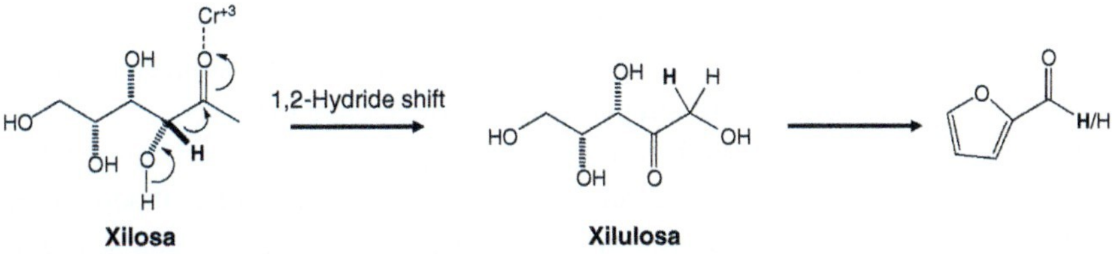

Figure 14.9 Mechanism of open-chain dehydration. Source: Binder et al. [33] Reproduced with permission of Wiley-VCH Verlag.

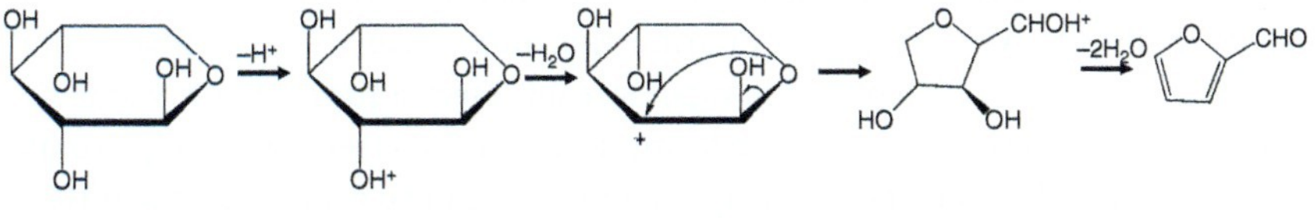

Figure 14.10 Mechanism of closed-chain dehydration. Source: Nimlos et al. [34] Reproduced with permission of ACS Publications.

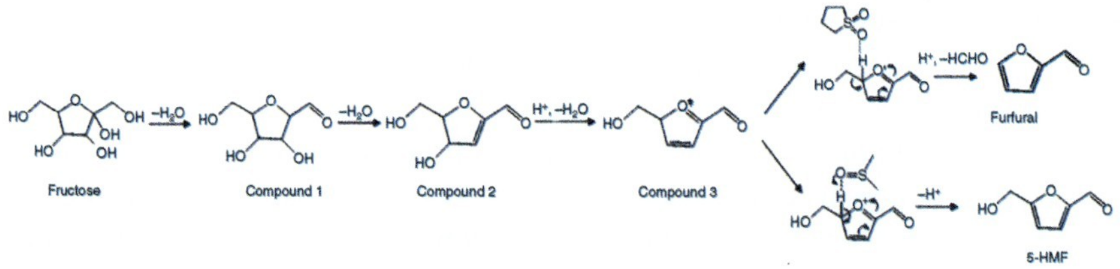

Figure 14.11 Mechanism of furfural synthesis from fructose. Source: Asakawa et al. [35] Reproduced with permission of Royal Society of Chemistry.

14.5.1.4 Purification

Producing furfural from bagasse and corncobs can generate other compounds such as HMF and formic acid. Furfural must be separated from impurities and purified to increase the desired concentration. Besides that, during furfural production, a resinification reaction and furfural condensation occur that cause decreasing in furfural concentration [36]. The resinification reaction is a reaction that occurs between furfural with furfural, while the condensation reaction occurs between furfural and intermediate compounds. Both of these reactions occur in the liquid phase of the product mixture. Therefore, the furfural produced must be immediately separated and purified. The furfural separation process from the mix is generally used by the evaporation method. Because in the vapor phase, it can avoid furfural resinification and condensation reactions [36]. Several purification processes in the vapor phase include steam striping, nitrogen striping, supercritical carbon dioxide, and reactive distillation.

Steam stripping is the process of separating furfural from the mixture using steam. This steam stripping process is the most commonly used in commercial-scale furfural production processes. The purification process is then continued with distillation. However, this steam stripping process requires a lot of energy and a high cost. Danon et al. reported that producing 1 tonne of furfural takes 25–35 tons of steam. In addition, the steam stripping process requires optimum conditions for the amount of steam used. The more steam used can reduce the catalyst's performance, but using too little steam causes the extraction process to be ineffective [37].

Nitrogen stripping can be used in the furfural separation process. High product purity, high furfural yield, reduced product dilution, and simplicity of separating the stripping agent from the catalyst are all benefits of utilizing nitrogen stripping [38]. In addition, using nitrogen stripping can reduce energy requirements and costs in commercial-scale furfural production. However, recycling nitrogen used requires high compression.

The vapor stream from the furfural reactor, which contained the vapor stream from pretreatment, was delivered for purification and recovery via a distillation column. Furfural was recovered in this vapor stream. First, a furfural-rich vapor side product with volatile components like formic acid removed through the top product stream is produced by the first azeotropic column to remove water from the furfural product. Next, the first column's furfural that has escaped via the top product stream is recovered using the second column. Finally, the second column's bottom product, which contains furfural, is delivered to a decanter, where the furfural-rich phase is collected for the third and final distillation column [39].

The furfural extraction process can be carried out in the liquid–liquid phase using organic solvents. Some solvents that can be used are toluene, butanol, MIBK and γ-valerolactone (GVL). The extraction process involves a biphasic system that can prevent the resinification reaction and furfural condensation. The furfural purification process on a commercial scale still uses distillation. Distillation is a purification process that is easy to operate and low in investment costs. However, the operational cost of the distillation process is still high because it requires high energy. Several studies have been carried out by combining distillation with distillation or with other methods such as extraction. Integrating distillation with the extraction process can reduce production costs by more than 40% [40].

14.5.2 Factors Affecting the Furfural Production

14.5.2.1 Catalyst

Catalysts are used in the furfural production process to accelerate the dehydration reaction of xylose to furfural. The furfural production process can occur without a catalyst, especially when using LCB as a substrate. This is because organic acids are produced in this process, which can be catalysts or namely autocatalytic [41].

Many studies have been conducted to develop catalysts that can increase furfural yield. In general, the catalysts used in the furfural production process are divided into homogeneous catalysts and heterogeneous catalysts. Sulfuric acid is a mineral acid as a homogeneous catalyst widely used commercially. The use of a sulfuric acid catalyst can hydrolyze and dehydrate lignocellulose to furfural. Hydrochloric acid can also be used as a catalyst. The disadvantages of using mineral acids are that they are corrosive, difficult to recover from mixed products, and cannot be reused [38]. Other than mineral acid, some organic acids like maleic acid and Lewis acids, such as $FeCl_3$, $CrCl_3$, and $AlCl_3$ can be used as a catalyst in furfural production. These two acids have different properties; Lewis acids tend to direct dehydration reactions, while mineral acids favor the breaking of glycosidic bonds in carbohydrate polymers [42].

Heterogeneous catalysts are also used in the furfural process. The advantage of using heterogeneous catalysts is that they can be removed from the product after processing, hence they can be reused. Several heterogeneous catalysts used in the furfural process include zeolite and carbon based. The use of zeolite is due to its stable nature. Carbon is widely used as a catalyst and catalyst support because it is easy to obtain, inexpensive, recyclable, and environmentally friendly. The physical properties of carbon, such as porosity and surface area, can be modified depending on the manufacturing process and its activation.

The performance of a catalyst in a process is influenced by several factors, including acidity and surface area. The acid type influences the catalyst's acidity in the furfural process. Lewis acids accept electrons, whereas Bronsted acids donate protons. Bronsted acid sites fast up the conversion of xylose to furfural, while Lewis acid sites favor the xylose–xylulose–furfural route rather than the direct conversion of xylose to furfural.

14.5.2.2 Solvent

The water-based solvent was the common solvent system to produce furfural from lignocellulose. The production of furfural using this solvent requires a catalyst such as mineral acid, like sulfuric acid. Gozan et al. produce furfural from oil palm empty fruit bunch with dilute sulfuric acid. Organic solvents can also be used to hydrolyze and dehydrate lignocellulosic to furfural. Some organic solvents that can be used are methanol, ethanol using sulfuric and formic acid catalysts, and GVL with catalysts [43]. GVL is a safe, biodegradable, and nontoxic solvent. GVL can increase the rate of furfural formation, reduce the occurrence of side reactions and produce more stable furfural.

Ionic liquids are salts consisting of large organic cations and inorganic or organic anions. These salts may exist in the liquid phase and often have low boiling points. In addition, ionic liquids have low volatility, allowing high-temperature processing, dissolving suitable biomass, chemical and thermal stability, increasing catalytic activity, and stabilizing reactive components [44]. Ionic liquids used in the furfural production process include 1-butyl-3-methylimidazolium chloride [BMIM]Cl with various metal chlorides (e.g. $CrCl_3$, $CuCl_2 \cdot 2H_2O$, $CrCl_3/LiCl$, $FeCl_3 \cdot 6H_2O$, and $AlCl_3$) [44]. The advantage of ionic liquids is that they can be engineered according to their intended use. One is a Bronsted acid, which can act as a solvent and a catalyst.

Deep eutectic solvent (DES) is a liquid formed by combining a hydrogen bond acceptor (HBA) and a molecular hydrogen bond donor (HBD) in a particular ratio. DES has properties like an ionic liquid, although the DES ion content is lower. DES started to be used in the furfural process because it was cheaper, safer, and easier to make. It was also more biodegradable and compatible with living organisms [45]. Zhang and Yu [46] produce furfural using DES, which is made by mixing choline chloride with citric acid. This DES solvent can act as a reaction medium and a Brønsted acid catalyst. DES derived from choline chloride is commonly used in furfural production because Cl^- ions can disrupt the hydrogen bonding network of biomass and can dissolve carbohydrates. DES made from mixing choline chloride with dicarboxylic acids, such as oxalic acid, malonic acid, succinic acid, and citric acid can produce furfural without a catalyst. Yield furfural can be increased with an additional catalyst to the process.

The resignification and furfural condensation reactions cause side products that cause low-yield furfural. The biphasic process uses two solvents forming two layers, namely the aqueous and organic phases can increase furfural yield. The biphasic process using H_2O–Methyl isobutyl ketone (MIBK) and H_2O–Toluene as a catalyst with H_2SO_4 as a corn husk can produce furfural with yields of 80.1% and 76.3%, while without the addition of organic solvents only produce below 64%. In this condition, furfural was dominant in the organic phase, indicating furfural would be extracted to the organic phase after being produced [47].

14.6 Techno-economical Aspect

Techno-economic analysis was used to see the feasibility of furfural production. Commercial furfural production has been carried out with several existing processes, both batch and continuous systems. Quaker Oats technology and China's batch technology are commercial technology based on a batch system. Both methods are almost the same: acid hydrolysis, extraction with steam, and purification by distillation. The batch process produces low yields of furfural. Even though it is economically feasible, it generates low profits. Yield enhancement is carried out using a continuous process system. Commercial examples of continuous processes include Quaker continuous, Westpro-modified Huaxia continuous technology, Rosenlew continuous technology, SupraYield® continuous technology, and Multi-turbine-column (MTC) process.

In the furfural manufacturing process, the global market is divided into Quacker batch, China batch, Rosenslew continuous process, and others (Table 14.3). By 2021, China's batch process will be the dominant process segment, accounting for 82% of global revenue. This method is mostly used in China, the world's largest producer. One of the oldest manufacturing processes is the Quaker batch process, while the others are modified versions of previous processes.

The technology of biomass-based furfural production is generally integrated with the concept of lignocellulosic biorefinery. Furfural is formed from hemicellulose, while lignocellulose still contains cellulose and lignin that can be utilized for another product. Ntimbani et al. used sugarcane bagasse to produce furfural, which was integrated with ethanol as a coproduct. Ethanol is produced from furfural residues due to ethanol obtained from cellulose. This process has a furfural yield of 69% (11.44 g/100 g of dry bagasse) and 77–95% ethanol (9.57–11.58 g/100 g of dry bagasse) with process conditions of 170 °C, 0.5% H_2SO_4. Techno-economic calculation of furfural compares one-stage and two-stage furfural production integrated with ethanol coproduction from sugarcane bagasse and harvest residues with Aspen Plus V8.8. Contrary to the one-stage furfural and ethanol coproduction biorefinery, the two-stage biorefinery required additional capital investment. However, lower MESPs than the one-stage furfural biorefinery integrated with ethanol coproduction resulted from increased ethanol productivity [49].

Compared to integrated two-stage furfural and ethanol coproduction biorefineries, the one-stage furfural production exhibited greater heating demands (2034 against 741–1124 kW t^{-1} feed) and lower ethanol yields (90 versus 150–214 kg^{-1} of feed). One-stage furfural production generated more cost than two-stage furfural production [39]. In situ, furfural extraction was performed using MIBK, while simultaneous biomass processing and furfural synthesis was accomplished using aqueous choline chloride (ChCl). To generate an internal rate of return of 10% with a feedstock capacity of 1096 kt/y, the selling price of ethanol was determined to be USD 2.55/gal, and the selling price of lignin was determined to be US$ 500/ton. The lowest selling price of furfural that could generate such an IRR was determined to be US$ 625/t. This pricing is considerably less than the furfural market's average price, which is roughly US$ 1350/ton [50].

Table 14.3 Industrial-scale furfural production process [48].

Furfural production method	Process reactor	Process operation	Furfural yield
Quaker Oats (1922) in America	Closed	H_2SO_4, stripping (153 °C) at high pressure	40%–52%
Chinese batch in China	Closed	H_2SO_4, 160–165 °C	50%
Agrifurane/Petrole Chimie	Closed	H_2SO_4, 177–161 at 6–9 bar	—
Suprayield (1990) by Dr. Karl Zeitsch	Closed	H_3PO_4/CH_3COOH, 170–230 °C at high pressure	70%
Vedernikov's	Closed	H_2SO_4	55–75%
Quaker Oats (1997)	Continuous	H_2SO_4, 184 °C at 11 bar	55%
Escher Wyss	Continuous	H_2SO_4, 170 °C	—
Rosenlew	Continuous	Acids formed in situ at 180 °C at 10 bar	60%
Supratherm (1988)	Continuous	H_2SO_4, 200–240 °C	—
Stake	Continuous	Acids formed in situ at 230 °C at 27.7 bar	66%
Biofine (1990) by Dr. Stephen W. Fitzpatrick	Continuous	H_2SO_4, 220 °C at 25 bar	70%
CIMV (2010) in France	Continuous	$HCOOH/CH_3COOH$, 230 °C, pressurized	—
Multi-Turbin-Column in TU Delft (2012)	Continuous	H_2SO_4/NaCl, stripping (200 °C)	83%

Techno-economic calculations of furfural integrated with levulinic acid and ethanol were carried out with SuperPro. The optimization of furfural production process using oil palm empty fruit bunches as raw material is conducted with varying production capacities for each product. The results of the calculation with the ratio of production capacity of ethanol at 87%, furfural at 7%, and levulinic acid at 3% produce the net present value (NPV) of US$ 53 939 000; this model also resulted in the internal return rate (IRR), and payback period almost 29.77%, and 4.52 years respectively [51]. These economic parameters concluded that the process is feasible.

14.7 Future Trends

The global market of furfural has been segmented into furfuryl alcohol, solvents, intermediate compounds, and others. In 2021, furfuryl alcohol became the largest segment, accounting for a maximum of more than 86% of global revenue. Therefore, the foundry industry's growth leads to the demand for furfuryl alcohol, thereby increasing the furfural market. Based on raw materials, the global market has been segmented into corncobs, bagasse, rice straw, and rice husks. Corncob raw materials segment leads the global market with a share of more than 70% in 2021. It is expected to maintain its dominant position throughout the forecast years due to the extensive use of raw materials in furfural production. The furfural yield from corncob is higher than other materials. Corncob is the most extensively used raw material in China, Thailand, Russia, and Spain because it is very versatile and is the main raw material used to manufacture chemical products. Dominican Republic, South Africa, India, and America use bagasse as raw materials.

By 2021, the Asia Pacific region will hold a market share of more than 70% in terms of value. Due to its numerous large manufacturing firms and easy access to raw materials, China is a significant user of furfural in the Asia Pacific region. Due in large part to the presence of profitable furfural manufacturers in these nations, China, India, Vietnam, and Thailand are anticipated to dominate the regional furfural market. The key reasons boosting the demand for furfural in the Asia Pacific are greater government efforts and growing investment in bio-based goods. Important markets for furfural include those in North America and Europe.

14.8 Conclusions

Global demand for furfural applications in manufacturing plastics, adhesives, nematicides, inks, antacids, fungicides, fertilizers, and food flavors. Furfural is an ideal material for production due to its qualities, including thermosetting, physical resilience, and resistance to corrosion. Furfural can also be used in decolorizing and refining processes. Furfural can be produced from LCB, such as corncob, sugarcane bagasse, and other biomass. Companies worldwide are investing in research and development to introduce new and innovative products made from furfural. Challenges for future processes are increasing furfural yield and optimizing furfural plant's energy efficiency. Sustainability and environmentally friendly methods are also considered in furfural process development. The utilization of nontoxic material that is environmentally friendly and easy to separate has been developed to increase the sustainability of the furfural production process.

References

1 Kirk, R.E. and Othmer, D. (2004). Furan derivatives: on. *Encycl. Chem. Technol.* 10: 237–250.
2 Yemiş, O. and Mazza, G. (2011). Acid-catalyzed conversion of xylose, xylan and straw into furfural by microwave-assisted reaction. *Bioresour. Technol.* 102: 7371–7378.
3 Galkin, K.I. and Ananikov, V.P. (2019). When will 5-hydroxymethylfurfural, the "sleeping giant" of sustainable chemistry, awaken? *ChemSusChem* 12: 2976–2982.
4 Galkin, K.I. and Ananikov, V.P. (2020). The increasing value of biomass: moving from C6 carbohydrates to multifunctionalized building blocks via 5-(hydroxymethyl) furfural. *ChemistryOpen* 9: 1135–1148.

5 Lewkowski, J. (2001). Synthesis, chemistry and applications of 5-hydroxymethylfurfural and its derivatives. *Arkivoc* 1: 17–54.

6 Mascal, M. (2015). 5-(Chloromethyl) furfural is the new HMF: functionally equivalent but more practical in terms of its production from biomass. *ChemSusChem* 8: 3391–3395.

7 Pedersen, M.J., Jurys, A., and Pedersen, C.M. (2020). Scalable synthesis of hydroxymethyl alkylfuranoates as stable 2, 5-furandicarboxylic acid precursors. *Green Chem.* 22: 2399–2402.

8 Mathew, A.K., Abraham, A., Mallapureddy, K.K., and Sukumaran, R.K. (2018). Lignocellulosic biorefinery wastes, or resources? In: *Waste Biorefinery* (ed. T. Bhaskar, A. Pandey, S.V. Mohan, et al.), 267–297. Elsevier.

9 Gonçalves, F.A.M.M., Santos, M., Cernadas, T. et al. (2022). Advances in the development of biobased epoxy resins: insight into more sustainable materials and future applications. *Int. Mater. Rev.* 67: 119–149.

10 Research and Market 2022 Report: Furfuryl Alcohol Market Size, Share & Trends Analysis Report by Application (Resins, Solvents, Corrosion Inhibitors), by End Use (Foundry, Agriculture, Food & Beverages), by Region, and Segment Forecasts, 2021–2028.

11 Audemar, M., Wang, Y., Zhao, D. et al. (2020). Synthesis of furfuryl alcohol from furfural: a comparison between batch and continuous flow reactors. *Energies* 13: 1002.

12 Lindsey R V and Prichard W W 1980 Oxidation of butadiene to furan, US Patent 4298531A, Granted November 3rd, 1981.

13 Bruice, P.Y. (2017). *Organic Chemistry*. England: Pearson Prentice Hall.

14 Serrano-Ruiz, J.C., Luque, R., Campelo, J.M., and Romero, A.A. (2012). Continuous-flow processes in heterogeneously catalyzed transformations of biomass derivatives into fuels and chemicals. *Challenges* 3: 114–132.

15 Mariscal, R., Maireles-Torres, P., Ojeda, M. et al. (2016). Furfural: a renewable and versatile platform molecule for the synthesis of chemicals and fuels. *Energy Environ. Sci.* 9: 1144–1189.

16 Pérez, H.I., Manjarrez, N., Solís, A. et al. (2009). Microbial biocatalytic preparation of 2-furoic acid by oxidation of 2-furfuryl alcohol and 2-furanaldehyde with *Nocardia corallina*. *Afr. J. Biotechnol.* 8.

17 Sudiyani, Y., Sembiring, K.C., and Adilina, I.B. (2014). Bioethanol G2: production process and recent studies. In: *Biomass and Bioenergy* (ed. K.R. Hakeem, M. Jawaid, and U. Rashid), 345–364. Springer.

18 Sudiyani, Y. and Muryanto, M. (2012). The potential of biomass waste feedstock for bioethanol production. In: *Proceeding of International Conference on Sustainable Energy Engineering and Application* (ed. B. Prawara), 5–10. Research Center for Electrical Power and Mechatronics.

19 Dahmen, N., Lewandowski, I., Zibek, S., and Weidtmann, A. (2019). Integrated lignocellulosic value chains in a growing bioeconomy: status quo and perspectives. *GCB Bioenergy* 11: 107–117.

20 Limayem, A. and Ricke, S.C. (2012). Lignocellulosic biomass for bioethanol production: current perspectives, potential issues and future prospects. *Prog. Energy Combust. Sci.* 38: 449–467.

21 Machado, G., Leon, S., Santos, F. et al. (2016). Literature review on furfural production from lignocellulosic biomass. *Nat. Resour.* 07: 115–129.

22 Gao, Z., Mori, T., and Kondo, R. (2012). The pretreatment of corn stover with *Gloeophyllum trabeum* KU-41 for enzymatic hydrolysis. *Biotechnol. Biofuels* 5: 1–11.

23 Duff, S.J.B. and Murray, W.D. (1996). Bioconversion of forest products industry waste cellulosics to fuel ethanol: a review. *Bioresour. Technol.* 55: 1–33.

24 Sari, D.R. (2018). The potential of woody waste biomass from the logging activity at the natural forest of Berau District, East Kalimantan. In: *IOP Conference Series: Earth and Environmental Science*, vol. 144, 12061. IOP Publishing.

25 Lloyd, T.A. and Wyman, C.E. (2005). Combined sugar yields for dilute sulfuric acid pretreatment of corn Stover followed by enzymatic hydrolysis of the remaining solids. *Bioresour. Technol.* 96: 1967–1977.

26 Sun, Y. and Cheng, J. (2002). Hydrolysis of lignocellulosic materials for ethanol production : a review. *Bioresour. Technol.* 83: 1–11.

27 Alvarez-vasco, C. and Zhang, X. (2013). Bioresource technology alkaline hydrogen peroxide pretreatment of softwood: hemicellulose degradation pathways. *Bioresour. Technol.* 150: 321–327.

28 Shirkavand, E., Baroutian, S., Gapes, D.J., and Young, B.R. (2016). Combination of fungal and physicochemical processes for lignocellulosic biomass pretreatment – a review. *Renewable Sustainable Energy Rev.* 54: 217–234.

29 Metkar, P.S., Till, E.J., Corbin, D.R. et al. (2015). Reactive distillation process for the production of furfural using solid acid catalysts. *Green Chem.* 17: 1453–1466.

30 Wyman, C.E., Decker, S.R., Himmel, M.E. et al. (2005). Hydrolysis of cellulose and hemicellulose. In: *Polysaccharides: Structural Diversity and Functional Versatility* (ed. S. Dumitriu), 994–1033. CRC Press.

31 Galbe, M. and Zacchi, G. (2002). A review of the production of ethanol from softwood. *Appl. Microbiol. Biotechnol.* 59: 618–628.

32 Verardi, A., De Bari, I., Ricca, E., and Calabrò, V. (2012). Hydrolysis of lignocellulosic biomass: current status of processes and technologies and future perspectives. In: *Bioethanol*, vol. 2012 (ed. M.A.P. Lima), 95–122. InTech Rijeka.

33 Binder, J.B., Blank, J.J., Cefali, A.V., and Raines, R.T. (2010). Synthesis of furfural from xylose and xylan. *ChemSusChem* 3: 1268–1272.

34 Nimlos, M.R., Qian, X., Davis, M. et al. (2006). Energetics of xylose decomposition as determined using quantum mechanics modeling. *J. Phys. Chem. A* 110: 11824–11838.

35 Asakawa, M., Shrotri, A., Kobayashi, H., and Fukuoka, A. (2019). Solvent basicity controlled deformylation for the formation of furfural from glucose and fructose. *Green Chem.* 21: 6146–6153.

36 Dulie, N.W., Woldeyes, B., Demsash, H.D., and Jabasingh, A.S. (2021). An insight into the valorization of hemicellulose fraction of biomass into furfural: catalytic conversion and product separation. *Waste Biomass Valorization* 12: 531–552.

37 Danon, B., Marcotullio, G., and de Jong, W. (2014). Mechanistic and kinetic aspects of pentose dehydration towards furfural in aqueous media employing homogeneous catalysis. *Green Chem.* 16: 39–54.

38 Agirrezabal-Telleria, I., Gandarias, I., and Arias, P.L. (2013). Production of furfural from pentosan-rich biomass: analysis of process parameters during simultaneous furfural stripping. *Bioresour. Technol.* 143: 258–264.

39 Ntimbani, R.N., Farzad, S., and Görgens, J.F. (2021). Techno-economics of one-stage and two-stage furfural production integrated with ethanol co-production from sugarcane lignocelluloses. *Biofuels, Bioprod. Biorefin.* 15: 1900–1911.

40 Nhien, L.C., Long, N.V.D., Kim, S., and Lee, M. (2017). Techno-economic assessment of hybrid extraction and distillation processes for furfural production from lignocellulosic biomass. *Biotechnol. Biofuels* 10: 1–12.

41 Zhang, S., Lu, J., Li, M., and Cai, Q. (2017). Efficient production of furfural from corncob by an integrated mineral-organic-Lewis acid catalytic process. *BioResources* 12: 2965–2981.

42 Zhang, Z. and Zhao, Z.K. (2010). Microwave-assisted conversion of lignocellulosic biomass into furans in ionic liquid. *Bioresour. Technol.* 101: 1111–1114.

43 Mohamad, N., Yusof, N.N.M., and Yong, T.L.-K. (2017). Furfural production under subcritical alcohol conditions: effect of reaction temperature, time, and types of alcohol. *J. Japan Inst. Energy* 96: 279–284.

44 Peleteiro, S., Rivas, S., Alonso, J.L. et al. (2016). Furfural production using ionic liquids: a review. *Bioresour. Technol.* 202: 181–191.

45 Haldar, D. and Purkait, M.K. (2021). A review on the environment-friendly emerging techniques for pretreatment of lignocellulosic biomass: mechanistic insight and advancements. *Chemosphere* 264: 128523.

46 Zhang, L. and Hongbing, Y. (2013). Conversion of xylan and xylose into furfural in biorenewable deep eutectic solvent with trivalent metal chloride added. *BioResources* 8 (4): 6014–6025.

47 Mittal, A., Black, S.K., Vinzant, T.B. et al. (2017). Production of furfural from process-relevant biomass-derived pentoses in a biphasic reaction system. *ACS Sustainable Chem. Eng.* 5: 5694–5701.

48 Cousin, E., Namhaed, K., Pérès, Y. et al. (2022). Towards efficient and greener processes for furfural production from biomass: a review of the recent trends. *Sci. Total Environ.* 847: 157599.

49 Ntimbani, R.N., Farzad, S., and Görgens, J.F. (2021). Furfural production from sugarcane bagasse along with co-production of ethanol from furfural residues. *Biomass Convers. Biorefin.* 12: 5257–5267.

50 Kubic, Jr, W.L., Yang, X., Moore, C.M., and Sutton, A.D. (2020). A Process for Converting Corn Bran to Furfural without Mineral Acids (Los Alamos National Lab.(LANL), Los Alamos, NM (United States)).

51 Muryanto, M., Putri, K.L., Srinophakun, P., and Gozan, M. (2021). Techno-economic evaluation of integrated levulinic acid-bioethanol plant design based on oil palm empty fruit bunches. In: *IOP Conference Series: Earth and Environmental Science*, vol. 926, 12064. IOP Publishing.

15

Levulinic and Formic Acids from Rice Straw and Sugarcane Bagasse

Jabosar R. H. Panjaitan[1] and Misri Gozan[2]

[1] Institut Teknologi Sumatera, Department of Chemical Engineering, Lampung, Indonesia
[2] Universitas Indonesia, Department of Chemical Engineering, Depok, 16424, Jawa Barat, Indonesia

15.1 Introduction

With the fast growth of the world's population, the demand for pharmaceuticals, cosmetics, fuel oil, and food has also increased significantly. Levulinic and formic acids (FA) can be used as an alternative material in various industrial production processes. Levulinic and FA reaction production steps are shown in Figure 15.1.

Levulinic acid (LA; $C_5H_8O_3$, 4-oxopentanoic acid, −acetylpropionic acid, and ketovaleric acid) is one of the most studied chemical from various carbohydrates, which is produced from various types of biomasses, such as agricultural waste and food waste. Because it maybe used as a chemical building block in many different sectors, LA is a flexible intermediate. As a flavoring and plant growth stimulator, LA is also widely employed in the culinary and agriculture industries [1]. Because of its carbonyl and carboxyl groups, which are responsible for a variety of functional, chemical properties and reactive keto acids, the LA serves as a versatile intermediate that can be transformed into a variety of chemical derivatives to produce herbicides, insecticides, antifreeze, polymers, dyes, battery electrolytes, oil additives, solvents, polymers, food additives, and additives for food and cosmetics [2].

LA has a structure that consists of a carboxylic acid, a ketone carbonyl group, and a short chain of fatty acids. Using an acid catalysis procedure that typically employs sulfuric acid, LA was first created in 1870 and commercialized from starch in 1940. The US Department of Energy's top 12 carbohydrate-based building blocks list includes LA. Because of this, LA is a useful chemical [1].

The simplest carboxylic acid, formic acid (FA; HCOOH, methanoic acid), is soluble in water and many alcohols, including acetone and ether. FA is a useful chemical product with various uses, including in the pharmaceutical, culinary, textile, chemical, and agricultural sectors. This is due to its strong acid and reducing capabilities. FA can also be used in decalcification, acidification, silage, and preservation of animal feed, formic salt, rubber chemicals, catalysts, and plasticizers. FA can be used as an intermediate chemical in organic synthesis and has an important role in

Chemical Substitutes from Agricultural and Industrial By-Products: Bioconversion, Bioprocessing, and Biorefining,
First Edition. Edited by Suraini Abd-Aziz, Misri Gozan, Mohamad Faizal Ibrahim, and Lai-Yee Phang.
© 2024 WILEY-VCH GmbH. Published 2024 by WILEY-VCH GmbH.

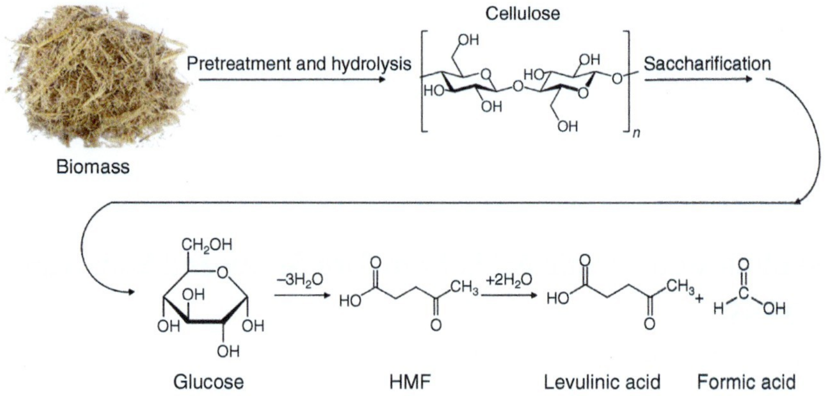

15 Levulinic and Formic Acids from Rice Straw and Sugarcane Bagasse

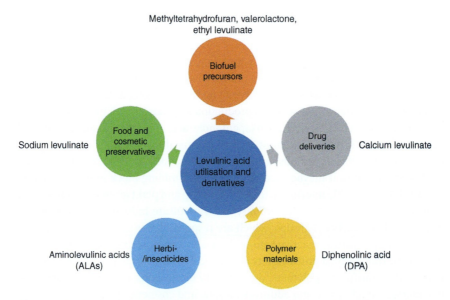

Figure 15.1 Reaction scheme of levulinic and formic acids from biomass. Source: Illustration and drawn by Gozan.

Figure 15.2 Levulinic and formic acids utilization. Source: Illustration and drawn Gozan.

producing hydrogen, alkanes, and other chemicals based on biological resources [3]. Levulinic and FAs utilizations are shown in Figure 15.2.

Recently, FA has been seen and proposed as a hydrogen storage component for eco-friendly fuel cell technologies. FA is a promising material because it can be decomposed into CO_2 and H_2 with the possibility of reversible FA retransformation [3].

Given the advantages and applications of levulinic and FAs in several sectors, there is a need for these chemicals, which are supplied by many businesses and nations.

Unlike other high-value biomass derivatives, which maybe derived by biological methods like fermentation, hydrogenation, or oxidation in the manufacturing

Table 15.1 Levulinic acid supplying countries and companies.

Country of producer	Vendor
United States	Advanced Biotech
	Aurochemicals
	Charkit Chemical
	Merck and Co.
	Synerzine
	The Sherwin Williams
	Vigon International
United Kingdom	Augustus Oils
	De Monchy Aromatics
China	Beijing Lys Chemicals
	Hefei TNJ Chemical Industry
	Shanghai M and U International Trade
	Simagchem
Italia	GFBiochemicals
India	Otto Chemie Pvt.
	RXChemicals
Japan	Tokyo Chemical Industry

Source: Mordor Intelligence [4]/Mordor Intelligence.

Table 15.2 Formic acid supplying countries and companies.

Country	Vendor
United States	BASF SE
	Eastman Chemical
	Haviland Enterprises
	NuGenTec
	Thermo Fisher Scientific
India	GNFC
	Rashtriya Chemicals and Fertilizers
Finland	Kemira Oyj
China	LUXI Group
Sweden	Perstorp Holding AB

Source: Modified from Mordor Intelligence [5].

process, LA can only be produced utilizing acid-catalyzed chemical procedures. Distillation, solvent extraction, reactive extraction, and absorption are usually used to separate and purify LA.

LA is currently produced by several countries and companies as vendors, as described in Table 15.1.

Fossil fuels are typically used in the manufacture or production of FA. The four steps in producing FA are methyl formate hydrolysis, hydrocarbon oxidation, formamide hydrolysis, and FA synthesis. The methyl formate-based process pathway is the most common of these four procedures [3].

FA is currently produced by several countries and several companies as vendors, as described in Table 15.2.

15.2 Potential of Biomass Source for the Production of Levulinic and Formic Acids

15.2.1 Rice Straw

Rice straw is produced more than 6 billion tons annually [6]. It is one of the primary raw materials utilized in biomass. Although some rice is produced in Mediterranean temperatures, rice is mostly cultivated in tropical and subtropical climates. In tropical climes, rice plants maybe grown yearly if enough water is available (such as in Southeast Asia). This implies that the straw generated might result in more than one harvest yearly. Table 15.3 below is a more detailed explanation of the composition of rice straw.

The biochemical makeup of rice straw is typical of lignocellulosic leftovers derived from agriculture, with an average content of 30–45% cellulose, 20–25% hemicellulose, 15–20% lignin, and a few minor organic compounds [10]. Rice straw has little nitrogen but a lot of inorganic substances, sometimes known as ash. Rice straw is possible feedstock for the bio-based economy.

15.2.2 Sugarcane Bagasse

In tropical nations, sugarcane (*Saccharum officinarum*) is frequently grown. There were over 1.84 billion tonnes of sugarcane produced worldwide in 2017. Sugar and alcohol production facilities often employ sugarcane. The plant does not, however, completely use sugarcane. After being employed in manufacturing, only around 30% of the pulpy fibrous waste is created [11, 12].

Bagasse, or sugarcane residue, is the fiber-containing material left behind after sugar is extracted. Bagasse, made from sugarcane, is a fibrous substance with cellulose as its major constituent. Bagasse made from sugarcane is manufactured in great numbers all over the world. A sort of substance that originates from the sugar industry is sugarcane bagasse. Typically, the paper industry uses sugarcane bagasse. Nevertheless, several recent studies have demonstrated that when subjected to specific mechanical and chemical processes, sugarcane bagasse can aid in the extraction

Table 15.3 Composition and characteristics of rice straw.

	Sources		
	[7]	[8]	[9]
Characteristic (% wet basis)			
Humidity	5.73	6.8	14.08
Fixed carbon	17.6	—	11.10
Volatile materials	61.4	80.1	60.55
Ash	21	9.6	13.26
Element composition (% dry basis)			
C	35.5	43.3	33.70
H	4.62	4.94	3.91
O	37.83	30.8	36.26
S	0.06	0.07	0.03
N	0.99	0.57	0.71
Cl	—	0.013	0.32
Ash (estimate)	29.23	—	18.67

of cellulose fibers, pure cellulose, nanocellulose fibers, and nanocellulose crystals. The components obtained from sugarcane bagasse may then be used to create various types of composite products and regenerate cellulose fibers.

Sugarcane bagasse shares many chemical properties with other plant cell walls. About 40–50% of sugarcane bagasse is made of cellulose, while 25–35% is

Table 15.4 Composition and characteristics of sugarcane bagasse.

Proximate analysis (%)	
Humidity	6.21
Fixed carbon	82.38
Volatile matters	2.94
Ash	8.47
Element composition (%)	
C	45.39
N	0.15
H	7.92
O	46.67
O/C	1.03
H/C	0.17

Source: Adapted from Samadi et al. [13].

hemicellulose [11]. Table 15.4 shows a more detailed explanation of the composition of bagasse.

Several factors affect the compositions of sugarcane bagasse [10], some of which are the use of fire or other methods to clean the pulp before cutting; Harvesting and packaging methods may result in larger or smaller dragging of dirt, grit, and vegetable residue, such as manual cutting methods, mechanical cutting, mincing and cutting to the end; the type of soil used when planting sugarcane (latosol, sandy soil, and other soil types) and the different procedures used for sugarcane cleaning.

15.3 Levulinic Dan Formic Acids Formation

Levulinic and FAs can be produced using biomass hydrolysis reaction. This reaction usually uses catalysts, such as acid or enzyme to speed up the reaction. Acid catalysts are often referred to as homogeneous catalysts. Heterogeneous catalysts, such as Mn/ZSM-5 and metal (IV) phosphate, can be used in place of homogeneous catalysts to facilitate the hydrolysis process [14]. Based on the existing literature, the amount of yield that can be produced from this reaction is 58–71%. The reaction mechanism of cellulose to LA also follows the mechanism in Figure 15.1.

Hierarchical MnO_x/ZSM-5 is used in research as a heterogeneous catalyst to convert delignified rice straws to LA. With the aid of a porous ZSM-5 framework, the conversion happened as a result of the interaction of the cellulose with either MnO_x in zeolites or Mn^{2+} ions in the solution. The findings of this study demonstrate that the ZSM-5 structure is kept intact in some way by the hierarchical system.

15.4 Pretreatment and Production Technologies

To produce FA and LA, the pretreatment procedure is crucial. Levulinic and formic acids pretreatment relies on the source of the material. Pretreatment for manufacturing FA and LA continually evolves as new methods are discovered. However, size reduction and delignification are the most typical pretreatments for manufacturing formic and LA.

15.4.1 Size Reduction

Size reduction is an initial step that is always carried out in the biomass pretreatment process. Several particle sizes have currently been established based on the primary processes that will be used to process the biomass, including pelletizing (particle size: 0.6–0.87 mm), gasification (particle size: 0.2–1.5 mm), pyrolysis (particle size: 0.25–2 mm), hydrolysis and fermentation (particle size: 0.03–10 mm), and briquette (particle size: 1.6–5.6 mm). Several equipments that can be used to carry out the size reduction process are a hammer mill, bead mill, ball mill, roller mill, and disk mill. The size reduction process required a lot of energy. The energy required to operate the size reduction machinery and the energy used to fracture the raw materials

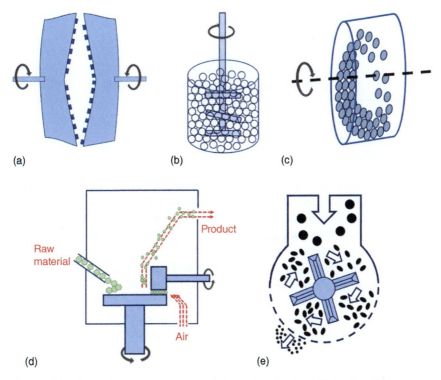

Figure 15.3 Size reduction equipment: (a) disc mill, (b) bead mill, (c) ball mill, (d) roller mill, and (e) hammer mill. Source: Illustration and drawn by Untung Nugroho Harwanto.

together makes up the total energy used for size reduction [15]. Several size reduction equipment can be seen in Figure 15.3.

15.4.2 Biological Pretreatment

Bagasse as biomass is mainly composed of polysaccharides and phenol-based compounds, especially cellulose. Lignin, cellulose, and hemicellulose are all present in bagasse, along with a few other readily soluble or sometimes referred to as ash-like chemicals. Treatment with white-rot fungus can lower levels of lignin concentration in lignocellulosic materials, including bagasse [16]. Some white-rot fungi often used for lignin biodegradation include *Pleurotus ostreatus, Phanerochaete sordila, Pycnoporus cinnabarinus, Sporotricum pulverulentum, Cyathus strecoreus, Pleurotus chrysosporium, Ceriperiopsis subvermispora, Lentinus edodes, Pleurotus eryngi,* and *Corolus versicolor.* The fungi *Ceriporiopsis subvermispora* and *L. edodes* are quite effective in degrading lignin and can also increase ethanol production from wood [17].

One study used bagasse samples treated with *L. edodes* for four weeks because the fungus effectively degraded lignin [18]. Lignin contained in bagasse can block or slow down the access of enzymes in breaking down polysaccharides in the hydrolysis process to increase the ethanol produced in the fermentation process [19]. Lignin

Table 15.5 Composition of sugarcane bagasse before and after treatment.

Bagasse	Composition (%)		
	Lignin	α-cellulose	Hemicellulose
Before pretreatment	24.2	52.7	17.5
After pretreatment	21.8	49.5	15.2

Source: Adapted from Itoh et al. [17].

peroxidase (LiP), manganese-dependent peroxidase (MnP), and laccase, which may breakdown lignin, are produced by white-rot fungus. [20]. These enzymes can oxidize the phenolic compounds in lignin so the bond will break. The hydrolysis will be optimal when more lignin is broken down, making the fermentation process turn it into ethanol more efficient. Bagasse treated with *L. edodes* for four weeks saw a decrease in the weight of lignin, holocellulose, and α-cellulose [18]. This happens because bagasse is a medium for growing mushrooms and a food source for the growth and development of these mushrooms.

White-rot fungi can disassemble complex lignin chains, which can block or slow down the access of enzymes in the hydrolysis process to increase the ethanol produced in the fermentation process [18, 19]. The analysis using *L. edodes* showed that the lignin content in bagasse was around 24.2% of the total bagasse. Holocellulose (cellulose and hemicellulose) in bagasse is around 70.2%. The α-cellulose and hemicellulose in bagasse are around 52.7% and 17.5%, respectively. After treatment with *L. edodes*, there was a decrease in bagasse weight. Loss of weight while treating white-rot fungi occurs in lignin or holocellulose. This treatment is said to be effective or has good selectivity if the fungus can degrade lignin more than the degradation of cellulose, which is characterized by a more significant loss of lignin weight compared to the loss of cellulose [21]. After treating *L. edodes* fungus for four weeks, bagasse composition decreased. The reduction in the composition of lignin holocellulose and α-cellulose is depicted in Table 15.5.

From Table 15.5, it can be said that white-rot fungi can degrade lignin bonds more than cellulose and hemicellulose. So, in this case, the fungus *L. edodes* has selectivity in degrading lignin in bagasse. In addition, the percentage loss of cellulose content was quite significant during the 4-week treatment. These results also show that the *L. edodes* fungus has a selectivity that is not too great during the 4-week treatment because quite a lot of cellulose is degraded. Lignin degradation by *L. edodes* fungus can occur during the process.

15.4.3 Delignification

Delignification is crucial in biomass processing to create several cellulose and hemicellulose-based derivative products. This stage aims to remove the lignin from the biomass so that the cellulose and hemicellulose maybe utilized to make specific goods.

Biological delignification is often carried out in the delignification process because this process uses environmentally friendly biological agents. Biological pretreatment has advantages, such as being environmentally friendly, having higher yield, operating at room temperature, producing few by-products, requiring little energy, not forming degradation products, and not needing a reactor with temperature and corrosion-resistant conditions [22]. Biological pretreatment functions as delignification involving bacteria and fungi. White-rot fungi, which are classified as Basidiomycetes fungi, can delignify biomass. The delignification process by white-rot fungi was influenced by the laccases enzyme produced by the fungus.

The most used chemical delignification method is alkaline delignification. Sodium hydroxide (NaOH), potassium hydroxide (KOH), and calcium hydroxide ($Ca(OH)_2$) are alkalines utilized in this method for delignification. Xylan hemicellulose and other components are cross-linked by intermolecular ester bonds during the saponification process of alkali, and the porosity of the hemicellulose rises when the cross-links are removed [23].

Alkaline delignification with low temperatures is one of the right choices for levulinic and FA production because, at low temperatures, cellulose degradation can be prevented. In this case, a soaked aqueous ammonia solution can be used. This process was usually carried out at room temperature for 14 hours with ±13% ammonia concentration. Several studies, such as Panjaitan and Gozan, used this method to examine FA production and its kinetic evaluation from empty oil palm fruit bunches with soaked aqueous ammonia solution pretreatment [24]. Gozan et al. studied the kinetics of LA and furfural synthesis after pretreating them with an aqueous ammonia solution [25].

Several studies on sugarcane bagasse and rice straw delignification have been carried out. Ozone was used to delignify sugarcane bagasse [26]. The alkaline hydrogen peroxide (AHP) method was reviewed in sugarcane bagasse delignification [27]. Sugarcane bagasse delignification was done using the steam explosion method followed by sodium hydroxide pretreatment [28]. An investigation was carried out on sugarcane bagasse and rice husk delignification using laccase enzymes and white-rot fungi [29] (Table 15.6).

15.5 Purification Technologies

Purifying levulinic and FAs from biomass hydrolysis can be done by various methods, but the most common methods are distillation and extraction. Nhien et al. [43] investigated LA purification with Aspen plus simulation using a distillation column with a dividing wall column (DWC), with the configuration as shown in Figure 15.4. Based on the study's results, it was found that the configuration of a distillation column with a DWC concept reduced energy requirements and annual operating costs by 16.4% and 20.6% compared to conventional multilevel distillation.

The extraction method can be used to purify LA. Kumar et al. [1] have reviewed and reported that various solvents (such as Diethyl carbonate, Isoamyl alcohol, and Methyl Isobutyl Ketone), as well as extractants (such as Amberlite LA-2 and

Table 15.6 Levulinic and formic acids production from sugarcane bagasse and rice straw.

No.	Raw material	Pretreatment method	Production method	Products	References
1	Sugarcane bagasse	Size reduction and dewaxing	Ionic liquid treatment (1-ethyl-3-methylimidazolium hydrogen sulfate)	Levulinic acid	[30]
2	Rice straw	Size reduction and solvent pre-extraction (ethanol/benzene)	Biphasic reactor system (HCL-DCM)	Levulinic acid	[31]
3	Sugarcane bagasse	—	Biofine process	Levulinic acid, furfural, electricity, and gamma valeractone	[32]
4	Empty fruit bunch and rice straw	—	Acid hydrolysis	Levulinic acid	[33]
5	Bamboo, Poplar, Eucalyptus, Pine, Bagasse, Straw, and Switchgrass	Size reduction	Acid hydrolysis using bifunctional catalyst (A15-Al$_2$(SO$_4$)$_3$) in a biphasic cosolvent (water–dimethoxymethane)	Levulinic acid	[34]
6	*Cicer arietinum*, cotton, *Pinus radiata*, and Sugarcane bagasse	Size reduction	Acid hydrolysis (HCl)	Levulinic acid	[35]
7	Sugarcane bagasse, rice straw, and soybean straw	Size reduction	3-step process (acid pretreatment, alkaline pretreatment, and catalytic depolymerization of cellulose)	Levulinic acid	[36]

8	Miscanthus, sweet sorghum bagasse, maize stover, and rice straw	Size reduction and delignification using SGL (Na_2CO_3 and Na_2S)	Acid hydrolysis using HCl	Levulinic acid	[37]
9	Sugarcane bagasse and miscanthus	Size reduction	Biofine process	Chemicals (levulinic acid, formic acid, furfural, bio-oil, biochar, graded oil, and ethyl levulinate)	[38]
10	Sugarcane fiber	Size reduction and alkali pretreatment	Acid hydrolysis	Levulinic acid, formic acid, and furfural	[39]
11	Rice straw	Size reduction	Acidic ionic liquid treatment	Levulinic acid	[40]
12	Sugarcane bagasse	—	Acid hydrolysis	Levulinic acid	[41]
13	Sugarcane bagasse	Acid pretreatment and alkali delignification	Acid hydrolysis	Levulinic acid	[42]

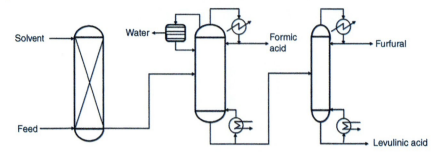

Figure 15.4 Purification of levulinic acid using distillation column with Dividing Wall Column. Source: Illustration drawn by Gozan based on Nhien et al. [43].

Tri-*n*-octylamine), could purify LA with up to 98% separation efficiency. The extraction method in LA purification was preferred due to the simpler process.

15.6 Economic Feasibilities

One of the crucial elements is a combination of levulinic and FA. These acids may all be used and employed in a variety of sectors. LA is one of the compounds that have a market due to several causes [44], including:

- A growing preference for LA over compounds made from petroleum.
- LA derivatives and other ecologically friendly solvents are in great demand in the global chemical sector. This is because petrochemical solvents pose a risk to both the environment and people.
- There maybe uses for ketals and LA esters in a variety of industries
- Regulatory bodies in nations like the United States and Canada support the creation of safer, more ecologically friendly substitutes, such as tough plasticizers made from LA.
- The market will be driven by growing demand for sustainable bio-plasticizers.

The cost of LA is now around US$27.58 and is estimated to reach US$93.65 million in 2029, growing at a CAGR of 14.17% between 2022 and 2029 [45]. From 2021 to 2026, the market for LA is projected to expand to 3436.9 tons, with a CAGR of 7.68% for market growth.

Another priceless chemical with enormous potential for utilization in major industries is FA. The primary driver of the growth in the FA market is its rising demand for preservatives [44]. FA is the best preservative for silage because of its superior preservation properties. In the manufacturing of pet food, silage is frequently employed. The need for animal food is rising, and FA is used more often as a preservative.

The rising demand for natural rubber is another factor driving the global market for FA [44]. FA is used to assure the quality of the finished rubber goods, including rubber sheets. Additionally, since FA helps the natural rubber industry's latex coagulation, employing it might reduce expenses. FA is one of the finest coagulants in manufacturing dry rubber because it can assure good uniformity in natural rubber products. The rubber industry chooses FA because it does not impair rubber's flexibility and boosts the color of rubber goods, in addition to benefits like improved performance and reduced pricing. This is another factor contributing to the market expansion for FA.

15.7 Case Studies

A study utilized a three-step process to produce LA from sugarcane bagasse [46]. Humic acids were also produced in this three-step process. The process consists mainly of pretreatment with sulfuric acid, delignification with sodium hydroxide, and acid-catalyzed conversion steps. Bagasse with a solid loading of 20.0% w/v was diluted in 1.0% (w/v) sulfuric acid and autoclaved for 80 minutes (at 121 °C). The resulting slurry was filtrated. After neutralization with NaOH (pH 6), delignification was performed for 90 minutes at 80 °C by adding 0.5% (w/v) NaOH. Acid-catalyzed conversions were made after the second neutralization until 190 °C. This study conducted six strategies to determine opportunities to produce LA and utilize the humins (HU) formed. Discussion regarding selectivity and economic feasibility was carried out for this biorefinery multistep strategy. A brief economic study is carried out by considering the process raw materials, other required process inputs, and the products obtained. Apart from LA and HU, which are included in product considerations, are FA and furfural. Earnings before interest, taxes, depreciation, and amortization (EBITDA) calculations were run to analyze the impact of a multistep strategy on LA production from sugarcane bagasse. The results showed that higher concentrations of LA and FA were obtained in scenarios that use H_2SO_4. LA and FAs (9.8 and 1.2 kg, respectively, per 100 kg dry bagasse) were obtained along with the inevitable production of HU (13.3 kg per 100 kg dry sugarcane bagasse). The biomass fractionation strategy in this biorefinery concept allows for separating other value-added components. Thus, this strategy allows for a smaller minimum selling price for LA, which is fundamental to many possible market applications [46].

HU is a challenging by-product as it has a chemical structure that is not yet suitable for many applications other than being burned as additional boiler fuel in biorefinery. This study [46] suggested that other potential uses of HU should also be explored to add more value to the LA production chain. Using EBITDA analysis, the best price scenario with a 30% margin is obtained: the LA price of US$98/t. This value is much better than the study's results using molasses as raw materials [33], which costs US$675/ton of product. A techno-economic analysis of LA production was studied from plantation waste biomass, namely empty fruit bunches or rice straw. Their research showed that 7–8 kg of biomass could produce 1 kg of LA [33]. Another economic evaluation of bioproducts (FA, LA, and 5-hydroxymethylfurfural) was using sugarcane bagasse as raw material with two pretreatment stages, namely hydrothermal and alkaline pretreatment. The Aspen simulation found that the process was run uneconomically, with the pretreatment and product purification stages being the most expensive stages, reaching 43.44% and 38.6% for equipment and investment costs [47]. All the studies using sugarcane bagasse [30, 43, 46], and molasse [48], as well as using *Sorghum bicolor* solid residue [49], showed the economic importance and the energy cost of the purification process, by distillation [46, 48, 49], solvent extraction [43, 50], or membrane [30]. Due to low cost and potential for recovery, biphasic solvent extraction will be an attractive approach for LA, FA, and other valuable chemical purification from those processes [50].

15.8 Conclusions

With the rising demand for energy, LA, and FA demand are predicted to continue to rise. LA and FA can be used as precursors for biofuels or as a source of hydrogen. Many cellulose-containing biomass can be used as potential raw materials, including rice straw and sugarcane bagasse. The reactions to forming LA and FA can be through acid and metal catalysts. Pretreatment is usually carried out using acid because of its relatively high reaction rate. The economics of LA and FA produced in a biorefinery system is feasible due to its low feedstock cost. However, the purification strategy is an important step in determining the economic value of the product.

References

1 Kumar, A., Shende, D.Z., and Wasewar, K.L. (2020). Production of levulinic acid: a promising building block material for pharmaceutical and food industry. *Mater. Today Proc.* 29 (3): 790–793.
2 Jeong, G. and Kim, S. (2021). Statistical optimization of levulinic acid and formic acid production from lipid-extracted residue of *Chlorella vulgaris*. *J. Environ. Chem. Eng.* 9 (2): 105142.
3 Harahap, A.F., Rahman, A.A., Sadrina, I.N., and Gozan, M. (2019). Production of formic acid from oil palm empty fruit bunch via dilute acid hydrolysis by response surface methodology. *IOP Conf. Ser.: Mater. Sci. Eng.* 673 (1): 012004.
4 Mordor Intelligence (2022). Levulinic acid market by end-user and geography – forecast and analysis 2022–2026. https://www.mordorintelligence.com/industry-reports/levulinic-acid-market
5 Mordor Intelligence (2022). Formic acid market by end-user and geography – forecast and analysis 2021–2025. https://www.mordorintelligence.com/industry-reports/formic-acid-market
6 Pinzi, S. and Dorado, M. (2011). Vegetable-based feedstocks for biofuels production. In: *Handbook of Biofuels Production, Processes and Technologies* (ed. R. Luque and J. Clark), 61–94. Woodhead Publishing Series in Energy.
7 Migo-Sumagang, M.V.P., van Hung, N., Detras, M.C.M. et al. (2020). Optimization of a downdraft furnace for rice straw-based heat generation. *Renewable Energy* 148: 953–963. https://doi.org/10.1016/j.renene.2019.11.001.

8 Sietske Boschma, D., Kees, I., and Kwant, W. (2013). Netherlands Programmes Sustainable Biomass. http://edepot.wur.nl/288866

9 Gummert, M., Hung, N.V., Chivenge, P., and Douthwaite, B. (2020). *Sustainable Rice Straw Management*. https://doi.org/10.1007/978-3-030-32373-8_1. Springer Nature.

10 NL Agency (2013). Rice straw and wheat straw – potential feedstocks for the biobased economy. NL Agency Report. 448025. Retrieved from 12 August 2022, from WUR. https://agris.fao.org/agris-search/search.do?recordID=NL2020010655

11 Mahmud, M. and Anannya, F. (2021). Sugarcane bagasse – a source of cellulosic fiber for diverse applications. *Heliyon* 7 (8): e07771.

12 Heniegal, A., Ramadan, M., Naguib, A., and Agwa, I. (2020). Study on properties of clay brick incorporating sludge of water treatment plant and agriculture waste. *Case Stud. Constr. Mater.* 13: e00397.

13 Samadi, S., Mohammadi, M., and Najafpour, G. (2016). Production of single cell protein from sugarcane bagasse by *Saccharomyces cerevisiae* in tray bioreactor. *Int. J. Eng. Trans. B* 29: 1029–1036.

14 Girisuta, B., Janssen, L.P.B., and Heeres, H.J. (2006). Green chemicals: a kinetic study on the conversion of glucose to levulinic acid. *Chem. Eng. Res. Des.* 84 (A5): 339–349.

15 Oyedeji, O., Gitman, P., Qu, J., and Webb, E. (2020). Understanding the impact of lignocellulosic biomass variability on size reduction process – a review. *ACS Sustainable Chem. Eng.* 8 (6): 2327–2343.

16 Samsuri, M., Prasetya, B., Hermiati, E., Idiyanti, T., Okano, K., Syafwina, Y.H. and Watanabe, T. (2005). Pretreatments for ethanol production from bagasse by simultaneous saccharification and fermentation. In Towards Ecology and Economy Harmonization of Tropical Forest Resources. Proceedings of the 6th International Wood Science Symposium, Bali (pp. 29-31).

17 Itoh, H., Wada, M., Honda, Y. et al. (2003). Bioorganosolve pretreatments for simultaneous saccharification and fermentation of beech wood by analysis and white rot fungi. *J. Biotechnol.* 103: 273–280.

18 Samsuri, M., Gozan, M., Mardias, R. et al. (2007). Utilisation of bagasse cellulose for ethanol production through simultaneous saccharification and fermentation by xylanase. *Makara J. Technol.* 11 (1): 4.

19 Sun, Y. and Cheng, J. (2002). Hydrolysis of lignocellulosic materials for ethanol production: review. *Bioresour. Technol.* 83: 1–11.

20 Ramos, J., Rojas, T., Navarro, F. et al. (2004). Enzymatic and fungal treatments on sugarcane bagasse for the production of mechanical pulp. *J. Agric. Food. Chem.* 52: 5057–5062.

21 Blanchette, R.A., Burnes, T.A., Eerdmans, M.M., and Akhtar, M. (1992). Evaluating isolates of *Phanerochaete chrysosporium* and *Ceriporiopsis subvermispora* for use in biological pulping process. *Holzforschung* 46: 105–115.

22 Moreno, A.D., Ibarra, D., Alvira, P. et al. (2015). A review of biological delignification and detoxification methods for lignocellulosic bioethanol production. *Crit. Rev. Biotechnol.* 35 (3): 342–354.

23 Singh, R., Shukla, A., Tiwari, S., and Srivastava, M. (2014). A review on delignification of lignocellulosic biomass for enhancement of ethanol production potential. *Renewable Sustainable Energy Rev.* 32: 713–728.

24 Panjaitan, J.R.H. and Gozan, M. (2017). Formic acid production from palm oil empty fruit bunches. *Int. J. Appl. Eng. Res.* 12 (14): 4382–4390.

25 Gozan, M., Panjaitan, J.R.H., Tristantini, D. et al. (2018). Evaluation of separate and simultaneous kinetic parameters for levulinic acid and furfural production from pretreated palm oil empty fruit bunches. *Int. J. Chem. Eng.* 2018: 1920180.

26 Hermansyah, C., Kasmiarti, G., and Panagan, A.T. (2021). Delignification of lignocellulosic biomass sugarcane bagasse by using ozone as initial step to produce bioethanol. *Polish J. Environ. Stud.* 30 (5): 4405–4411.

27 Niju, S. and Swathika, M. (2019). Delignification of sugarcane bagasse using pretreatment strategies for bioethanol production. *Biocatal. Agric. Biotechnol.* 20: 101263.

28 Ratti, R.P., Delforno, T.P., Sakamoto, I.K., and Varesche, M.B.A. (2015). Thermophilic hydrogen production from sugarcane bagasse pretreated by steam explosion and alkaline delignification. *Int. J. Hydrogen Energy* 40: 6296–6306.

29 Matei, J.C., Soares, M., Bonato, A.C.H. et al. (2020). Enzymatic delignification of sugar cane bagasse and rice husks and its effect in saccharification. *Renewable Energy* 157: 987–997.

30 Liang, X., Wang, J., Guo, Y. et al. (2021). High-efficiency recovery, regeneration and recycling of 1-ethyl-3-methylimidazolium hydrogen sulfate for levulinic acid production from sugarcane bagasse with membrane-based techniques. *Bioresour. Technol.* 330: 124984.

31 Kumar, S., Ahluwalia, V., Kundu, P. et al. (2018). Improved levulinic acid production from agri-residue biomass in the biphasic solvent system through synergistic catalytic effect of acid and products. *Bioresour. Technol.* 251: 143–150.

32 Kapanji, K.K., Haigh, K.F., and Gorgens, J.F. (2021). Techno-economics of lignocellulose biorefineries at South African sugar mills using the biofine process to co-produce levulinic acid, furfural, and electricity along with gamma valeractone. *Biomass Bioenergy* 146: 106008.

33 Isoni, V., Kumbang, D., Sharratt, P.N., and Khoo, H.H. (2018). Biomass to levulinic acid: a techno-economic analysis and sustainability of biorefinery processes in Southeast Asia. *J. Environ. Manage.* 214: 267–275.

34 Feng, J., Tong, L., Xu, Y. et al. (2020). Synchronous conversion of lignocellulosic polysaccharides to levulinic acid with synergic bifunctional catalysis in a biphasic cosolvent system. *Ind. Crops Prod.* 145: 112084.

35 Victor, A., Pulidindi, I.N., and Gedanken, A. (2014). Levulinic acid production from *Cicer arietinum*, cotton, *Pinus radiata* and sugarcane bagasse. *RSC Adv.* 84 (4): 44706–44711.

36 Lopes, E.S., Rivera, E.C., Gariboti, J.C.J. et al. (2020). Kinetic insights into the lignocellulosic biomass-based levulinic acid production by a mechanistic model. *Cellulose* 27: 5641–5663.

37 Pulidindi, I.N. and Kim, T.H. (2018). Conversion of levulinic acid from various herbaceous biomass species using hydrochloric acid and effects of particle size and delignification. *Energies* 11: 621.

38 Hayes, D. (2013). Report on optimal use of DIBANET feedstocks and technologies. https://www.celignis.com/celig_files/DIBANET_D5.3_FINAL_PUBLIC.pdf (accessed 15August 2022).

39 Rackemann, D.W. (2014). Production of levulinic acid and other chemicals from sugarcane fiber. *Queensland University of Technology (QUT)*. https://eprints.qut.edu.au/72741

40 Liu, L., Li, Z., Hou, W., and Shen, H. (2018). Direct conversion of lignocellulose to levulinic acid catalysed by ionic liquid. *Carbohydr. Polym.* 181: 778–784.

41 Miranda, J.C.C., Ponce, G.H.S.F., Neto, J.M., and Concha, V.O.C. (2019). Simulation and feasibility evaluation of a typical levulinic acid (LA) plant using biomass as substrate. *Chem. Eng. Trans.* 74: 901–906.

42 Lopes, E.S., Dominices, K., Lopes, M. et al. (2017). A green chemical production: obtaining levulinic acid from pretreated sugarcane bagasse. *Chem. Eng. Trans.* 57: 145–150.

43 Nhien, L.C., Long, N.V.D., and Lee, M. (2015). Design and optimization of the levulinic acid recovery process from lignocellulosic biomass. *Chem. Eng. Res. Des.* 107: 126–136.

44 Technavio (2021). Formic acid market by end-user and geography – forecast and analysis 2021–2025, report IRTNTR43378. https://www.technavio.com/report/formic-acid-market-industry-analysis

45 DataBridge Market Research (2022). Global levulinic acid market – industry trends and forecast to 2029, Report SKU-52420. https://www.databridgemarketresearch.com/reports/global-levulinic-acid-market#:~:text=Data%20Bridge%20Market%20Research%20analyses,period%20of%202022%20to%202029.

46 Lopes, E.S., Leal Silva, J.F., Rivera, E.C. et al. (2020). Challenges to levulinic acid and humins valuation in the sugarcane bagasse biorefinery concept. *Bioenergy Res.* 13: 757–774.

47 Junior, M.M.J., Rodrigues, F.A., Costa, M.M., and Guirardello, R. (2022). Economic evaluation for bioproducts production from carbohydrates obtained from hydrolysis of sugarcane bagasse. *Chem. Eng. Trans.* 92: 703–708.

48 Lopes, E.S., Silva, J.F.L., Nascimento, L.A.D. et al. (2021). Feasibility of the conversion of sugarcane molasses to levulinic acid: reaction optimization and techno-economic analysis. *Ind. Eng. Chem. Res.* 60 (43): 15646–15657.

49 Gozan, M., Ryan, B., and Krisnandi, Y. (2018). Techno-economic assessment of levulinic acid plant from *Sorghum bicolor* in Indonesia. *IOP Conf. Ser.: Mater. Sci. Eng., A* 345: 012012.

50 Cousin, E., Namhaed, K., Pérès, Y. et al. (2022). Towards efficient and greener processes for furfural production from biomass: a review of the recent trends. *Sci. Total Environ.* 847: 157599.

16

Cellulase as Biocatalyst Produced from Agricultural Wastes

Wichanee Bankeeree[1], Suraini Abd-Aziz[2], Sehanat Prasongsuk[1], Pongtharin Lotrakul[1], Syahriar NMM Ibrahim[1], and Hunsa Punnapayak[1]

[1] *Chulalongkorn University, Plant Biomass Utilization Research Unit, Department of Botany, Faculty of Science, Bangkok, 10330, Thailand*
[2] *Universiti Putra Malaysia, Department of Bioprocess Technology, Faculty of Biotechnology and Biomolecular Sciences, Serdang, 43400 UPM, Selangor, Malaysia*

16.1 Introduction

Cellulases belong to the glycosyl hydrolase (GH) enzyme, which catalyzes the degradation of β-1,4-glycosidic linkage presented in linear glucose polymer, especially cellulose in the plant cell wall. These enzymes play an essential role not only in ecological carbon recycling but also in a wide range of industrial applications, such as mixing in detergent to improve the softness of cotton fabric, fiber digesting to improve the quality of animal feed, de-inking of recycled paper, polishing of textiles, and clarifying of fruit juice. Because of their broad applicability, cellulases have been regarded as the third most demanded industrial enzymes in the global retail market [1]. In recent decades, there has been growing interest in cellulases since the gradual depletion of fossil resources has encouraged numerous studies on finding alternative resources and/or alternative fuels. Ethanol has been considered as the alternative and renewable biofuel that is firstly produced from sugarcane, corn, and cassava. However, the cost competition between food and fuel is a major challenge for the bioethanol business, which requires cheap monosaccharides for subsequent biological and/or chemical conversion.

Consequently, in the second-generation biofuels, cost-free lignocellulosic biomass (LCB), such as agricultural wastes and forest residues, was considered an economical feedstock for monosaccharide preparation. Although several techniques are available for LCB hydrolysis, environmental pollution, and undesirable by-products are also points of concern. Therefore, enzymatic hydrolysis using cellulases is a promising technique due to their high selectivity and environmentally friendly condition. With this scenario, the market demand for cellulases has constantly increased, with values at US$9.9 and US$10.6 billion in 2019 and 2020, respectively [1].

Chemical Substitutes from Agricultural and Industrial By-Products: Bioconversion, Bioprocessing, and Biorefining,
First Edition. Edited by Suraini Abd-Aziz, Misri Gozan, Mohamad Faizal Ibrahim, and Lai-Yee Phang.
© 2024 WILEY-VCH GmbH. Published 2024 by WILEY-VCH GmbH.

The global cellulase market is estimated to reach US$23 billion by 2025, expanding at a compound annual growth rate (CAGR) of 5.5% from 2018 to 2025 [2]. The high cost of cellulases due to their high market demand will restrict the expansion of the cellulosic bioethanol industry. Hence, new strategies for enhancing cellulase production have become essential. This chapter includes the recent research on cellulase-producing strains and the application of agricultural waste as a feedstock to enhance cellulase production with high activity and specificity.

16.2 Cellulases Diversity

As cellulases are one of the most demanded industrial enzymes, the fundamental knowledge of these enzymes is also essential for further research. Although cellulases catalyze only a single type of β-1,4-glycosidic linkage, these enzymes are diverse in catalytic functions, crystal structures, and amino acid sequences. The reason might be due to the complex structure of natural cellulose in the plant cell wall, which contains both amorphous and crystalline regions. In addition, the molecular interactions between cellulose and other matrices, such as hemicellulose and lignin, have also been identified as important factors [2].

16.2.1 Functional Types of Cellulases

Cellulases are categorized in different ways and typically divided into three groups (i) endoglucanases (EGs) (EC 3.2.1.4; 1,4-β-D-glucan-4-glucanohydrolase or carboxymethyl cellulase), (ii) exoglucanases (EC 3.2.1.91; cellobiohydrolase and EC 3.2.1.74; cellodextrinase) and (iii) β-glucosidases (BGL) (EC 3.2.1.21) based on their catalytic functions and specific substrates. EGs preferably attack the amorphous region of cellulose and randomly cleave the internal bond to generate oligosaccharides with varying lengths, while cellobiohydrolases (CBHs; also known as exoglucanases) processively act with crystalline portions to liberate cellobiose as the major product from cellulose end chains. Lastly, BGL are not active with cellulose chains, while they can hydrolyze short-chain oligosaccharide (3–4 units) and cellobiose to glucose at the nonreducing ends. EGs can act synergistically with CBHs to hydrolyze the crystalline cellulose in plant cell walls when CBHs can disrupt the crystalline regions to make them more accessible for EGs, and EGs can create new sites for CBHs attack. This synergistic reaction could result in 5–10 times more hydrolysis activity than a single enzyme usage. The combination of different CBHs that can attack different ends of the cellulose chain (reducing or nonreducing ends) also presented the synergistic action. The final products from CBH and/or EG, such as cellobioses and cellodextrins, can inhibit their free enzyme activities by binding to an allosteric site, called feedback inhibition. Therefore, the presence of BGL is highly required for the efficient hydrolysis of these products to glucose.

16.2.2 Cellulase Structures

The structures of cellulase also vary based on their catalytic types and organism sources, which generally comprise multimodule including (i) structural and (ii) functional domains. Among functional domains, catalytic and carbohydrate-binding modules (CBM) play an essential role in cellulose hydrolysis. CBM is responsible for the specific adsorption of the enzyme through van der Waals interactions between aromatic amino acids, such as tyrosine or tryptophan, onto a hydrophobic patch of cellulose surfaces that leads to disrupt the crystalline structure and liberate the individual cellulose chains [3]. CBM is usually located in the active site at the N- and/or C-terminus of a catalytic module; therefore, this tight association of substrate with the catalytic module provides an efficient way for enzymatic degradation [4]. CBM is classified into 45 families based on their ligand specificity, amino acid similarity, and structural characteristics [5]. All fungal cellulases are in CBM1 (with 30 amino acids), and the first discovered family was found from CBHI and CBHII of *Trichoderma reesei* [5]. The CBM2 and CBM3 families mostly contain the CBMs of aerobic and anaerobic bacterial cellulases, respectively.

In contrast, other CBM families are diverse and might present an affinity with other substrates together with cellulose. In some cases, multiple CBMs in the same or different families are found within a cellulase. The multiple CBMs show a significantly higher affinity with crystalline cellulose than a single module [6]. Catalytic modules can be classified into three types according to the topographical arrangement, including (i) cleft/groove, (ii) tunnel, and (iii) crater/pocket, also found to correlate with their functions as shown in Figure 16.1 [7]. The cleft or groove shape is usually found in EGs, where the active site is open and allows random binding with internal cellulose chains. In contrast, the closed active site as the tunnel is occurred by covering single or multiple polypeptide loops that uniquely exhibit in CBH or exoglucanases to access only at the end of the cellulose chain. The last topography of crater or pocket shape is encountered in BGL that recognize nonreducing end of cellobioses and cellodextrins [7].

According to the Carbohydrate-Active Enzyme (CAZy) database, cellulases were classified into 14 GH families [2]. The fungal EGs are mostly classified into GH5,

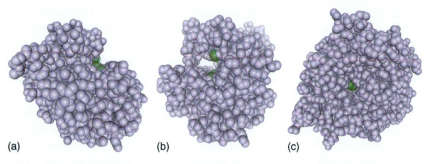

Figure 16.1 Crystal structure of cellulase with the different shapes of catalytic modules, including groove (a), tunnel (b), and pocket (c) shapes. Source: Wichanee Bankeeree (author) using the PyMOL program based on the amino acid sequences of *Trichoderma reesei* (PDB ID: 3QR3 and 4I5U, respectively) and *Thermoanaerobacterium saccharolyticum* (PDB ID: 7E5J).

GH6, GH7, GH9, GH12, GH45, and GH74 families, while exoglucanases are found in GH6, GH7, and GH48 families. In family GH6, all known exoglucanases attack cellulose at the nonreducing end, while exoglucanase members in family GH7 and GH48 attack at the reducing end of the cellulose chain [8]. BGL are generally classified in family GH1 and GH3. However, some data from previous studies have challenged this distinction. For example, EGs with active-site cleft produced from *Aspergillus ochraceus* (AS-HT-Celuz A; [9]) and *Bacillus subtilis* (EG5C; [10]) were also reported to hydrolyze the crystalline cellulose. This endo-/exo-function has been defined as the processive EG. This type of enzyme is usually found in the cellulolytic system, which may not contain exoglucanase. Most known processive EGs are in the largest cellulase family, GH5 and GH9, and enzymes are generally produced from plants, animals, bacteria, and a few brown-rot fungi [11]. Only Cel48F from *Clostridium cellulolyticum* has been reported to belong to the GH48 family [12].

16.2.3 Catalytic Mechanisms

The catalytic mechanisms of cellulase or other GH enzymes are generally divided into hydrolysis with (i) retention and (ii) inversion according to the change of anomeric carbon stereochemistry. Both mechanisms rely on carboxylic groups of two amino acids, i.e. aspartic acid and glutamic acid, located in the catalytic domain at the opposite site. The retaining cellulases, such as the GH7 family, operate the hydrolysis via a double displacement mechanism through glycosylation and deglycosylation. In glycosylation, once cellulose is in the catalytic panel, the anomeric carbon of cellulose is attacked by the nucleophilic carboxylate to form a glycosyl-enzyme intermediate with the aid of another carboxylate that serves as an acid residue for protonation of glycosidic oxygen. Subsequently, deglycosylation occurs by attacking the water molecule (or a new glycosyl hydroxyl group in transglycosylation) that leads to the deprotonation at base residue, breaking of glycosyl-enzyme intermediate, restoring of both carboxylates and generating of product with the same β-configuration to the substrate. Some retaining cellulases, such as GH1, GH3, and GH9 families, also exhibit transglycosylation activity, while it was no report about this activity from inverting cellulases [13]. The hydrolysis of inverting cellulases such as GH6, GH8, and GH45 operate with a single displacement. One carboxylate would be a base residue to remove a proton from a water molecule to increase its nucleophilicity. In contrast, another acid residue donates a proton to the anomeric carbon of the substrate and generates a product with a different stereochemistry.

On the basic knowledge of the structure and catalytic mechanism, some techniques, such as protein and metabolic engineering, have been proposed to enhance cellulase's desired properties and performance. However, the modified cellulases are typically more specialized to the tested substrate, such as soluble cellulose. At the same time, it remains a rare success for natural lignocelluloses because of the wide distribution of plant biomass. Detailed information, such as substrate characteristics and enzyme affinity, is still necessary for case-by-case studies.

16.3 Cellulase-producing Microorganisms

16.3.1 Aerobic Microorganisms

A wide range of aerobic and anaerobic microorganisms are capable of producing cellulase, of which bacteria and fungi are regarded as potential producers. Cellulases from different resources are mainly expressed in two different systems. The noncomplex system is the mixture of extracellular cellulases typically produced by aerobic bacteria and fungi. The well-known studies of noncomplexed cellulase systems have been conducted on soft-rot fungus, *T. reesei*, which contains at least five EGs (EGI, EGII, EGIII, EGIV, and EGV), two CBHs (CBHI and CBHII) and two BGL (BGLI and BGLII). With conventional mutagenesis, *T. reesei* mutants with high cellulase production have been reported, and mutant strain RUT-C30 is now being used commercially [14].

16.3.2 Anaerobic Microorganisms

In anaerobic fungi and bacteria, multienzyme complexes called cellulosomes have been observed that are comprised of (i) dockerin (Doc) modules enclosing cellulases or other accessory enzymes and (ii) cohesin (Coh) modules containing structural proteins. Multiple Coh modules are generally linked to form a scaffoldin containing Doc modules for binding to other scaffoldins and CBMs for attacking the recognized sites on the plant cell wall. Some cellulosomes can be released as cell-free complexes, while most of them anchor on the bacterial surface through an affinity interaction between type-II Doc modules located at the C-terminal of primary scaffoldin and one of the type-II Coh modules, such as scaffoldin dockerin-binding protein A (SdbA), open reading frame 2 (Orf2p), and outer layer protein B (OlpB). These type-II Cohs typically contain S-layer homology (SLH) module and function as an anchor protein for binding to the bacterial cell surface. The prototype of cellulosome was first described in *Clostridium thermocellum* in which the interaction between cellulosome integrating protein A (CipA; with nine Cohs) and OlpB contained seven Cohs resulted in a large cellulosome, accommodating up to 63 enzymes (Figure 16.2) [15]. The first report of cellulosome-like structures in anaerobic fungi was found in *Neocallimastix frontalis* [16].

Different types and GH families of cellulases have been discovered within a cellulosome, and this incorporation can lead to efficient use for cellulose degradation. However, a broader range of enzymes is required to assist in the degradation of raw plant material. Therefore, synthetic cellulosomes consisting of native systems with specific accessory enzymes have been created to produce efficient enzymatic cocktails that enable a faster degradation of given substrates. There are several techniques for reconstructing cellulosomes, and the interaction between CohI and DocI has garnered much attention. Li et al. [17] reported the fusion of BGL from *Caldicellulosiruptor* sp. F32 (CaBGLA) with type-I Coh domain for coordinating in *Cl. thermocellum* cellulosomes when almost all BGL from cellulolytic bacteria are the cytoplasmic enzymes. The recombinant cellulosome significantly increased the

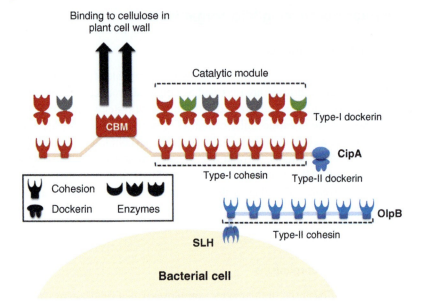

Figure 16.2 Schematic of the bacterial cellulosome. CBM is a carbohydrate-binding module, CipA is cellulosome integrating protein A, OlpB is outer layer protein B, and SLH is the S-layer homology module. Source: Wichanee Bankeeree (author).

total cellulolytic activity and produced glucose at a rate two times greater than the mixture of cellulosomes and free CaBGLA. The 3.5-fold boost in relative activity was also reported by the incorporation of bifunctional xylanolytic (Xyn10Y) and mannan-degrading (Man26A) enzymes into *Cl. thermocellum* CipA8 [18]. The synergistic reaction of xylanase, mannanase, and cellulase in the enzyme library of cellulosome could be due to the degradation of hemicellulose, which improves the penetration of enzymes into complex cellulose. To overcome lignocellulosic conversion, mono-copper polyphenol oxidase (laccase-like Tfu1114) originated from *Thermobifida fusca* was firstly fused to DocI from *Cl. cellulolyticum* at either the N or C terminus; however, the expression of these proteins was not detected. Since xylanase (XynT6) from *Geobacillus stearothermophilus* has been found to express very effectively in the cellulosome system, this protein was used as a solubility tag by fusing at the N-terminus of Doc-bearing laccase. Then, the resulting bifunctional chimera was successfully expressed and could be used simultaneously to hydrolyze raw lignocellulosic material [19].

Although the proximity of cellulase and related enzymes in cellulosomes can enable synergistic reactions and increase the cellulose hydrolysis yield, the industrial production of cellulosomes is still restricted due to the requirement of anaerobic conditions. Extensive studies have been carried out to express the recombinant cellulosomes in other host strains that can be cultivated aerobically, such as *Escherichia coli* and *Lactobacillus* sp. For instance, the recombinant cellulosomes that contained cellulolytic enzymes (Cel5D and Cel9/48A) from a hyperthermophilic bacterium, *Caldicellulosiruptor bescii*, were constructed along with thermophilic Coh–Doc modular from *Cl. thermocellum* (T-t), *Cl. clariflavum* (V-v), and *Archaeoglobus fulgidus* (G-g) and then cloned into *E. coli*. The cellulase activities at 75 °C of the recombinant cellulosomes were 15- to 25-fold higher than that of the native *Cl. thermocellum* cellulosome after incubation for 24 and 72 hours, respectively [20]. The key finding of this study is the stability interactions between specific Coh–Doc pairs from thermophilic bacteria and archaea that could function at high temperatures for extended periods. Lactic acid bacteria (LAB) are one of the appropriate host cells for expressing recombinant scaffoldins from a mesophilic cellulolytic bacterium, such as *Clostridium cellulovorans* and *Clostridium papyrosolvens*, because of their similarity in cell membrane component (Gram-positive) and GC content, which implied the similar codon usage. For example, the recombinant scaffolding CbpA from *Cl. cellulovorans*, which consist of three EG, one CBD (X2), and three SLH domains, could successfully introduce in *Lactococcus lactis* and serve as surface-displayed cellulosome [21]. There have also been reports of yeast hosts expressing modified cellulosomes, while the minicellulosomes containing a few enzymes with 12 so far have been identified [22]. Recently, the large cellulosome complex with 63 enzymes on the cell surface of *Kluyveromyces marxianus* has been reported. Two genes encoding for CipA and OlpB were integrated into the yeast genome. The recombinant yeast effectively brokedown the cellulosic substrates and could release 3.09 and 8.61 g/l of ethanol from avicel and phosphoric acid-swollen cellulose, respectively [23].

16.4 Cellulase Properties

Some cellulase characteristics, such as thermostable cellulases, are assumed to have a higher economic value because cellulose swells efficiently at high temperatures. Furthermore, thermostable cellulases also have the advantages of shorter hydrolysis times, lower contamination risk, easier recovery of volatile products, and lower cooling costs after thermal pretreatment. Consequently, isolation and screening of new microbes with desirable traits that can be used for industrial purposes are essential. Thermostable CMCases have been recently reported from *Aspergillus aculeatus* PN14, in which more than 65% of its activity was maintained at 90 °C for 90 minutes. The optimal temperature of these enzymes ($\geq$80%) was in the ranges of 60–90 °C, at pH 3.0–11.0 [24], which might be suitable for application in paper manufacturing and lignocellulosic material conversion. On the contrary, cold-active cellulases also have the potential for biotechnological processes, such as waste bioremediation and pretreatment in cold climates, healthcare and pharmaceutical approaches, food and beverage industries, and textile and molecular biology. Examples of recent isolates of cellulase-producing microbes with cold-active properties are *Nocardiopsis dassonvillei* [25] and *Exiguobacterium sibiricum* K1 [26]. The CMCases produced from these isolates were active ($\geq$60%) in temperatures of 4–50 °C, at pH 4.0–9.0. These new isolates might act as important sources of cellulase genes with potential properties for further research. In recent decades, around 55,200 cellulase genes within 13 GH families from fungi, bacteria, and archaea have been reported. The distinct variation across different lineages was observed according to the large-scale genome analysis [27].

With the advent of genetic manipulation, the individual or combination of efficient cellulase-encoding genes could be mutated to improve their properties and/or the level of expression before being transformed into other host cells, such as bacteria (*E. coli* and *Bacillus* spp.), yeast (*Pichia pastoris, Saccharomyces cerevisiae, Hansenula polymorpha, Kluyveromyces lactis*, and *Yarrowia lipolytica*), and filamentous fungi (*Aspergillus* spp., *T. reesei, Rhizopus oryzae*, and *Penicillium chrysogenum*) with the different techniques [28]. Some of the successfully engineered strains have been patented, such as recombinant BGL from *Talaromyces piceae* (TpBGL3A) [29] and *A. aculeatus* (AaBGL1) [30] with production yields of more than 2000 U/ml.

16.5 Strategies to Improve Cellulase Production

Once the effective cellulase-producing microorganisms have been obtained, the next step for reducing cellulase cost is to optimize the production processes. Commercial cellulases are typically produced using pure cellulose or its derivatives, such as carboxymethyl cellulose (CMC), as a carbon source, while they are costly substrates. Therefore, alternative substrates such as cellulose-rich biomass are required for replacement.

16.5.1 Utilization of Agricultural Waste

Agriculture is an important sector of the global economy that impacts the growth of several countries. The increment could evidence in world production of wheat (783.92 million tonnes; MT), corn (1172.58 MT), barley (147.72 MT), oat (24.21 MT), sorghum (60.32 MT), rice (507.99 MT), copra (5.86 MT), palm (79.19 MT), and soybean (389.77 MT) in each year [31]. Expanding the agricultural or agro-industrial sector may also result in environmental pollution due to disorderly waste management practices, such as dumping, burning, or burying. Instead of being eliminated, these wastes are rich sources of cellulose (20–40% w/w), along with hemicellulose (20–30%) and lignin (1–15%), which can be used as renewable sources to produce various bio-products, including cellulase [32]. With this approach, the carbon source of the production medium is replaced by raw agricultural wastes, and several literatures successfully produced cellulase, as shown in Table 16.1.

The gram-negative bacterium *Klebsiella* sp. PRW-1 has been reported to produce EG, CBH, and BGL with the efficient activities of 34.62 ± 0.01, 2.43 ± 0.02, and 3.00 ± 0.17 U/ml, respectively, by using sugarcane trash as a sole carbon source. However, the yields of these cellulases from production medium containing sorghum husks, corn straw, or sugarcane bagasse were lower than that of CMC [41]. Raw material's variable structure and insolubility might explain this phenomenon because the water-soluble CMC can result in a shorter period of cellulase production when it is not necessary to breakdown the recalcitrant lignin. Therefore,

Table 16.1 Cellulase production by different bacteria and fungi using various agricultural wastes as a carbon source.

Agricultural waste	Microorganism	Cellulase type[a]	Activity	Harvesting time (days)	Optimum condition	References
Sugarcane trash	Enterobacter sp.	FPase	0.2 U/ml	8	pH 5.0, 50 °C	[33]
		CMCase	12.2 U/ml			
		BGL	37.1 U/ml			
Rice bran	Bacillus carboniphilus	CMCase	4040.5 U/ml	2	pH 9.0, 50–60 °C	[34]
Grass straw	Bacillus tequilensis	CMCase	242.5 U/ml	14	pH 5.0, 40 °C	[35]
Banana peduncle	Acinetobacter indicus	CMCase	1.2 U/ml	2	pH 7.0, 37 °C	[36]
Copra meal	Aspergillus tubingensis	CMCase	750.0 U/g[b]	5	pH 5.0, 65 °C	[37]
Rice straw	Penicillium oxalicum	FPase	1.4 U/ml	7	pH 5.0, 60 °C	[38]
		CMCase	2.0 U/ml			
		BGL	2.7 U/ml			
Orange peel	Emericella variecolor	FPase	31 U/g	4	pH 5.0, 50 °C	[39]
		CMCase	145.0 U/g			
		BGL	157.0 U/g			
Sugarcane bagasse	Aureobasidium pullulans	FPase	3.1 U/ml	2.5	pH 5.5, 50 °C	[40]
		CMCase	8.7 U/ml			

a) FPase, CMCase, Avicelase, and BGL referred to total cellulase activity, endoglucanase activity, exoglucanase activity, and β-glucosidase activity.
b) U/g dry weight of the substrate.

some physical, mechanical, and/or chemical pretreatments have been proposed to partially remove lignin and increase cellulose exposure. Similar results were also found in *Enterobacter* sp. SUK-Bio in which EG, CBH, and BGL activities were varied within different carbon sources. Although sugarcane trash and wheat straw induced lower CBH activities (0.85 ± 0.01 and 0.82 ± 0.08 U/ml, respectively) than that of CMC (1.72 ± 0.02 U/ml), they could assist in producing the BGL with the activities of 15.90 ± 0.65 and 21.20 ± 0.71 U/ml, respectively, which it could not detect in the medium containing CMC. In addition, the activities of xylanase (33.26 ± 0.35 U/ml) and glucoamylase (26.26 ± 0.22 U/ml) were observed in the presence of sorghum husk as the sole carbon source [33]. Rice bran also could be used to produce thermo-alkalophilic CMCases by *Bacillus carboniphilus* CAS3 and *Bacillus licheniformis* AU01. Both enzymes presented the maximum activities (188 U/ml) at 50 °C, pH 9.0, with their stability of more than 60% of their initial activity at 80 °C, pH range of 7.0–11.0 for 24 hours [34]. From the gut fluid of a giant African land snail, *Bacillus tequilensis* G9 was isolated and has the potency to produce cellulases by using grass straw as a sole carbon source after incubation for 14 days. The maximum EG activity of 242.5 ± 23.6 U/ml was observed at 40 °C, pH 5.0 [35]. Only after two days of incubation at 37 °C, banana peduncle has been reported as desired feedstock for CMCase production (1.2 U/ml) from newly isolated *Acinetobacter indicus* KTCV2 [36].

Fungi, including soft-rot, brown-rot, and white-rot, provide the ability to produce cellulase with excellent yields as the natural decomposer, so this microorganism is commercially used for cellulase production due to the secretion of multiple enzymes. Among all filamentous fungi, three genera of *Penicillium*, *Trichoderma*, and *Aspergillus* are the most well-known models for cellulase production from laboratory to industrial scales [1]. Oil palm trunk is usually an unutilized waste that has been reported to induce cellulase production from *Aspergillus fumigatus* SK1 through solid-state fermentation (SsF) with the CMCase, FPase, and BGL activities of 76, 199, and 48 U/g dry-weight substrates, respectively. Under the same condition, *A. fumigatus* SK1 also produced xylanase at a high concentration (418.70 U/g) compared to the strong xylanase producer (2.7-fold), *Aspergillus niger* MS80, which was cultivated in pure birchwood xylan as substrate [42]. It indicated that the untreated feedstock might facilitate the production of other auxiliary enzymes for lignocellulosic degradation. High titers of mannanase (1023 U/g) along with thermostable CMCase (750 U/g), BGL (71 U/gds), xylanase (167 U/g) and α-galactosidase (54 U/g) were generated from *A. tubingensis* NKBP-55 after cultivation with copra meal, a mannan-rich waste, for five days. This crude enzyme could generate glucose, xylose, mannose, and mannobiose from the enzymatic hydrolysis of sugarcane bagasse and rice straw, indicating the enzyme's efficiency. Therefore, this cocktail enzyme might be viable for the saccharification of LCB due to its cooperativity [37]. Differences in cellulase profiles produced by different substrates were evidenced within the same fungal strain. Corn stover was the most effective carbon source to produce CMCase from *P. oxalicum* GZ-2, while the highest level of FPase was found in a medium containing rice straw [38]. Thus, selecting a suitable substrate was critical for the successful production of cellulases.

16.5.2 Production Processes

Several attempts to improve the production titers of cellulase have been reported through various approaches, including the use of better bioprocess technologies. Cellulase and other related enzymes like mannanase, xylanases, and pectinases are often produced through two types of fermentation, including (i) submerged fermentation (SmF) and (ii) solid-state fermentation (SsF), which have numerous benefits and some drawbacks to each approach. SmF is a preferred procedure on an industrial scale that is also used by Novozymes and Dyadic, two global leaders in enzyme production, to produce commercial cellulase from *T. longibrachiatum* and *Myceliophthora thermophila* [39]. The reason is the ease of variable regulation, end-product recovery, and reproducibility. In addition, several types of bioreactors, such as stirred tanks, bubbles, and airlift reactors, are available for large-scale production, and the economic scale has been reported to be 20–200 m^3 [40]. Several studies have been conducted to assess the techno-economic of cellulase production in different types of SmF via process simulation. With these scenarios, the fed-batch approach is advised over the batch process due to the higher economic benefits while the unit production costs of both processes are comparable. It was evidenced by cellulase production from *T. reesei* using batch and fed-batch processes (170 metric tonnes/batch). The shorter payback time (1.88 years) and the higher internal rate of return (IRR) (36.95%), positive net present value (NPV) (US$140,328,000), and certainty of achieving the base-case value (69.91% of US$20.71/kg) were observed from a fed-batch approach compared to a batch process (2.12 years, 33.36%, US$132,043,000, and 47.80% of US$21.83/kg, respectively) [43]. The main drawback of a batch process is catabolite repression due to glucose accumulation from cellulose hydrolysis at the late stages, which can repress the genes related to cellulase production. In contrast, a fed-batch process involves continuously or periodically supplying nutrients in a bioreactor to maintain a specific nutrient concentration, increasing the specific growth rate and overproduction of some required metabolites. However, optimizing the various factors in a fed-batch system for increased product concentration remains a difficult approach. One of the critical operational variables is the availability of oxygen, which has a significant impact on the growth rate and the ability to produce enzymes of aerobic producers. When high-speed agitation has been used to promote aeration and mass transfer within the reactor, it can also be harmful to filamentous fungal hyphae.

In SsF, microorganisms are cultivated on solid substrates, especially agricultural waste, which maybe more favorable for filamentous fungi because it resembles a natural habitat for their growth. The fungal morphology during the SsF process was discovered to be distinct from that of an SmF process, in which a long mycelial mat is formed to penetrate inside substrate materials and the interspace between solid particles to absorb available nutrients and oxygen. The total water for microbial growth in SsF is typically derived from the moisture content of substrate material in the absence or near absence of free water. With this fact, downstream processing from SsF compared to SmF is easier and less expensive since enzymes are secreted at high concentrations due to the lack of free liquid, as illustrated in Figure 16.3.

Because of the lower electricity consumption in the downstream process, particularly in the lyophilization step, the average of all environmental impacts of complete cellulase production by *T. reesei* via SsF using coffee husk as a substrate has been reported to be 13% lower than that of the SmF process [44]. The economic assessments of cellulase production by *Cl. thermocellum* using paper pulp as a substrate also indicated that the unit cost of production using SsF (US$15.67/kg cellulase) is 2.5-times lower than that of using SmF (US$40.36/kg cellulase) with a certainty of 99.6% (9959 out of 10,000 cases) [45]. This finding suggests that SsF is a valid economic process for producing cellulase enzymes while minimizing environmental impact. Therefore, SsF has gained traction as an economic process in the last two decades. However, the fermentable process should be chosen based on the final product and its subsequent applications. SmF might be better for products that require high purity in the pharmaceutical sector because the purification process

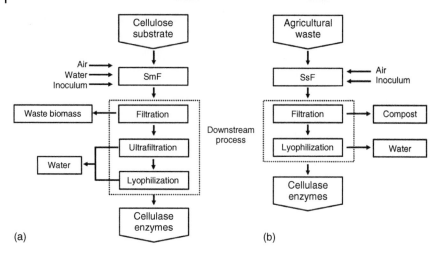

Figure 16.3 Flowchart of cellulase production using solid-submerged fermentation (SmF; a) and solid-state fermentation (SsF; b). Source: Wichanee Bankeeree (author).

can be more controlled and defined. On the contrary, SsF is more appropriate for high-volume, low-value products like cellulases used in manufacturing biofuels.

16.5.3 Consolidated Bioprocessing

In general, bioconversion of lignocellulosic feedstock requires a series of steps, which include (i) pretreatment to remove and separate the recalcitrant structure of biomass, (ii) production of lignocellulolytic enzymes, (iii) hydrolysis of pretreated biomass to generate sugars, and (iv) conversion of obtained sugars into valuable products via potential fermenting microorganisms. The conventional biorefinery process is the separate hydrolysis and fermentation (SHF). In contrast, simultaneous saccharification and fermentation (SSF) have been developed to reduce contamination risk and prevent enzyme feedback inhibition. However, the inclusion of external enzymes is necessary for both processes. Consolidated bioprocessing (CBP) has been proposed as an alternative process that integrates all steps into a single reactor to further reduce investment expenses and simplify the process. With this process, co-treatment (CT) and consolidated bio-saccharification (CBS) are the CBP-derived strategies when CBS separates fermentation from the CBP process due to specific fermentation conditions. The success of CBS relies on the ability of microorganisms to utilize substrates, produce enzymes, and convert them into high-efficiency value-added products, which can be accomplished using either a native single microbe, a genetically modified microbe, or a consortium. For example, the ethanol production from olive mill waste using *Fusarium oxysporum* isolated from the Meknes region in Morocco. This isolate had high and diverse cellulase activities and the ability to produce and tolerate ethanol. The ethanol yield (2.47 g/l) was equal to the yield with monosaccharides [46]. The co-culture of *B. tequilensis* G9 and *S. cerevisiae* has also been reported to produce bioethanol with a high yield from grass straw [35]. Itaconic acid is a versatile building block for renewable polyesters with enhanced functionality that could produce by sequential cocultivation of cellulase producer *T. reesei* RUT-C30 and itaconic acid-producing yeast, *Ustilago maydis*, in the medium containing recalcitrant α-cellulose (270 g/l) as a sole carbon source. This remarkable yield (33.8 g/l) was comparable with that of SHF, while the cost of the enzyme had just a minor impact [47]. Although several studies have proven the concept of CBP, the commercial process has been limited due to low conversion efficiency. Therefore, the natural exploration or genetic modification of microorganisms is still ongoing to obtain high product yields of consolidated productions between cellulase enzymes and expected products.

16.6 Techno-economic Analysis to Produce Biofuels

The techno-economic analysis (TEA) is a procedure that employs software modeling to provide critical information on economic viability based on technical and financial input parameters. The methodology in brief initiated by developing a flow diagram of all processes and then determining the mass and energy balance to estimate total capital investment (TCI) and total production cost (TPC). Then, based on the predefined scenarios, other economic variables such as the NPV, return on investment (ROI), payback period (PBP), and IRR are calculated from the cash flow analysis and used as indicators to determine profitability. Therefore, this information assists in the decision-making process for investments and research. Numerous studies over the past decades have outlined potential techniques for converting LCB to biofuels and value-added chemicals using TEA and their profitability was also affected by cellulase production.

A case study to depict the TEA of second-generation ethanol from sugarcane bagasse using cellulase enzymes produced on-site versus commercial has been proposed by Carpio et al. [48]. The on-site enzyme production plant in this study was composed of two separate units from *T. reesei* and *Aspergillus awamori* for supplying high levels of BGL. However, low cellulase activity (1 FPU/ml) was achieved from the mixture of *T. reesei* CBH and *A. awamori* BGL at a ratio of 1 : 4. As a result, more than 80% of the operating cost (US$270 MM) was from the on-site enzyme production process. The impact of cellulase activity on the cost of enzyme production was further analyzed, and the hypothetical scenario revealed that a single unit of enzyme production with cellulase activity of around 10 FPU/ml would result in a 20-time reduction in operating cost. To enhance the titer of cellulase, a promising fungus for commercial cellulase production, *M. thermophila*, was considered as a model for cellulase production by SsF using sugarcane bagasse as a substrate [49]. The enzyme production plant was integrated with first- or second-generation ethanol processes and the economic prospects were investigated. The highest cellulase activity obtained from SsF with substrate availability at 12,000 tons was around 12 FPU/gds with the values of IRR, ROI, and PBP at 11.9%, 8.8%, and 5.4 years, respectively. Although PBP in this study was acceptable by less than 10 years, the IRR and ROI were under the reference values ($\leq$13% and $\leq$30%, respectively).

The appropriate cellulase activity in this study was 18 FPU/gds that correspond to IRR, ROI, and PBP at 20%, 30%, and three years, respectively. Therefore, alternative approaches, such as enriching the substrate and genetically modifying the microorganism, could be considered to increase the enzyme activity to obtain more acceptable economic values. The economics of converting sugarcane bagasse to ethanol using a high-solid loading SsF (15% w/v) under different scenarios, including on-site/commercial enzyme supply and enzyme loading, was assessed through the minimum ethanol selling price (MESP) [50]. The synergistic action between on-site (5–30 U/WIS) and purchased enzymes (10 U/WIS) has been reported to decrease the MESP by up to 25% compared to the use of purchased enzymes (20 Baht/l) as caused by xylanase from *Scheffersomyces stipites*. The integration of the SsF process with on-site/purchased enzymes and the *S. stipites/S. cerevisiae* consortia resulted in the lowest MESP of 15.7 Baht/l (US$1.66/gal), which was 6% less than the market selling price (US$1.76/gal). It could be concluded that the processing platform in this study is crucial for the development of a low-cost biorefinery industry.

16.7 Conclusions

Cellulases are highly in-demand industrial enzymes that can be used in various industrial applications. Therefore, several techniques have been proposed to reduce the cost of these enzymes, to lower the cost of final products in biorefinery. Numerous effective cellulase-producing strains with the desirable characteristics of cellulase have been explored and used as models for genetic manipulation. Several designed cellulases and synergistic enzymes have been successfully generated while it is still ongoing. Besides this, another strategy is the utilization of agricultural waste as a sole carbon source to produce these enzymes through SsF and SmF. This approach also has the potential to solve the problems of waste management and environmental pollution. However, several efforts still need to be considered to further reduce cellulase costs.

References

1 Singh, A., Bajar, S., Devi, A., and Pant, D. (2021). An overview on the recent developments in fungal cellulase production and their industrial applications. *Bioresour. Technol. Rep.* 14: 100652. https://doi.org/10.1016/j.biteb.2021.100652.

2 Jayasekara, S. and Ratnayake, R. (2019). Microbial cellulases: an overview and applications. In: *Cellulose* (ed. A.R. Pascual and M.E.E. Martin), 83–104. London: IntechOpen.

3 Pérez, S. and Samain, D. (2010). Structure and engineering of celluloses. In: *Advances in Carbohydrate Chemistry and Biochemistry*, vol. 64 (ed. D. Horton), 25–116. Cambridge: Academic Press.

4 Pérez, S. and Tvaroška, I. (2014). Carbohydrate–protein interactions: molecular modeling insights. In: *Advances in Carbohydrate Chemistry and Biochemistry*, vol. 71 (ed. D. Horton), 9–136. Cambridge: Academic Press.

5 Van Tilbeurgh, H., Tomme, P., Claeyssens, M. et al. (1986). Limited proteolysis of the cellobiohydrolase I from *Trichoderma reesei*. *FEBS Lett.* 204 (2): 223–227.

6 Sidar, A., Albuquerque, E.D., Voshol, G.P. et al. (2020). Carbohydrate binding modules: diversity of domain architecture in amylases and cellulases from filamentous microorganisms. *Front. Bioeng. Biotechnol.* 8 (871): 1–15.

7 Xu, Q., Luo, Y., Ding, S.Y. et al. (2011). Multifunctional enzyme systems for plant cell wall degradation. In: *Comprehensive Biotechnology*, 2e (ed. M. Moo-Young), 15–25. Burlington: Academic Press.

8 Poidevin, L., Feliu, J., Doan, A. et al. (2013). Insights into exo- and endoglucanase activities of family 6 glycoside hydrolases from *Podospora anserina*. *Appl. Environ. Microbiol.* 79 (14): 4220–4229.

9 Asha, P., Divya, J., and Bright Singh, I.S. (2016). Purification and characterisation of processive-type endoglucanase and β-glucosidase from *Aspergillus ochraceus* MTCC 1810 through saccharification of delignified coir pith to glucose. *Bioresour. Technol.* 213: 245–248.

10 Wu, B., Zheng, S., Pedroso, M.M. et al. (2018). Processivity and enzymatic mechanism of a multifunctional family 5 endoglucanase from *Bacillus subtilis* BS-5 with potential applications in the saccharification of cellulosic substrates. *Biotechnol. Biofuels* 11 (1): 20.

11 Wu, S. and Wu, S. (2020). Processivity and the mechanisms of processive endoglucanases. *Appl. Biochem. Biotechnol.* 190 (2): 448–463.

12 Parsiegla, G., Reverbel, C., Tardif, C. et al. (2008). Structures of mutants of cellulase Cel48F of *Clostridium cellulolyticum* in complex with long hemithio-cellooligosaccharides give rise to a new view of the substrate pathway during processive action. *J. Mol. Biol.* 375 (2): 499–510.

13 Wang, X., Wu, Y., and Zhou, Y. (2017). Transglycosylation, a new role for multifunctional cellulase in overcoming product inhibition during the cellulose hydrolysis. *Bioengineered* 8 (2): 129–132.

14 Li, C., Lin, F., Li, Y. et al. (2016). A β-glucosidase hyper-production *Trichoderma reesei* mutant reveals a potential role of cel3D in cellulase production. *Microb. Cell Fact.* 15 (1): 151.

15 Leibovitz, E., Ohayon, H., Gounon, P., and Béguin, P. (1997). Characterization and subcellular localization of the *Clostridium thermocellum* scaffoldin dockerin binding protein SdbA. *J. Bacteriol.* 179 (8): 2519–2523.

16 Wilson, C.A. and Wood, T.M. (1992). The anaerobic fungus *Neocallimastix frontalis*: isolation and properties of a cellulosome-type enzyme fraction with the capacity to solubilize hydrogen-bond-ordered cellulose. *Appl. Microbiol. Biotechnol.* 37 (1): 125–129.

17 Li, X., Xiao, Y., Feng, Y. et al. (2018). The spatial proximity effect of beta-glucosidase and cellulosomes on cellulose degradation. *Enzyme Microb. Technol.* 115: 52–61.

18 Leis, B., Held, C., Andreeßen, B. et al. (2018). Optimizing the composition of a synthetic cellulosome complex for the hydrolysis of softwood pulp: identification of the enzymatic core functions and biochemical complex characterization. *Biotechnol. Biofuels* 11 (1): 220.

19 Davidi, L., Moraïs, S., Artzi, L. et al. (2016). Toward combined delignification and saccharification of wheat straw by a laccase-containing designer cellulosome. *Proc. Natl. Acad. Sci. U.S.A.* 113 (39): 10854–10859.

20 Kahn, A., Moraïs, S., Galanopoulou, A.P. et al. (2019). Creation of a functional hyperthermostable designer cellulosome. *Biotechnol. Biofuels* 12 (1): 44.

21 Tarraran, L., Gandini, C., Luganini, A., and Mazzoli, R. (2021). Cell-surface binding domains from *Clostridium cellulovorans* can be used for surface display of cellulosomal scaffoldins in *Lactococcus lactis*. *Biotechnol. J.* 16 (8): 2100064.

22 Fan, L.-H., Zhang, Z.-J., Mei, S. et al. (2016). Engineering yeast with bifunctional minicellulosome and cellodextrin pathway for co-utilization of cellulose-mixed sugars. *Biotechnol. Biofuels* 9 (1): 137.

23 Anandharaj, M., Lin, Y.-J., Rani, R.P. et al. (2020). Constructing a yeast to express the largest cellulosome complex on the cell surface. *Proc. Natl. Acad. Sci. U.S.A.* 117 (5): 2385–2394.

24 Nargotra, P., Sharma, V., Sharma, S. et al. (2022). Purification of an ionic liquid stable cellulase from *Aspergillus aculeatus* PN14 with potential for biomass refining. *Environ. Sustainability* 5 (3): 313–323.

25 Sivasankar, P., Poongodi, S., Sivakumar, K. et al. (2022). Exogenous production of cold-active cellulase from polar *Nocardiopsis* sp. with increased cellulose hydrolysis efficiency. *Arch. Microbiol.* 204 (4): 218.

26 Kumari, S., Kumar, A., and Kumar, R. (2022). A cold-active cellulase produced from *Exiguobacterium sibiricum* K1 for the valorization of agro-residual resources. *Biomass Convers. Biorefin.* https://doi.org/10.1007/s13399-022-03031-w.

27 Liu, L., Huang, W.-C., Liu, Y., and Li, M. (2021). Diversity of cellulolytic microorganisms and microbial cellulases. *Int. Biodeterior. Biodegrad.* 163: 105277.

28 Juturu, V. and Wu, J.C. (2014). Microbial cellulases: engineering, production and applications. *Renewable Sustainable Energy Rev.* 33: 188–203.

29 Wu, J., Xia, W., Xu, X., and Huang, Y. (2021). Preparation and use of thermophoric β-glucosidase. WO2021217902A1, filed 17 July 2020 and issued 4 November 2021.

30 Sumitani, J., Yutaro, B., Shuji, T., and Kawaguchi, T. (2022). Mutant β-glucosidase. US Patent 2021/0024969A1, filed 21 February 2019 and issued 28 January 2021.

31 USDA Foreign Agriculture Service. (2022). World Agricultural Production. https://www.fas.usda.gov/data/world-agricultural-production

32 Benocci, T., Aguilar-Pontes, M.V., Zhou, M. et al. (2017). Regulators of plant biomass degradation in ascomycetous fungi. *Biotechnol. Biofuels* 10 (1): 1–25.

33 Waghmare, P.R., Patil, S.M., Jadhav, S.L. et al. (2018). Utilization of agricultural waste biomass by cellulolytic isolate *Enterobacter* sp. SUK-Bio. *Agric. Nat. Resour.* 52 (5): 399–406.

34 Annamalai, N., Rajeswari, M.V., and Balasubramanian, T. (2014). Enzymatic saccharification of pretreated rice straw by cellulase produced from *Bacillus carboniphilus* CAS 3 utilizing lignocellulosic wastes through statistical optimization. *Biomass Bioenergy* 68: 151–160.

35 Dar, M.A., Pawa, K.D., Rajput, B.P. et al. (2019). Purification of a cellulase from cellulolytic gut bacterium, *Bacillus tequilensis* G9 and its evaluation for valorization of agro-wastes into added value byproducts. *Biocatal. Agric. Biotechnol.* 20: 101219.

36 Karlapudi, A.P., Venkateswarulu, T.C., Srirama, K. et al. (2019). Purification and lignocellulolytic potential of cellulase from newly isolated *Acinetobacter indicus* KTCV2 strain. *Iran. J. Sci. Technol. Trans. A: Sci.* 43 (3): 755–761.

37 Prajapati, B.P., Kumar Suryawanshi, R., Agrawal, S. et al. (2018). Characterization of cellulase from *Aspergillus tubingensis* NKBP-55 for generation of fermentable sugars from agricultural residues. *Bioresour. Technol.* 250: 733–740.

38 Liao, H., Fan, X., Mei, X. et al. (2015). Production and characterization of cellulolytic enzyme from *Penicillium oxalicum* GZ-2 and its application in lignocellulose saccharification. *Biomass Bioenergy* 74: 122–134.

39 Srivastava, N., Srivastava, M., Manikanta, A. et al. (2017). Production and optimization of physicochemical parameters of cellulase using untreated orange waste by newly isolated *Emericella variecolor* NS3. *Appl. Biochem. Biotechnol.* 183 (2): 601–612.

40 Vieira, M.M., Kadoguchi, E., Segato, F. et al. (2021). Production of cellulases by *Aureobasidium pullulans* LB83: optimization, characterization, and hydrolytic potential for the production of cellulosic sugars. *Prep. Biochem. Biotechnol.* 51 (2): 153–163.

41 Waghmare, P.R., Kshirsagar, S.D., Saratale, R.G. et al. (2014). Production and characterization of cellulolytic enzymes by isolated *Klebsiella* sp. PRW-1 using agricultural waste biomass. *Emir. J. Food Agric.* 44–59.

42 Ang, S.K., Shaza, E.M., Adibah, Y. et al. (2013). Production of cellulases and xylanase by *Aspergillus fumigatus* SK1 using untreated oil palm trunk through solid state fermentation. *Process Biochem.* 48 (9): 1293–1302.

43 Taiwo, A.E., Tom-James, A., Falowo, O.A. et al. (2022). Techno-economic analysis of cellulase production by *Trichoderma reesei* in submerged fermentation processes using a process simulator. *S. Afr. J. Chem. Eng.* 42: 98–105.

44 Zhuang, J.A., Marchant, M.E., Nokes, S.J., and Strobel, H. (2007). Economic analysis of cellulase production methods for bio-ethanol. *Appl. Eng. Agric.* 23 (5): 679–687.

45 Catalán, E. and Sánchez, A. (2020). Solid-state fermentation (SSF) versus submerged fermentation (smf) for the recovery of cellulases from coffee husks: a life cycle assessment (LCA) based comparison. *Energies* 13 (11): 2685.

46 Nait M'Barek, H., Arif, S., Taidi, B., and Hajjaj, H. (2020). Consolidated bioethanol production from olive mill waste: wood-decay fungi from Central Morocco as promising decomposition and fermentation biocatalysts. *Biotechnol. Rep.* 28: e00541.

47 Schlembach, I., Hosseinpour Tehrani, H., Blank, L.M. et al. (2020). Consolidated bioprocessing of cellulose to itaconic acid by a co-culture of *Trichoderma reesei* and *Ustilago maydis*. *Biotechnol. Biofuels* 13 (1): 207.
48 Carpio, R.R., Secchi, S.G., Barros, R.O. et al. (2022). Techno-economic evaluation of second-generation ethanol from sugarcane bagasse: commercial versus on-site produced enzymes and use of the xylose liquor. *J. Cleaner Prod.* 369: 133340.
49 Mendes, F.B., Ibraim Pires Atala, D., and Thoméo, J.C. (2017). Is cellulase production by solid-state fermentation economically attractive for the second-generation ethanol production? *Renew. Energy* 114: 525–533.
50 Khajeeram, S. and Unrean, P. (2017). Techno-economic assessment of high-solid simultaneous saccharification and fermentation and economic impacts of yeast consortium and on-site enzyme production technologies. *Energy* 122: 194–203.

17

Conversion of Glycerol Derived from Biodiesel Production to Butanol and 1,3-Propanediol

Prawit Kongjan[1], Alissara Reungsang[2,3,4], and Sureewan Sittijunda[5]

[1]*Prince of Songkla University, Faculty of Science and Technology, Department of Science, Charoen Pradit Rd, Pattani, 94000, Thailand*
[2]*Khon Kaen University, Faculty of Technology, Department of Biotechnology, Mittraphap Rd, Khon Kaen, 40002, Thailand*
[3]*Khon Kaen University, Research Group for Development of Microbial Hydrogen Production Process from Biomass, Mittraphap Rd, Khon Kaen, 40002, Thailand*
[4]*Academy of Science, Royal Society of Thailand, Si Ayutthaya Rd, Bangkok, 10700, Thailand*
[5]*Mahidol University, Faculty of Environment and Resource Studies, Phuttamonthon 4 Rd, Nakhon Pathom, 73170, Thailand*

17.1 Introduction

Crude glycerol or glycerine is a waste commonly generated during biodiesel production [1, 2]. The top five biodiesel producers are the United States (USA), Indonesia, Brazil, Argentina, and Thailand, but its production is rapidly increasing worldwide [3]. During the transesterification process, 10 kg of biodiesel produces approximately 1 kg of crude glycerol as a by-product [1, 2]. The high annual production of this by-product decreases the price of crude and refined glycerol [4, 5]. Owing to its negative environmental impact, the disposal of crude glycerol is a major challenge in the biodiesel industry [2]. An industrial-scale process for refining crude glycerol to produce pure glycerol remains elusive. Refined glycerol is a suitable raw material for diverse industries, including cosmetics, pharmaceuticals, food and beverages, personal, and healthcare [2, 4, 5]. In 2014, approximately 65% of worldwide pure glycerol was obtained from refining biodiesel-derived crude glycerol [6]. However, this process is uneconomical for small- and medium-sized biodiesel enterprises. Therefore, the direct bioconversion of crude glycerol into valuable products can enhance the sustainability of the biodiesel industry and reduce the environmental impact of crude glycerol disposal.

Recently, the direct utilization of crude glycerol for the production of biofuels and valuable products, such as hydrogen (H_2), methane (CH_4), ethanol (C_2H_5OH), butanol (C_4H_9OH), 1,3-propanediol (1,3-PDO; $CH_2(CH_2OH)_2$), 2,3-butanediol (2,3-BDO; ($CH_3CHOH)_2$), citric acid ($C_6H_8O_7$), acrolein (C_3H_4O), and polyhydroxyalkanoates (PHAs) has been investigated [1, 2, 5, 7, 8]. Among these chemicals used in several industries, the production of butanol and 1,3-PDO from crude

Chemical Substitutes from Agricultural and Industrial By-Products: Bioconversion, Bioprocessing, and Biorefining,
First Edition. Edited by Suraini Abd-Aziz, Misri Gozan, Mohamad Faizal Ibrahim, and Lai-Yee Phang.
© 2024 WILEY-VCH GmbH. Published 2024 by WILEY-VCH GmbH.

glycerol was the focus of the present study. 1,3-PDO, for example, is a monomer for synthesizing polymers, such as polyurethane and polytrimethylene terephthalate (PTT) [1, 5, 8, 9]. Conversely, butanol is used as a gasoline and diesel additive [10–12], whereas its utilization is limited as a building block for chemical production. Properties of the feedstock, metabolic paths, and downstream processes involved in converting crude glycerol to butanol and 1,3-PDO have also been inadequately characterized.

The present chapter focuses on the microbially-assisted production of butanol and 1,3-PDO from crude glycerol. First, the characteristics and impurities in crude glycerol are examined, followed by biological processes that generate butanol and 1,3-PDO, as well as the uses of these products, especially as raw materials for the production of other chemicals. Finally, a downstream strategy for purifying and recovering these compounds is proposed, also challenges and perspectives are highlighted.

17.2 Crude Glycerol Characteristics and Impurities

Biodiesel is produced via the transesterification of triglycerides, such as waste cooking, vegetable, algal oils, and animal fats, using alcohol (usually methanol) in the presence of a catalyst [13]. Such catalyzed processes can be categorized into homogenous, heterogeneous, and enzymatic. The process is termed homogeneous if the catalyst and reactants are in the same phase during transesterification, whereas the process is heterogeneous if the catalyst and reactants are in distinct phases [13]. Enzymes are suitable for the transesterification of waste cooking oil because these are unaffected by the water content of the oil feedstock and the lengths of free fatty acid (FFA) chains [13, 14]. Therefore, enzymes are preferred as catalysts for novel or advanced production that aims to enhance the characteristics of biodiesel. Homogenous catalysis (e.g. alkali-catalyzed transesterification) is most commonly used for the production of biodiesel because of its high yield, low corrosion of equipment, and high reaction rate [13, 14]. However, the drawback of this catalysis approach is the potential reaction between FFAs and alkali catalysts, which produces soap. During production, owing to density and polarity differences, biodiesel forms a separate phase on top of crude glycerol [13, 14].

Crude glycerol (Figure 17.1) is a mixture containing glycerol and impurities such as soap, water, methanol, salt, and non-glycerol organic matter [2, 14]. However, its composition varies depending on the properties of feedstocks, the transesterification reaction, and posttreatment.

In general, the glycerol content ranges from 20% to 80% (w/w), whereas the soap content varies from 0% to 25%, and the methanol content is from <1% to 28% (w/w) (Table 17.1). During alkali-catalyzed transesterification, soap and crude glycerol are produced concurrently. The utilization of a 50 : 50% (w/w) mixture of chicken fat and soybean oil during transesterification using potassium hydroxide (KOH) as the catalyst yields 23.2% soap in the crude glycerol (Table 17.1).

Figure 17.1 Crude glycerol discharged after biodiesel production process.

Table 17.1 Summary of the feedstock, catalysts, and crude glycerol compositions.

		Crude glycerol compositions % (w/w)				
Feedstock	Catalyst	Glycerol	Soap	Methanol	Non-glycerol-organic matter	References
Seed oils[a]	3.8–4.2% NaOCH$_3$	62.5–76.6	NA	NA	NA	[14]
Jatropha oil	NA	18.0–22.0	29.0	14.5	11.0–21.0	[15]
Palm oil	NA	80.5	NA	0.5	<2.0	[16]
Canola oil	NaOH[b]	56.5	15.3	28.3	NA	[17]
Chicken fat/soybean oil (50 : 50% w/w)	KOH[b]	62.3	23.2	14.4	NA	[17]
Palm oil	NA	78.03	NA	NA	NA	[1]
Waste cooking oil	KOH[b]	34.1	NA	NA	NA	[8]

NA, not available.
a) Seed oils included the following: Ida Gold Mustard, Pacific Gold Mustard, rapeseed, canola, crambe, soybean, and waste cooking oils.
b) The concentration or ratio of the catalyst was unavailable.

Both heterogeneous and homogenous catalytic transesterification processes also yield crude glycerol containing high (5.0–7.0% w/w) amounts of salts [18]. In contrast, enzymatic catalytic transesterification produces high-purity crude glycerol [2, 13, 18]. Owing to impurities, crude glycerol varies in color from light yellow to dark brown, and its state at room temperature ranges from liquid to wax [19]. Impurities also affect the conversion of crude glycerol into valuable products via biological processes. During the production of butanol and 1,3-PDO, for example, the inhibition of microbial activity by FFAs depends on the number of double bonds and lengths of the acids [20]. Owing to its two double bonds, linoleic acid, for example, is characterized by two kinks. Consequently, it forms a prominent interface with the cell membrane, which significantly impedes diffusion [21]. In fact, because of the two kinks in linoleic acid, diffusion can be completely inhibited [22]. In addition, soap can limit the diffusion of the substrate into microorganisms [20, 23]. The cell membrane function alteration and bacterial growth and activity impedance of methanol are concentration dependent. Venkataramanan et al. [20], for example, reported that a methanol concentration of 2.5–5 g/l produced no effect on bacterial activity. In contrast, bacterial growth and metabolism diminished at methanol concentrations ≥10 g/l [20, 23]. Therefore, the treatment of crude glycerol is often necessary to improve its purity, even though the cost of purification is sometimes uneconomical.

17.3 Bioconversion of Crude Glycerol into Butanol and 1,3-Propanediol

Glycerol can be converted into valuable products, such as 1,3-PDO, ethanol, butanol, 2,3-BDO, PHA, lipids, and docosahexaenoic acid (DHA) using microorganisms (e.g. bacteria, fungi, and microalgae), and the associated reactions involve diverse pathways [1, 2, 5, 8, 11, 12, 14, 18]. Depending on the desired product, bioconversion can be accomplished under anaerobic, microaerobic, or aerobic conditions [1, 2, 5, 8, 11, 12, 14, 18].

The conversion of crude glycerol to 1,3-PDO can occur via anaerobic and microaerobic fermentation using bacterial strains, such as *Clostridium butyricum*, *Citrobacter* sp., *Enterobacter aerogenes*, *Klebsiella oxytoca*, *Klebsiella pneumoniae*, and *Lactobacillus diolivorans* [2, 5, 18]. *Klebsiella pneumoniae* and *C. butyricum* are commonly used as inocula to produce 1,3-PDO because of their high yields, simple productivity, and high substrate tolerance [2, 5, 18]. Figure 17.2 shows the natural mechanism for the production of 1,3-PDO. This mechanism involves both oxidative and reductive pathways. In the oxidative pathway, glycerol is converted to pyruvate via glycolysis, and this process produces energy. Conversely, in the reductive pathway, glycerol is converted to 1,3-PDO [2, 5, 8, 18]. The bacteria use approximately 5–10% of glycerol to produce biomass. In the reductive pathway, glycerol is converted to 3-hydroxypropionaldehyde (3-HPA) glycerol via dehydration using dehydratase, and this is then transformed using 1,3-propanediol dehydrogenase to 1,3-PDO [5, 9, 24]. In this pathway, nicotinamide adenine dinucleotide (NADH), which is generated in the oxidative pathway, is consumed. Therefore, in addition to microbes, the availability of NADH depends on the operation or fermentation conditions [5, 9, 24]. Consequently, the 1,3-PDO yield reflects the stoichiometries of the reduction and oxidation reactions (Table 17.2).

Klebsiella pneumoniae, for example, produces high concentrations of 1,3-PDO (49.3–66.3 g/l) under microaerobic conditions, whereas *C. butyricum* and *Enterobacter* sp. produce low concentrations under anaerobic conditions. This is because under microaerobic conditions, microbes, especially *K. pneumoniae*, gain energy from the conversion of substrates via respiration and fermentation to maximize cell

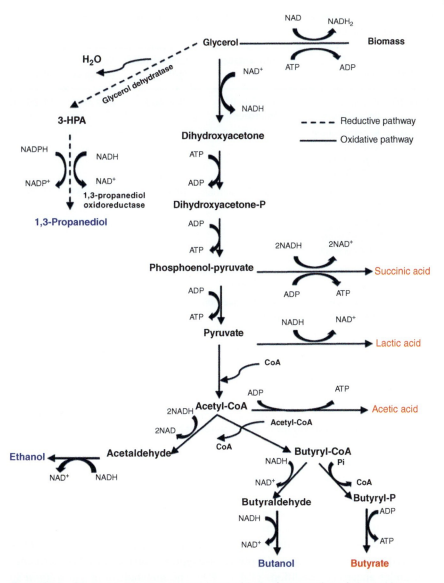

Figure 17.2 Biochemical pathways associated with the fermentation of glycerol by microorganisms. Text in red represents products associated with acids, whereas that in blue is related to solvents. Source: Adapted from Vivek et al. [5].

growth and the generation of products [33]. In contrast, under anaerobic conditions, microbes simply convert substrates to products to balance the $NAD^+/NADH$ ratio [33]. Moreover, operations using the fed-batch can reduce the inhibition of substrates relative to the batch mode [4].

In the oxidative pathway, glycerol is converted to pyruvate via glycolysis, and this is followed by the transformation of pyruvate to acetyl-CoA [5, 23]. Subsequently, acetyl-CoA is converted to butyryl-CoA using butyryl-CoA dehydrogenase,

Table 17.2 Summary of data from the diverse bioconversion of crude glycerol into 1,3-PDO and butanol.

Products	Inoculum	Fermentation mode	Concentration (g/l)	Yield (mol/mol)	References
1,3-PDO	Klebsiella pneumoniae K. pneumoniae TUAC 01 K. pneumoniae ATCC 15380	Fed-batch/microaerobic	49.3–66.3	0.43–0.85	[15, 25, 26]
1,3-PDO	Clostridium butyricum C. butyricum DS30 C. butyricum NCIMB 8082	Batch/anaerobic condition	14–67	0.56–0.67	[27–29]
1,3-PDO	Enterobacter sp. MU01	Batch/anaerobic condition	0.70	0.24	[1]
Butanol	Clostridium pasteurianum	Batch/anaerobic condition	6–29.8	0.24–0.55[a]	[10, 30, 31]
Butanol	C. pasteurianum DSM 525 (immobilized cell)	Repeated batch/anaerobic	—	0.25–0.26[a]	[32]
Butanol	C. pasteurianum DSM 525	Batch/anaerobic	4.18–8.95	0.26–0.35[a]	[25]

a) The reported unit is g/g and converted into mol/mol (a value obtained from the molecular weight of glycerol divided by the molecular weight of butanol).

which is then transformed to butyraldehyde (a precursor of butanol) by NAD- and NADP-dependent butyraldehyde dehydrogenase. Finally, butyraldehyde is converted to butanol using butanol dehydrogenase (Figure 17.2). Butyric acid and CO substantially increase the activity of butyraldehyde dehydrogenase. To produce butanol, acetyl-CoA is converted using the acetone–butanol–ethanol (ABE) pathway (a solvent production process). Clostridia are typically involved in ABE fermentation under anaerobic conditions (Table 17.2). The metabolism of *Clostridium* sp. comprises acidogenic and solventogenic phases. First, these microorganisms convert glycerol to organic acids, especially butyrate and acetate [10–12], and following a build-up of acids, the assimilation of organic acids into the solvent by microorganisms begins. In practice, butanol fermentation is impeded by its accumulation in the fermentation broth, which is usually described as "end-product inhibition" [10–12]. The threshold level of butanol for *Clostridium pasteurianum* is approximately 17 g/l [34], whereas that for *Clostridium acetobutylicum* is 11–12 g/l [35].

17.4 Purification and Recovery of 1,3-Propanediol and Butanol

17.4.1 1,3-Propanediol

17.4.1.1 Ion-exchange Resin-based Separation

Ion-exchange resin-based separation is considered a sustainable process because of its referable capacity and low energy consumption. In this process, 1,3-PDO is initially separated from the fermentation broth by removing biomass via centrifugation at 10,000 rpm at 4 °C. Organic macromolecules, cell debris, and proteins are then prepared using a mixture of chitosan or polyacrylamide, activated charcoal, and kieselguhr. The fermentation broth is added to 2 M HCl to adjust its pH to 5 before vacuum filtration. The filtered fermentation broth is then distilled under vacuum at approximately 95 °C to remove water and ethanol. The solution obtained that is rich in 1,3-PDO, 2,3-BDO, fractions of glycerol, and organic acids (citric, succinic, lactic, and acetic acids) is passed through a column packed with acidic exchange resins to recover the 1,3-PDO. Resin beds are pretreated using a 2 M HCl–NaOH solution, 2 M HCl is added to the resins subsequently for reconversion to the H-form, and these beds are rinsed using deionized water. The H-form resin is highly efficient for separating 1,3-PDO from fermentation broth [26]. In the three sequential steps of solute diffusion across the liquid film surrounding the particle, diffuses occurs through the polymeric matrix of the resin, and the adsorption of 1,3-PDO on the H-form resins involves chemical reactions with functional groups in the resin matrix.

Consequently, ethanol (75%) is commonly employed to elute 1,3-PDO [27]. Mitrea et al. [26] reported that after treatment of the fermentation broth using biomass removal, flocculation, vacuum filtration, and vacuum distillation processes, a clear broth for the H-cationic resin separation is obtained (Figure 17.3). Consequently, 1,3-PDO with a purity of 91% in ethanol can be obtained using a 7-time vacuum concentrator.

17.4.1.2 Hydrodistillation-based Separation

High purity 1,3-PDO (99.3%) can be obtained from fermentation broth using the following steps sequentially: (i) ultrafiltration to remove biomass and proteins, (ii) vacuum distillation associated with the continuous addition of glycerol to eliminate water, inorganic salts, and organic acids, (iii) alkaline hydrolysis of the 1,3-PDO esters generated during vacuum distillation, and (iv) distillation–deodorization to obtain purified 1,3-PDO (Figure 17.4). More than 99% of cells and proteins in the fermentation broth can be removed using a hollow fiber membrane during ultrafiltration. A rotary evaporator operated at 80 °C and a 200–500 mbar vacuum were used to concentrate the permeate obtained from ultrafiltration. The concentrate, which mainly contained 1,3-PDO (47.3 wt%), glycerol (0.75 wt%), acetate

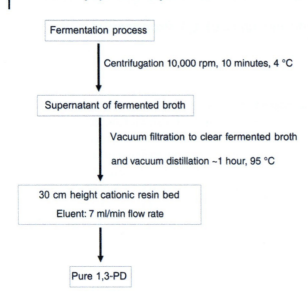

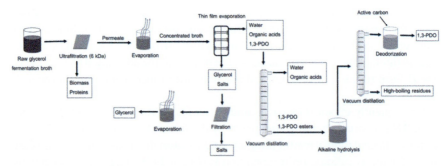

Figure 17.3 Block flow diagram for the ion-exchange resin-based 1,3-PDO separation. Source: Adapted and redrawn from Mitrea et al. [26].

Figure 17.4 Process flow diagram for the distillation-based 1,3-PDO separation from fermentation broth. Source: Adapted and redrawn from Zhang et al. [28].

(10.1 wt%), butyrate (4.8 wt%), water, and components of the residual medium, was placed in a thin-film evaporator to remove salts via the continuous addition of glycerol to suspend salts and other heavy impurities. The thin-film evaporator, operated at 20 mbar and 140–180 °C, caused water, free organic acids, and 1,3-PDO to evaporate. Glycerol-containing suspended salts discharged from the bottom of the thin-film evaporator were passed through the filtration unit to recover glycerol. The desalted distillate obtained from the thin-film evaporator was then subjected to a vacuum distillation (20 mbar and 120 °C) to eliminate water and most organic acids from the top of the column. However, 1,3-PDO esters of butyric and acetic acids can be generated during vacuum distillation. Therefore, the product at the bottom of the column was hydrolyzed using NaOH. The hydrolysis product, which contained 1,3-PDO (86.9 wt%), glycerol (4.66 wt%), sodium acetate (2.61 wt%), and sodium butyrate (2.16 wt%) was distilled at 140 °C under 20 mbar. The distillate was deodorized by mixing with 2 wt% activated carbon for three hours. Finally, the

1,3-PDO present in the raw fermentation broth was recovered at approximately 76% [28], and its purity was 99.63%.

17.4.2 Butanol Separation

Traditional distillation for the recovery of butanol from fermentation broth characterized by a low concentration of butanol is an expensive process. In addition, ABE fermentation usually involves solvent-toxicity problems. In situ separation processes, such as membraneless gas stripping and membrane pervaporation, are practical and economically feasible options to tackle these issues. However, because products recovered from gas stripping and pervaporation contain ABE, distillation is further applied to remove acetone and ethanol out for butanol purification [29].

17.4.2.1 In Situ Gas Stripping

Gas stripping is an important technique for separating butanol from fermentation broth. It involves the passage of broth through a fermenter (Figure 17.5) in the presence of carrier gas (e.g. N_2 or CO_2), which enables the selective removal of volatile components [30]. Gas stripping exhibits advantages relative to other removal processes owing to its simple and inexpensive operation and resistance to fouling or clogging because of biomass production. During butanol fermentation, the carrier gas and the vapor-containing butanol condense in the low-temperature condenser. The carrier gas passing through the condenser is then recycled to the fermenter. Advantages of this method in butanol recovery include the absence of negative effects on microbial cells, inexpensive operations, simplicity, and negligible loss of nutrients and intermediates in the culture broth. Product removal via gas stripping is facilitated by the volatility of ABE. The stripping gas is introduced in the fermenter via a rotary shaft, whereas volatile gases are condensed and recovered from the condenser.

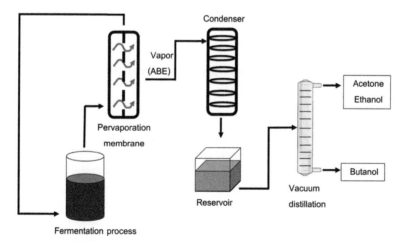

Figure 17.5 Process flow diagram for the in situ gas stripping process for butanol separation. Source: Adapted and redrawn from Ezeji et al. [30].

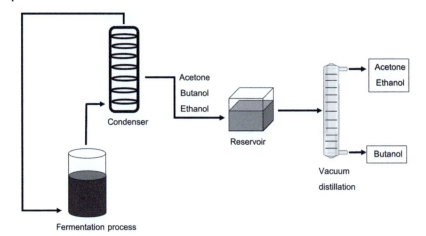

Figure 17.6 Process flow diagram for the in situ pervaporation steps for butanol separation during fermentation. Source: Adapted and redrawn from Qureshi and Blaschek [32].

17.4.2.2 In Situ Pervaporation

Pervaporation enables the separation of butanol from fermentation broth via a selective (solid/liquid) membrane in a vapor, followed by recovery via condensation (Figure 17.6). This efficient separation technique does not harm microorganisms either. Mass transport through pervaporation membranes involves a sorption–diffusion mechanism, wherein sorption of the liquid phase occurs at the feed side, followed by transport through the membrane, and then desorption as a vapor phase at the permeate side. The efficiency of pervaporation depends on membrane properties, such as stability, selectivity, and permeation flux [29, 31]. Pervaporation for the separation of butanol from fermentation broth has been achieved by adding silicone, silicalite, polypropylene, and an immobilized liquid (polypropylene-oleyl alcohol) in membrane pores. However, the modified system is unstable because the oleyl alcohol diffuses out of the polypropylene membrane [32].

17.5 Applications of 1,3-Propanediol and Butanol

Figure 17.7 shows that 1,3-PDO is a substance used in many industrial processes. However, it is principally employed in the textile industry, especially in the carpet-making sector [5, 9, 18]. The primary polymer involved is PTT, produced from the polycondensation of terephthalic acid and 1,3-PDO [4, 5, 28]. Due to its reducing nature, 1,3-PDO is a suitable monomer for synthesizing biodegradable polymers to produce plastics, adhesives, and laminates. It also reacts with dicarboxylic acids, such as succinic and lactic acids, to produce polymers and copolymers that exhibit diverse properties [4, 5, 28]. In addition, 1,3-PDO is utilized in the production of personal care, food, and pharmaceutical/medicine (e.g. vitamin H and immunosuppressive treatments) products, as well as perfumes, cosmetics, detergents, resins, solvents, engine coolants, and insect repellents

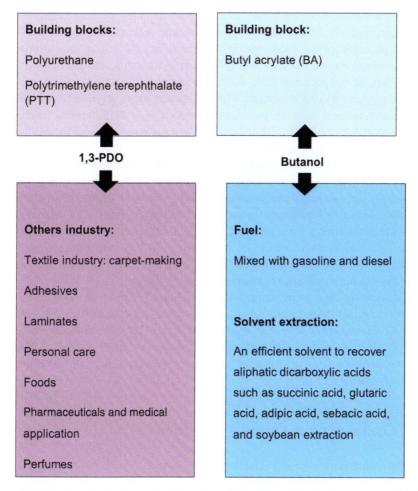

Figure 17.7 Diagram showing applications of 1,3-PDO and butanol as raw materials for chemicals and fuel, as well as the associated industries.

[2, 5, 18, 36]. Owing to its high reactivity, 1,3-PDO is suitable for polymerization and condensation reactions to produce biodegradable polymers that can replace synthetic plastics, thereby helping prevent ecological imbalances caused by the release of toxic pollutants. On a global scale, the market for 1,3-PDO is categorized based on its application in the production of polyurethane, PTT, personal care, detergent, and other products. In 2020, for example, the worldwide market for polyurethane applications was worth US$80.9 million. The increasing demand for high-performance and bio-based polymer textiles, footwear, and automotive applications will certainly elevate the revenue associated with the 1,3-PDO market. Bio-based polyurethane coatings, waterproof breathable films, TPU outsoles, and synthetic leather covers can also allow market expansion.

Butanol is alcohol that comprises five isomers, and it is sometimes referred to as butyl alcohol. It is dominantly utilized as an additive for gasoline, diesel, and other petroleum-based fuels that power internal combustion engines (spark and compression ignition engines). This is because, compared with ethanol, it is less corrosive to engines and reduces fuel consumption and emissions [37]. Butanol is a favorite solvent for recovering aliphatic dicarboxylic acids like succinic and glutaric acid, among others. In the presence of an esterification catalyst, butanol can also produce butyl acrylate (BA) via its reaction with methyl acrylate or acrylic acid [38]. BA is used in architectural coatings, paints, inks, textiles, sealants, adhesives, leather finishes, and paper. In addition, BA is emerging as a preferred material for synthesizing thermoplastics, such as ethylene acrylate copolymers (EAC) [38]. Using BA enhances materials' toughness, flexibility, aesthetic properties, and molding characteristics. This material is also resilient to climatic conditions, especially in direct sunshine [38]. A process for the bioconversion of butanol to volatile short-chain esters, along with ABE fermentation, has been developed to reduce the inhibitory impact of butanol during the fermentation process. Short-chain esters, especially those with <10 carbon atoms, are used to impart flavors in food, cosmetics, and pharmaceutical products [12]. Butyl butyrate is produced from the reaction of butanol and butyric acid in the presence of lipase as the catalyst [39].

Consequently, it has a fruity scent, and it is commonly employed in the industry to provide sweet fruity flavors comparable to those of pineapples. In 2022, the predicted market worth of volatile short-chain esters was US$2.44 billion, compared with US$1.83 billion in 2014 [40]. The increased demand is attributed to the food processing, personal care, cosmetics, and chemical synthesis sectors, which require biosynthesized esters.

17.6 Challenges and Future Perspective

Regarding industrial applications, the enormous potential for the bioconversion of crude glycerol to 1,3-PDO and butanol is hindered by obstacles, such as low productivity and product yield, as well as poor performance because of impurities [2]. For instance, Kongjan et al. [1] found that the highest 1,3-PDO production of 0.70 g/l was obtained from crude glycerol using the isolated strain *Enterobacter* sp. MU-01, which is relatively low compared to pure glycerol. In addition, the microbial production of 1,3-PDO is quite low, comparable to the chemical synthesis from ethylene oxide and other petroleum-based chemicals. However, producing 1,3-PDO from renewable resources maybe more environmentally friendly and less costly than chemical processes [41]. Innovative strategies to promote applications include the improvement of metabolism pathways, process design, impurity removal and utilization, and product inhibition, especially for butanol.

Regarding 1,3-PDO, the development of a metabolic approach to keep microorganisms in the exponential phase, while providing them with reductants and energy for the active biotransformation of glycerol to 1,3-PDO is in progress. The inactivation or elimination of the acetaldehyde dehydrogenase enzyme responsible for acetaldehyde production in *K. pneumoniae*, for example, increased the 1,3-PDO yield from pure glycerol by 0.7 mol/mol [42]. Moreover, an operation that favors the reductive over the oxidative pathway can enhance product separation. Considering an operation at pH values of 6.5–7.5 [43], which are favorable for the production of 1,3-PDO, a decrease in the pH to a value <6.5 stimulates the production of 2,3-BDO [44].

Owing to the toxicity of butanol in broth during the ABE fermentation process, a strategy that integrates butyrate and butanol removal has been proposed to improve the butanol yield. The combination of butanol removal via vacuum membrane distillation and the addition of butyrate can enhance the butanol concentration by 29.8 g/l, with *C. pasteurianum* CH_4 and pure glycerol as the inoculum and carbon source, respectively [11]. Gas stripping–pervaporation was employed to improve the butanol concentration by up to 75.5 g/l using glucose as the carbon source. Additionally, synthetic fermentation gas containing 20% hydrogen and 80% carbon dioxide could be used as stripping gas to recover butanol from the synthetic ABE with a high butanol concentration of 80–170 g/l [45]. However, research to further evaluate in situ butanol recovery from crude glycerol is necessary.

In general, chemical catalytic transesterification for biodiesel production is associated with the abundant soap as an impurity in crude glycerol. Enzyme catalytic transesterification reduces the formation of impurities, whereas pretreatments, such as the addition of an acid, remove impurities before using a substrate. However, technologies that involve the utilization of impurities in crude glycerol and the production of 1,3-PDO and butanol are required. The chemical conversion of crude glycerol to polyglycerol combined with the reductive production of 1,3-PDO in the ABE process, for example, can reduce the toxic effects of salt and soap [46]. Salts in crude glycerol impart base properties that are necessary for the conversion of crude glycerol to polyglycerol. Moreover, a long chain polyglycerol can be formed because of the high proportion of soap (12.50%) [46]. Considering that glycerol in crude glycerol is utilized in this strategy, an equilibrium is necessary between chemical conversion for the removal of impurities and the biological production of 1,3-PDO or butanol.

17.7 Conclusions

In addition to produce pure glycerol that can be utilized as a raw material in the food and beverages, pharmaceuticals, cosmetics, and textile industries, crude glycerol can be converted into diverse chemical building blocks via biological processes. Several factors, especially impurities in crude glycerol, influence biological pathway's production, yield, and productivity. To ensure its suitability for industrial applications, strategies, such as engineering the metabolism of microorganisms, impurity removal simplification, process design optimization, integration of ABE in butanol recovery, and development of technology that utilizes impurities in crude glycerol for the production of valuable products are required. The development of unique in situ downstream processes for recovering useful products and reducing associated inhibitory impacts is desired.

Acknowledgment

The author would like to acknowledge financial support from the National Research Council of Thailand, Thailand Science Research and Innovation (TSRI); Senior Research Scholar (Grant No. RTA6280001), Mahidol University, and Biodiversity-Based Economy Development Office (BEDO).

References

1 Kongjan, P., Jariyaboon, R., Reungsang, A., and Sittijunda, S. (2021). Co-fermentation of 1,3-propanediol and 2,3-butanediol from crude glycerol derived from the biodiesel production process by newly isolated *Enterobacter* sp.: optimization factors affecting. *Bioresour. Technol. Rep.* 13: 100616.
2 Luo, X., Ge, X., Cui, S., and Li, Y. (2016). Value-added processing of crude glycerol into chemicals and polymers. *Bioresour. Technol.* 215: 144–154.
3 Narin T. (2017). Thailand industry outlook 2017–19 biodiesel. https://www.krungsri.com/bank/getmedia/2a1ede06-1b73-4282-99d9-3ef7c787714a/IO_Biodiesel_201705_EN.aspx (accessed 13 April 2022).
4 Lee, C.S., Aroua, M.K., Daud, W.M.A.W. et al. (2015). A review: conversion of bioglycerol into 1,3-propanediol via biological and chemical method. *Renewable Sustainable Energy Rev.* 42: 963–972.
5 Vivek, N., Pandey, A., and Binod, P. (2017). Production and applications of 1,3-propanediol. In: *Current Developments in Biotechnology and Bioengineering* (ed. A. Pandey, S. Negi, and C.R. Soccol), 719–738. Elsevier.
6 Radiant Insight Inc. (2015). Glycerol market size, price trend, research report 2022. https://www.radiantinsights.com/research/glycerol-market (accessed 20 April 2022).
7 Garlapati, V.K., Shankar, U., and Budhiraja, A. (2016). Bioconversion technologies of crude glycerol to value added industrial products. *Biotechnol. Rep.* 9: 9–14.
8 Sittijunda, S. and Reungsang, A. (2020). Valorization of crude glycerol into hydrogen, 1,3-propanediol, and ethanol in an up-flow anaerobic sludge blanket (UASB) reactor under thermophilic conditions. *Renew. Energy* 161: 361–372.
9 Kaur, G., Srivastava, A.K., and Chand, S. (2012). Advances in biotechnological production of 1,3-propanediol. *Biochem. Eng. J.* 64: 106–118.
10 Karimi, K., Tabatabaei, M., Sárvári Horváth, I., and Kumar, R. (2015). Recent trends in acetone, butanol, and ethanol (ABE) production. *Biofuel Res. J.* 2 (4): 301–308.
11 Lin, D.S., Yen, H.W., Kao, W.C. et al. (2015). Bio-butanol production from glycerol with *Clostridium pasteurianum* CH_4: the effects of butyrate addition and in situ butanol removal via membrane distillation. *Biotechnol. Biofuels* 8 (1): 168.
12 Xin, F., Dong, W., Jiang, Y. et al. (2018). Recent advances on conversion and co-production of acetone-butanol-ethanol into high value-added bioproducts. *Crit. Rev. Biotechnol.* 38 (4): 529–540.

13 Babadi, A.A., Rahmati, S., Fakhlaei, R. et al. (2022). Emerging technologies for biodiesel production: processes, challenges, and opportunities. *Biomass Bioenergy* 163: 106521.

14 Tan, H.W., Abdul Aziz, A.R., and Aroua, M.K. (2013). Glycerol production and its applications as a raw material: a review. *Renewable Sustainable Energy Rev.* 27: 118–127.

15 Hiremath, A., Kannabiran, M., and Rangaswamy, V. (2011). 1,3-Propanediol production from crude glycerol from jatropha biodiesel process. *New Biotechnol.* 28 (1): 19–23.

16 Liu, Y.P., Sun, Y., Tan, C. et al. (2013). Efficient production of dihydroxyacetone from biodiesel-derived crude glycerol by newly isolated *Gluconobacter frateurii*. *Bioresour. Technol.* 142: 384–389.

17 Pyle, D.J., Garcia, R.A., and Wen, Z. (2008). Producing docosahexaenoic acid (DHA)-rich algae from biodiesel-derived crude glycerol: effects of impurities on DHA production and algal biomass composition. *J. Agric. Food Chem.* 56 (11): 3933–3939.

18 Yang, F., Hanna, M.A., and Sun, R. (2012). Value added uses for crude glycerol a byproduct of biodiesel production. *Biotechnol. Biofuels* 5 (1): 13.

19 Ciriminna, R., Pina, C.D., Rossi, M., and Pagliaro, M. (2014). Understanding the glycerol market. *Eur. J. Lipid Sci. Technol.* 116 (10): 1432–1439.

20 Venkataramanan, K.P., Boatman, J.J., Kurniawan, Y. et al. (2012). Impact of impurities in biodiesel-derived crude glycerol on the fermentation by *Clostridium pasteurianum* ATCC 6013. *Appl. Microbiol. Biotechnol.* 93 (3): 1325–1335.

21 Furusawa, H. and Koyama, N. (2004). Effect of fatty acids on the membrane potential of an alkaliphilic *Bacillus*. *Curr. Microbiol.* 48: 196–198.

22 Desbois, A.P. and Smith, V.J. (2010). Antibacterial free fatty acids: activities, mechanisms of action and biotechnological potential. *Appl. Microbiol. Biotechnol.* 85 (6): 1629–1642.

23 Sittijunda, S. and Reungsang, A. (2012). Media optimization for biohydrogen production from waste glycerol by anaerobic thermophilic mixed cultures. *Int. J. Hydrog. Energy* 37 (20): 15473–15482.

24 Drożdżyńska, A., Pawlicka, J., Kubiak, P. et al. (2014). Conversion of glycerol to 1,3-propanediol by *Citrobacter freundii* and *Hafnia alvei* newly isolated strains from the Enterobacteriaceae. *New Biotechnol.* 31 (5): 402–410.

25 Gallardo, R., Alves, M., and Rodrigues, L. (2014). Modulation of crude glycerol fermentation by *Clostridium pasteurianum* DSM 525 towards the production of butanol. *Biomass Bioenergy* 71: 134–143.

26 Mitrea, L., Leopold, L.F., Bouari, C., and Vodnar, D.C. (2020). Separation and purification of biogenic 1,3-propanediol from fermented glycerol through flocculation and strong acidic ion-exchange resin. *Biomolecules* 10 (12): 21601.

27 Wang, S., Dai, H., Yan, Z. et al. (2014). 1,3-Propanediol adsorption on a cation exchange resin: adsorption isotherm, thermodynamics, and mechanistic studies. *Eng. Life Sci.* 14 (5): 485–492.

28 Zhang, C., Sharma, S., Wang, W., and Zeng, A.P. (2021). A novel downstream process for highly pure 1,3-propanediol from an efficient fed-batch fermentation of raw glycerol by *Clostridium pasteurianum*. *Eng. Life Sci.* 21 (6): 351–363.

29 Lee, S.Y., Park, J.H., Jang, S.H. et al. (2008). Fermentative butanol production by Clostridia. *Biotechnol. Bioeng.* 101 (2): 209–228.

30 Ezeji, T.C., Qureshi, N., and Blaschek, H.P. (2003). Production of acetone, butanol and ethanol by *Clostridium beijerinckii* BA101 and in situ recovery by gas stripping. *World J. Microbiol. Biotechnol.* 19 (6): 595–603.

31 Azimi, H., Thibault, J., and Tezel, F.H. (2019). Separation of butanol using pervaporation: a review of mass transfer models. *J Fluid Flow Heat Mass Transfer* 6: 9–38.

32 Qureshi, N. and Blaschek, H.P. (2001). Recovery of butanol from fermentation broth by gas stripping. *Renewable Energy* 22 (4): 557–564.

33 Yen, H.W., Li, F.T., and Chang, J.S. (2014). The effects of dissolved oxygen level on the distribution of 1,3-propanediol and 2,3-butanediol produced from glycerol by an isolated indigenous *Klebsiella* sp. Ana-WS5. *Bioresour. Technol.* 153: 374–378.

34 Biebl, H. (2001). Fermentation of glycerol by *Clostridium pasteurianum* batch and continuous culture studies. *J. Ind. Microbiol. Biotechnol.* 27 (1): 18–26.

35 Branduardi, P., De Ferra, F., Longo, V., and Porro, D. (2014). Microbial *n*-butanol production from *Clostridia* to non-*Clostridial* hosts. *Eng. Life Sci.* 14 (1): 16–26.

36 Xu, B. and Ma, C. (2019). Advances in the production of 1,3-propanediol by microbial fermentation. *AIP Conf. Proc.* 2110 (1): 10842.

37 Zhen, X., Wang, Y., and Liu, D. (2020). Bio-butanol as a new generation of clean alternative fuel for SI (spark ignition) and CI (compression ignition) engines. *Renewable Energy* 147: 2494–2521.

38 Ajekwene, K. (2020). Properties and applications of acrylates. In: *Acrylate Polymers for Advanced Applications* (ed. Á. Serrano-Aroca and S. Deb), 35–46. Intech Open.

39 Stergiou, P.Y., Foukis, A., Filippou, M. et al. (2013). Advances in lipase catalyzed esterification reactions. *Biotechnol. Adv.* 31 (8): 1846–1859.

40 Grand View Research Inc. (2016). *Report on Fatty Acid Ester Market*. San Francisco CA: Grand View Research Inc.

41 Białkowska, A.M. (2016). Strategies for efficient and economical 2,3-butanediol production: new trends in this field. *World J. Microbiol. Biotechnol.* 32 (12): 200.

42 Zhang, G.L., Ma, B.B., Xu, X. et al. (2006). Fast conversion of glycerol to 1,3-propanediol by a new strain of *Klebsiella pneumoniae*. *Biochem. Eng. J.* 32: 93–99.

43 Sattayasamitsathit, S., Methacanon, P., and Prasertsan, P. (2011). Enhance 1,3-propanediol production from crude glycerol in batch and fed-batch fermentation with two-phase pH-controlled strategy. *Electron. J. Biotechnol.* 14: 1–12.

44 Wong, C.L., Yen, H.W., Lin, C.L., and Chang, J.S. (2014). Effects of pH and fermentation strategies on 2,3-butanediol production with an isolated *Klebsiella* sp. Zmd30 strain. *Bioresour. Technol.* 152: 169–176.

45 Kongjan, P., Tohlang, N., Khaonuan, S. et al. (2022). Characterization of the integrated gas stripping condensation process for organic solvent removal from model acetone-butanol-ethanol aqueous solution. *Biochem. Eng. J.* 182: 108437.

46 Din, N.S.M.N.M., Idris, Z., Yeong, S., and Hassan, H. (2013). Preparation of polyglycerol from palm biodiesel crude glycerin. *J. Oil Palm Res.* 25: 289–297.

18

Sustainability of Chemical Substitutes from Agricultural and Industrial By-products

Lai-Yee Phang[1], Suraini Abd-Aziz[1], Misri Gozan[2,3], and Mohamad F. Ibrahim[1]

[1]*Universiti Putra Malaysia, Faculty of Biotechnology and Biomolecular Sciences, Department of Bioprocess Technology, Serdang, 43400, Selangor, Malaysia*
[2]*Universitas Indonesia, Kampus UI, Bioprocess Engineering Program, Faculty of Engineering, Department of Chemical Engineering, Depok, 16424, Indonesia*
[3]*Research Center for Biomass Valorization, Universitas Indonesia, Faculty of Engineering, Kampus UI, Depok, 16424, Indonesia*

18.1 Introduction

The chemical industry experienced enormous challenges and pressures (as a result of legislation, social, market, or others) for change at all stages in the lifecycle or supply chain of chemical products in the last years of the twentieth century. This is mainly consequential of the concerns over human safety and destructive attitude toward the environment [1]. Since then, scientists, industrialists, and politicians have been looking for suitable chemicals and greener substitutes for many important chemicals. One of the approaches to the mechanisms for the product substitution is the route being based on chemicals derived from alternative, renewable resources, such as biomass, agricultural and industrial by-products. The use of agricultural and industrial by-products has gained attention in recent years since these by-products can be converted into various value-added chemical substitutes, using different technologies.

Producing chemical substitutes from agricultural and industrial by-products could positively impact the community and environment, such as reducing environmental problems, improving the social-economic, decreasing dependence on depleted or nonrenewable resources, and strengthening the chemical supply. Figure 18.1 shows the routes of chemical substitution. A twin-route approach to the chemical products substitution mechanism was proposed, in which one route focuses on existing commercial chemicals that can substitute the substances of concern, while another route will be based on chemicals derived from renewable resources.

Apart from adopting the mechanisms for product substitution based on existing commercial chemicals, chemicals derived from biomass and/or agro-industrial by-products becomes another important approach. A few criteria should be included in developing the production of a sustainable chemical using biomass or

Chemical Substitutes from Agricultural and Industrial By-Products: Bioconversion, Bioprocessing, and Biorefining, First Edition. Edited by Suraini Abd-Aziz, Misri Gozan, Mohamad Faizal Ibrahim, and Lai-Yee Phang.
© 2024 WILEY-VCH GmbH. Published 2024 by WILEY-VCH GmbH.

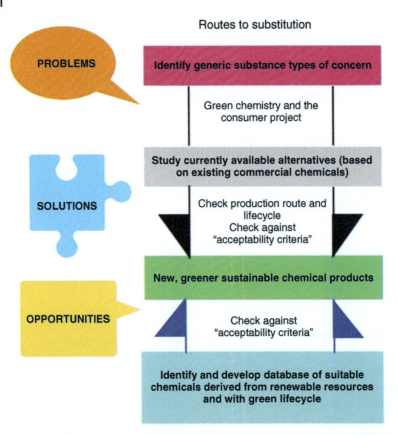

Figure 18.1 Routes to the substitution of chemical products. Source: Adapted from Clark [1].

industrial by-products, (i) the feedstock to have excellent long-term availability in the region and, most importantly, not compromise food production; (ii) process step yield not less than 70% (excluding any fermentation step) and (iii) obtainable at not less than 99% purity or in a suitable formulation (where appropriate) [1].

Favorable and promising agribusiness models increase awareness and promotion of agricultural and industrial residues as sustainable resources in the future. These chemical substitutes derived from agricultural and industrial by-products hold great potential to resolve global environmental crises. However, this option depends on various factors, such as availability, policy, cost of the resources, the capital cost of the process equipment and facilities, and markets for the chemical substitutes. In addition, the sustainability of the biomass-based chemical industry will be influenced by the viability of the conversion processes, the location and weather-dependent replacement time of the different feedstocks, energy requirements, waste generation during production, and their remediation in the natural environment [2].

Using biomass sources effectively and efficiently in the recycling processes is essential for constructing a sustainable society with low carbon, circulation, and natural symbiosis. Lignocellulose residues from agriculture and forestry residues are being extensively studied or reported biomass in the valorization model since

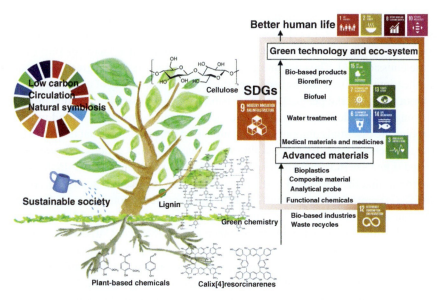

Figure 18.2 SDGs illustration of a sustainable society for bio-industry and recycling plant wastes leaching to the development of advanced materials and utilities to green and eco-technologies for better human life. Source: Kobayashi and Nakajima [3]/with permission of Elsevier.

they are abundant in the resources. As shown in Figure 18.2, the bio-industry and recycling plant wastes leaching to develop advanced materials and chemicals can be promoted and realized via sustainable development goals (SDGs) [3]. In 2015, the United Nations Member States developed SDGs with the motto of "The SDGs are the blueprint for achieving a better and more sustainable future for all" to realize the connection with nature and the person [4]. The SDG 9 focuses on building resilient infrastructure, promoting inclusive and sustainable industrialization, and fostering innovation. It covers the whole process, including conversion, transformation, and modification of waste materials into advanced materials and green products. More SDGs are achieved along the conversion or transformation process of the waste materials through the application of green and eco-technologies for better human life.

Life cycle analysis (LCA) is necessary to fully assess the sustainability of a given process. In order to have an accurate comparison of different procedures in this analysis, a sufficient amount of data from these procedures is required. Thus, it is time-consuming for the data collection process. Lignocellulosic biomass is the most promising and reputed feedstock considering its availability and low cost. Hence, this feedstock often serves as the research or study model for developing a sustainable concept in a circular bioeconomy. The carbohydrate fraction of this lignocellulosic biomass can be converted into sugars, which act as a primary carbon source in the microbial biocatalysis system to produce various chemical substances. The chemical process that produces biochemical products from microorganisms or enzymes is called bioprocess. In a review of sustainable homogenous cellulose modification, an environmental factor (E-factor) and basic toxicity information were used to compare various approaches to provide a quick and useful sustainability assessment [5]. E-factor evaluates the amount of waste generated by the process. Therefore, an ideal sustainable process should achieve an E-factor of zero (zero waste emission) [6]. The environmental, health, and safety (ESH) green metrics is another approach that can be used to evaluate the sustainability of a process [7].

From the industrial perspective, the feasibility of an industrial production process is strongly dependent on the overall process economics. Cost analysis and process mapping are essential components in an industrial process design that influence the overall process economics. Process simulation, which applies various software tools, is used to analyze individual unit operations and their relationships within the overall process that the developed mathematic model can represent. Aspen Plus, Aspen HYSYS (HYprotech SYStem), and SuperPro Designer are the software tools designed for bioprocess simulation. Alnur Auli et al. [8] reviewed and compared the application of the three bioprocess simulators, Aspen Plus, HYSYS, and SuperPro Designer, for xylitol production from hemicellulosic biomass. A bioprocess scheme composed of several unit operations including the fermentation process, is simulated using the selected software tool that involves experimental data collection. Moreover, the techno-economic assessment can also be performed by using this simulation software. In this chapter, the techno-economic assessment for (i) cellulase production using sugars derived from oil palm empty fruit bunches (OPEFB) and (ii) biofertilizer production from molasses, based on Aspen Plus, are described. The environmental and social impacts of the bioprocesses of these two products are discussed.

18.2 Sustainable Development Strategies, Policies and Regulations in Indonesia and Malaysia

18.2.1 Indonesia's National Sustainable Development Strategy

The agricultural, plantation, and forest areas in Indonesia produce crops for consumption and sale. Meanwhile, most agricultural, plantation, and forest by-products remain in the fields. Indonesia produces 146.7 million tons of biomass annually, with the potential as a renewable energy source of 49.81 GW (gigawatt) [9]. Renewable energy (RE) is the Indonesian government's main priority, specifically in improving legislation and establishing strategies to pursue SDGs. However, the potential uses of biomass in other industries, such as food, oleochemicals, and cosmetics have created competing interests in developing a circular economy for various green products [10]. The Indonesian government has issued several sustainable development policies and regulations. Roadmap SDGs Indonesia launched in 2019 defines issues and projections of main SDGs indicators in each goal as an important tool to guide all stakeholders on the directions and targets of the Indonesian 2030 agenda (Indonesia Secretariat for SDGs). It consists of policy direction for every indicator or goal, SDGs interlinkages, and SDGs' financing needs. The second pillar for the Indonesian Government's National Economic Recovery is economic diversification to build a more resilient and sustainable economy in which low-carbon development with the tagline of Green Recovery is an option. The Ministry of National Development Planning, Indonesia, has redesigned and launched six major strategies for Indonesia's economic transformation to achieve the 2045 Indonesia Vision. One of the six major strategies focuses on the transition to Green Economy. The basic concept of a Green Economy consists of decarbonization, resource efficiency, and environmental improvement. Investing in sustainable management to promote more sustainable consumption and production activities (circular economy), that reduce the potential waste generation in the whole supply chain, is one of the several strategic sectors in developing the green economy [11]. Green finance promotes green technology innovation and microenterprise, which leads to achieving the SDGs via environmental and economic sustainability [12].

18.2.2 Malaysia's Policies and Regulations for Sustainable Development

Malaysia has abundant agricultural residues and industrial by-products, which can be sustainably converted into chemical substitutes. These products include food and pharmaceutical ingredients, fine specialty and platform chemicals, and polymers. Bioeconomy in Malaysia [13]. The 12th Malaysia Plan (2021–2025) with the objective of "A Prosperous, Inclusive, Sustainable Malaysia," highlighted the greenhouse gas emissions to be reduced to 45% of GDP by 2030 in line with Paris Agreement. One of the three themes in the 12th Malaysia Plan is advancing sustainability toward guaranteeing continuous economic growth while protecting the environment and continuing Malaysia's commitments to global targets. The main policies driving Malaysia's biorefinery and bioeconomy development are National Biotechnology Policy (NBP) and Green Technology Policy. In Malaysia, one of the biomass contributors is the palm oil industry, and other contributors include the rubber industry, rice mills, wood industry, and coconut and sugarcane plantations.

The total amount of biomass produced by Malaysia is more than 160 million tons annually [14]. The NBP launched in 2005, is the strongest boost for Malaysia's biotechnology and bioeconomy development. The policy identified three main biotechnology areas (agricultural biotechnology, healthcare biotechnology, and industrial biotechnology) that are critical for the development of the bio-business. Besides, Malaysia implemented its program entitled "Bioeconomy Transformation Program" in 2013 to develop the national bio-based industry. The NBP 2.0 is launched in 2022 to achieve high-technology nation status by 2030. This policy focuses on agriculture and food security, healthcare and well-being, and biotechnology in industrialization and circular economy. Moreover, the government has established the Malaysia Sustainable Development Goal Foundation (MySDG Foundation) and Malaysia Sustainable Development Goals Trust Fund (MySDG Fund) to facilitate the nation's sustainable development agenda. The Malaysia Government has allocated an initial fund of RM20 million as part of Malaysia's five-year national development plans, which provides the overarching policy framework for Malaysia to achieve the SDGs in the medium- and long-term [15].

18.3 Case Study 1: Techno-economic Analysis for the Production of Cellulase

18.3.1 Cost Analysis Using SuperPro Designer

The microorganisms identified so far involved in the production of cellulase and related enzymes are bacteria, some fungi, and actinomycetes. However, due to the slow growth of the fungus, the cost of cellulase production is high for this process. On the other hand, bacterial cultures grow rapidly and have short generation time and other beneficial characteristics. Therefore, bacteria have good potential to produce cellulase and other enzymes for industrial applications [16].

Biomass from oil palm industry is an example of natural resource potential in many Asian countries, especially Indonesia and Malaysia. Palm oil processing forms several by-products and residues with economic potential [17, 18]. At 45.80% dry weight, OPEFB are solid residues with high cellulose content. In this case, the high cellulose content of empty fruit branches (EFB) was used as a substrate for bacterial culture to produce cellulase. Using OPEFB as an alternative substrate, the cost of cellulase production on an industrial scale can be reduced. Its industrial application requires cost-effective enzymes on a large scale. One common solution is the expression of individual cellulase enzymes using recombinant DNA [5]. *Escherichia coli* and *Bacillus* sp. are the two most commonly used microbes to express recombinant proteins [19, 20]. The feasibility of producing recombinant cellulase from *E. coli* was then evaluated in this case study using the SuperPro Designer Simulator. The production of recombinant *endo*-β-1,4-glucanase is based on kinetic studies carried out and also on previous kinetic studies [21] with the process scheme shown in Figure 18.3. And the characteristics of OPEFB processed by three different procedures are shown in Table 18.1. Pretreated EFB was fermented using *E. coli* BPPTCC-EgRK2 obtained from BPPT Indonesia.

SuperPro Designer was used to calculate mass and energy balance and estimate various economic parameters using data, such as equipment purchase, direct and indirect costs. To evaluate the most suitable cellulase production process, comparing yield values, batch duration, pretreatment performance, and economic parameters is necessary.

The alkaline or base pretreatment process can fulfill 1% of the Indonesian market share by using 8.5 tons of OPEFB as a substrate, smaller than the acid–base and steam explosion pretreatment, which require 19 and 29 tons, respectively. Base pretreatment also gave the highest product yield, 0.03, compared to acid–base and vapor explosion, which gave yields of 0.021 and 0.008. As for the delignification efficiency, the acid–base sequence gave the highest lignin removal by removing 90% of lignin from OPEFB [23]. The steam explosion does not remove lignin from OPEFB but provides high hemicellulose removal, which increases permeability and improves enzymatic digestibility [24]. However, this process also produces fermentation inhibitors, such as levulinic acid (0.11 g/l), formic acid (0.21 g/l), hydroxymethyl furfural (0.71 g/l), and furfural (0.23 g/l) [24]. Similar results were also achieved by other studies [25].

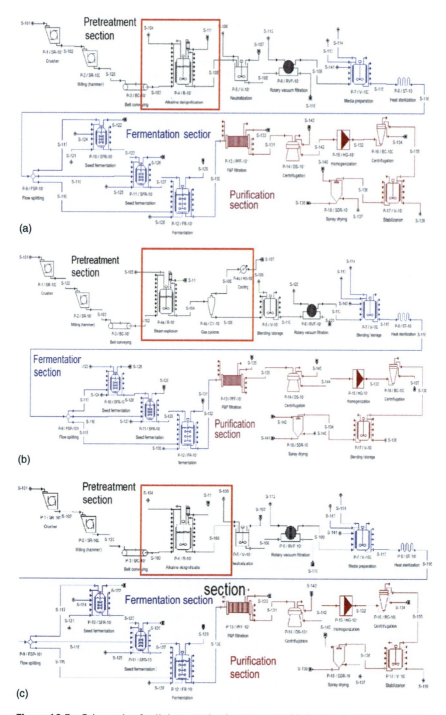

Figure 18.3 Schematic of cellulase production process with (a) alkaline (base) pretreatment; (b) vapor explosion pretreatment and (c) sequential acid–base pretreatment. Source: Adapted from Surya et al. [22].

Table 18.1 Summary of simulation effects of different pretreatment methods on project profitability.

Parameter	Alkaline/Base (CHEMEX)	Steam explosion	Acid–base
TKKS feed/batch (Ton)	8.5	29	19
Yield p/s	0.03	0.008	0.021
Unit number	18	20	21
Inhibitor fermentation	—	Present	—
Removal efficiency			
• Celluloses	19%	30%	41%
• Hemicellulose	67%	82%	86%
• Lignin	80%	0%	90%
Total capital investment (US$)	35,195,000	53,227,000	83,383,000
Operating cost (US$)	23,094,000	34,948,000	41,692,000
Internal rate of return (IRR)	14.61	N/A	N/A
Net present value (9.6%) (US$)	7,118,000	−73,411,000	−114,013,000
Payback period (PBP)	4.84	N/A	N/A

Source: Surya et al. [22]/IOP Publishing/CC BY 3.0.

Total capital investment (TCI) is calculated based on the purchase and installation costs of each equipment, piping, instrumentation, facility dependent, engineering, and construction costs. Table 18.1 shows each simulation's TCI and annual operating costs. The TCI values for alkaline pretreatment, steam explosion pretreatment, and acid–base pretreatment were US$35,195,000, US$53,227,000, and US$83,383,000, respectively. The high capital investment in steam and acid–base explosion pretreatment is due to the number of units. Steam pretreatment has 20 different units, while acid–base uses 21 units with multiple rotary vacuum filtration units, which results in higher capital investment in equipment and installation costs. Raw material costs, labor costs, and waste treatment costs strongly influence annual operating costs for the production of recombinant cellulase. Acid–base pretreatment has the highest operating cost at US$41,692,000 because it is energy intensive. Steam explosions also provide high operating costs due to low conversion rates, resulting in high raw material and waste treatment costs. The factory's revenue comes only from cellulase sales with the revenue of US$30,800,000. The calculation in the simulation is based on annual cellulase production of 154 tons, which is 1% of the targeted market share [22]. Net present value (NPV) is the key to determine the feasibility of a project. The simulation shows that alkaline pretreatment gives the highest NPV of US$7,118,000. The steam explosion pretreatment and acid–base pretreatment processes give negative NPV values due to higher operational costs.

The project's feasibility is also evaluated on its sensitivity to key variables, such as selling prices, raw material costs, and labor costs. Projects that are heavily influenced by external factors are considered riskier. Figure 18.4 shows changes in labor costs,

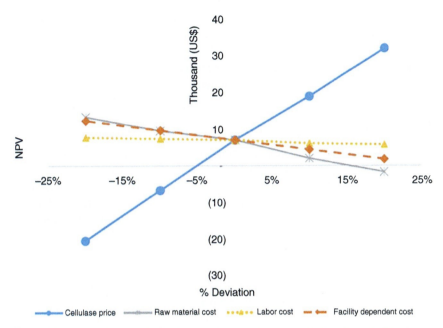

Figure 18.4 Sensitivity analysis of recombinant cellulase production from OPEFB. Source: Surya et al. [22]/IOP Publishing/CC BY 3.0.

raw material costs, selling prices, and facility costs on the NPV value. It was found that the selling price of cellulase had the effect on NPV.

Three different pretreatment processes were compared to find the most feasible process. The production of recombinant cellulase using OPEFB treated with alkali gave the highest NPV, Internal rate of return (IRR), and Payback period (PBP) with values of US$7,118,000, 14.61%, and 4.84 years, respectively.

18.3.2 Environmental Impact Analysis

Using alkaline as a pretreatment method in producing cellulase has both advantages and disadvantages to the environment. NaOH is widely used in cellulase production because it is an electrolyte solution easily soluble and ionized in polar solvents. However, the usage of NaOH might pollute the environment. This is because Na^+ ions can replace Ca^{2+} or Mg^{2+} (soil nutrients), causing damage to the structure and pores of the soil, decreasing the cycle or movement of water and air molecules in the soil, and reducing nutrient transfer in the soil. Besides, the soil pH becomes higher in the presence of NaOH. The increasing soil pH has a detrimental effect on soil physical properties, which in turn will interfere the plant growth [26, 27]. The selection of KOH as a substitute for NaOH can be made because it is based on the similarity of the properties of KOH with NaOH. NaOH and KOH have hygroscopic properties (absorb water vapor). NaOH and KOH also have high solubility in water. At a temperature of 25 °C, the solubility of NaOH in water is 1000 g/l and the solubility of KOH in water is 1210 g/l. NaOH and KOH also have the property of being easily ionized into their ions [28]. In addition, potassium is one of the main macronutrients needed for plant growth. The high concentration of K^+ and low concentration of Na^+ in the soil will reduce the stress level of plants due to the effect of dissolved salt ions in the soil [28]. Therefore, replacing NaOH with KOH is expected to produce more environmentally friendly waste that can benefit the environment.

18.3.3 Social Aspect

The enzyme industry has developed rapidly and occupies an important position in the industrial sector. Public awareness of environmental problems is getting higher, as well as pressure from experts and environmentalists, making enzyme technology an alternative to replace various chemical processes in the industrial sector [29]. Considering social aspects is very important in choosing the location of an industry, especially for the palm-based industries, which have various dimensions of social issues [30]. Issues that need to be avoided include social conflicts, land conflicts, marginalization of indigenous peoples and their livelihoods, conflicts in employment, and trade, to loss of biodiversity [31, 32]. The location of the cellulase enzyme production project discussed is determined based on several parameters: infrastructure readiness, availability of raw materials, availability of labor, and ease of trade. Let us only consider the availability of raw materials. The volume of raw materials required for this cellulase enzyme plant is relatively low. So, the enzyme plant can be built in every palm-producing area, where palm-producing areas with a capacity of 45 tons of fresh fruit per day can still meet the needs of the factory's EFB.

Another factor to be taken into account is infrastructure readiness. In this case study [22] the Cilegon Industrial-Estate Area (KIEC) was chosen, located in the province of Banten [33]. Banten Province produced more than 42 kilo tons of palm oil solid waste, the majority of which were EFB and 56.8% of this production came from state-owned oil palm plantations, with a potential value of TKS waste of 24,130.78 tons. Government plantations promote a stable supply of raw materials throughout the enzyme production period. This location with infrastructure readiness also has a very adequate human resource development study with an educated workforce to work in this region. The readiness of industrial areas near oil palm plantation areas is considered to have the best social readiness that can improve the population's welfare and are in line with local regional development priorities.

18.4 Case Study 2: Techno-economic Analysis for the Production of Biofertilizer

18.4.1 Cost Analysis Using SuperPro Designer

Plant growth-promoting bacteria (PGPB) is a term used to define soil bacteria that can increase plant growth and productivity if carried out under adequate environmental conditions. Living and thriving in soil and plant roots, PGPB, under optimal conditions, can increase crop yields by 5–30% [34]. *Azospirillum brasilense* is a gram-negative, aerobic, nitrogen-fixing, and naturally occurring bacterium commonly found on plant root surfaces and soil, which fixes about 20–40 kg of atmospheric nitrogen per hectare [35]. They can also produce extracellular growth-regulating hormones (phytohormones) important for plant development, such as auxin, vitamin cytokines, and gibberellins [36].

Biofertilizers are active products or microbial inoculants of bacteria, algae, and fungi, combined or separately, increasing nutrient availability in plants, and thereby increasing crop yields and productivity. They can add almost all the nutrients normally consumed by plants through the natural processes of atmospheric nitrogen fixation, phosphorus dissolution, and plant growth stimulation through the synthesis of growth-promoting substances [34]. The biofertilizer production plant was established near a sugar factory. Molasses was the main raw material for consumption in the industrial fermentation stage because it was supplied at a constant rate and low prices. It is also a substrate that is very good for the microorganisms used. Sugarcane drops can be stored in the factory and used at any time by the biofertilizer production plant. The use of molasses (molasses) as a raw material for producing liquid biological fertilizers through a submerged fermentation process has been patented for industrial application [37]. In the chemical process industry (CPI), the simulation approach is an important and indispensable tool mostly used for designing, assessing, optimizing, and analyzing projects, systems, and processes [38–41]. Thus, the feasibility of producing fertilizer using molasses (molasses) with the bacterial microorganism *A. brasilense* was evaluated in this study using the SuperPro Designer Simulator software.

The liquid biofertilizer production process consists of three stages, as shown in Figure 18.5. In the first step, bacterial propagation, bacteria are cultured in different volumes of flasks containing a certain culture medium until the desired cell concentration is reached. Then, the liquid culture containing the live cells (pre-inoculum) was transferred (inoculated) to a larger volume flask containing the culture medium. After the seed fermentation stage is complete, the cell suspension contained in the seed fermenter is inoculated into the industrial fermenter. The fermented products are harvested in a cylindrical vertical tank with stirring and cooling. The factory will have a production capacity of 44 tons of liquid biofertilizer per year, with the duration of each production batch being 109 hours/batch. Hence, the total number of production batches required per year to meet the production capacity is 78 batches/year, where the plant's annual operation is approximately 7900 hours/year.

The main items that affect the Total Plant Direct Cost (TPDC) are the purchase of equipment and piping, while construction and engineering are the main items that affect the value of the Total Plant Indirect Cost (TPIC). Thus, the total Direct Fixed Capital Cost obtained is US$3,700,000. Raw material items have little effect on operating costs, which is only 4.01% of the total [42]. The main substances and chemicals consumed in the process have relatively low purchase costs. The unit production cost calculated for one bottle of 1.5 l liquid biofertilizer formulated is US$24,009, with a total income of US$985,000 per year.

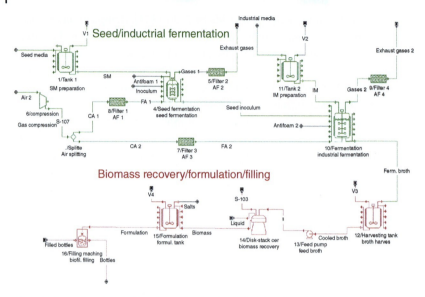

Figure 18.5 Simplified flow chart of a liquid biofertilizer production plant. Source: Redrawn and modified from Pérez Sánchez [42].

Table 18.2 Parameters of profitability values.

Parameter	Value
Net present value (NPV)	US$716,000
Net profit	US$422,000
Gross margin	24.97%
Internal rate of return (IRR)	2.55%
Payback period (PBP)	6.70 yr
Return on investment (ROI)	14.93%

Source: Adapted from Pérez Sánchez et al. [42].

Based on the profitability parameters in Table 18.2, the net profit obtained is US$422,000 with a gross margin percentage of 24.97%. The NPV value obtained is US$716,000. An NPV value greater than 0 indicates that this investment will generate a profit greater than the minimum value. Then the IRR value obtained based on the calculation of the entire cumulative cash flow of the biofertilizer factory is 2.55%. The PBP is 6.7 years.

Meanwhile, the return on investment (ROI) value obtained is 14.93%. This investment can be profitable for investors with higher interest rates than MARR. According to the four parameters calculated in the profitability analysis, the study concluded that this biofertilizer factory-made from molasses is economically feasible.

Some of the assumptions used in the calculations in this case study must be deepened. In this case study, the price of molasses, ammonium sulfate, and sucrose was supplied at zero cost by a sugar factory near the biofertilizer factory. At the same time, other raw materials were obtained at a fairly low price. Molasses as a by-product at zero cost is acceptable. However, the price of ammonium sulfate and sucrose must be calculated properly. Even so, this study has presented an approach that shows that this biofertilizer factory is economically feasible to continue.

18.4.2 Environmental Impact Analysis

Studies show that plant-associated microbes can promote plant growth in natural and extreme conditions. Apart from being directly involved in photosynthesis and forming amino acids that occur without causing environmental damage, studies also show that biofertilizers can play a role in bioremediation [43]. The mechanism that is thought to occur by plant growth-promoting rhizobacteria (PGPR) is the production of a deaminase, which reduces ethylene levels in plants. In this mechanism, the ability of host plants to cope with stress due to heavy metal poisoning increases. The study also shows that biofertilizers have the ability to biofortify biofertilizers in the amelioration of abiotic stress.

In the case presented here [44], *A. brasilense* is one of the most widely used microorganisms for biofertilizers. *Azospirillum brasilense* can fix nitrogen, solubilize phosphorus, and secrete plant growth-promoting hormones such as auxins, vitamin cytokines, and gibberellins. *Azospirillum brasilense* also can increase plant growth yield by 35% and has a high resistance to changes in environmental conditions. Apart from the bacteria, the molasses residue in the biofertilizer might have a positive effect. Molasses is a by-product of the sugar industry with the characteristics of having a high sugar content [44]. The presence of sugar content can be a source of energy for soil microorganisms that are carrying out fermentation. Soil fertility increases when molasses is applied in crop cultivation [45].

18.4.3 Social Aspect

In the case raised in this subsection, biofertilizer made from molasses is very beneficial for the growth of corn (*Zea mays*). This biofertilizer is very likely to be useful for other plants. A study using the innovative symbiosis energy model at the Manado Slaughterhouse, Indonesia, demonstrated the environmental and social impacts before and after the pilot plant was built [46]. It was found that villagers could earn an additional income of around US$400–500 per year from the sale of chili peppers by using the biofertilizer produced by the pilot mill. The introduction of technology to village communities also brought new knowledge and perspectives on the use of animal waste. The social impact of this research is not only having sustainable alternative energy but also aiming to improve the quality of life and a clean environment for rural communities. The combination of biogas and biofertilizers provides social benefits for households and also for women to be more productive in supporting households to earn additional income, especially for women who get new income from nonagricultural activities [47] in addition to education and job creation.

18.5 Challenges and Market Opportunities

There are several challenges associated with the use of chemical substitutes made from agricultural and industrial by-products, which are the availability and reliability of by-products as feedstocks for the production of chemical substitutes, the cost of producing chemical substitutes from by-products, the technical challenges associated with the production of chemical substitutes from by-products, and the regulatory challenges associated with the use of chemical substitutes made from agricultural and industrial by-products.

18.5.1 Availability and Reliability of By-products as Feedstocks

Agricultural by-products are often seasonal and may not be available in consistent quantities throughout the year, whereby the chemical industries continuously rely on the consistent supply of feedstock materials. This can make it difficult to consistently produce chemical substitutes in sufficient quantities. In addition, the availability of agricultural by-products can be affected by factors, such as crop yields, market demand, and transportation costs. Industrial by-products, such as waste streams from manufacturing processes, may also vary in quantity and quality depending on the specific industry and production processes.

18.5.2 Production Cost

Although most of the agricultural and industrial by-products are considered waste and needed to be treated before discharge, some of the by-products are being sold depending on factors, such as crop yields, market demand, and transportation costs. Other factors include the cost of energy, labor, and other inputs required for the production process [48]. Since the technology for the conversion of agricultural and industrial by-products to chemical substitutes is still a new technology, the capital cost to setup the technology is high and might be risky.

18.5.3 Technical Challenges

One challenge is the development of effective processes for converting by-products into the desired chemical substitutes. This may require the development of new technologies or the optimization of existing processes. For example, the conversion of agricultural by-products into chemical substitutes may require the development of processes for breaking down the by-products into their constituent components and then synthesizing the desired chemicals. Similarly, the conversion of industrial by-products into chemical substitutes may require the development of processes for purifying the by-products and separating out the desired components [48]. Another technical challenge can be the scale-up of production processes. It maybe necessary to develop processes that can be easily scaled up to meet the demand for chemical substitutes. This can be particularly challenging while working with by-products, as the quality and quantity of these by-products may vary. There maybe technical challenges associated with ensuring the quality and consistency of the chemical substitutes produced from by-products [49]. This may require the development of testing and quality control processes to ensure that the substitutes meet the necessary specifications.

18.5.4 Market Opportunities

There are a few market opportunities for chemical substitutes made from agricultural and industrial by-products. These substitutes can be used as alternatives to traditional chemical products in a variety of applications, and there is increasing demand for sustainable alternatives as consumers and businesses, which seek out more environmentally friendly products. One potential market opportunity is in the replacement of traditional chemicals in the production of plastics, adhesives, and personal care products. Chemical substitutes made from agricultural by-products maybe used in the production of biodegradable plastics, which can be an environmentally friendly alternative to traditional petroleum-based plastics [50]. These substitutes may also be used as ingredients in personal care products, such as shampoos, lotions, or toothpaste.

Another market opportunity for chemical substitutes made from by-products is sustainable agriculture. Chemical substitutes made from agricultural by-products, such as plant residues or animal manure, maybe used as sustainable alternatives to traditional synthetic fertilizers and pesticides. These substitutes can help to improve soil fertility and plant growth and support the trend toward sustainable agriculture practices. There are market opportunities for chemical substitutes made from agriculture by-products for the urban farming application. As the demand for locally grown products increases, there maybe an increase in demand for organic fertilizers and other products made from by-products.

In the industrial sector, chemical substitutes made from industrial by-products may be used as replacements for traditional chemicals in the production of paints, coatings, and other industrial products. There maybe market opportunities for chemical substitutes made from by-products in industries that generate large quantities of by-products, such as the pulp and paper industry, the food and beverage industry, wine industry, and many more [51]. Using these by-products as feedstocks for the production of chemical substitutes can help to reduce waste streams and generate additional revenue streams for these industries [52].

18.6 Conclusions

The sustainability of chemical substitutes made from agricultural and industrial by-products depends on several factors. One important factor is the environmental impact of the production process. Chemical substitutes made from agricultural and industrial by-products may have a lower environmental impact than those made from virgin materials, as they can reduce the demand for resources and energy required to extract and process raw materials. However, the production of chemical substitutes still requires energy and may produce waste or emissions, so it is important to consider the overall environmental impact of the process. Another factor to consider is the availability and reliability of the by-products being used as feedstocks for the chemical substitutes. If the supply of by-products is limited or unreliable, it may not be possible to consistently produce the chemical substitutes in sufficient quantities. Finally, the economic feasibility of using by-products as feedstocks for chemical substitutes is also an important factor. The cost of producing chemical substitutes from by-products should be competitive as an alternative substitution for feedstock sustainability. Overall, the sustainability of chemical substitutes made from agricultural and industrial by-products will depend on a combination of environmental, logistical, and economic factors. It is important to carefully consider these factors to make important decisions about the use of these substitutes.

References

1 Clark, J.H. (2007). Green chemistry for the second generation biorefinery—sustainable chemical manufacturing based on biomass. *J. Chem. Technol. Biotechnol.* 82 (7): 603–609.
2 Mika, L.T., Cséfalvay, E., and Németh, Á. (2018). Catalytic conversion of carbohydrates to initial platform chemicals: chemistry and sustainability. *Chem. Rev.* 118 (2): 505–613.
3 Kobayashi, T. and Nakajima, L. (2021). Sustainable development goals for advanced materials provided by industrial wastes and biomass sources. *Curr. Opin. Green Sustainable Chem.* 28: 100439.
4 Guandalini, I., Sun, W., & Zhou, L. (2019). Assessing the implementation of Sustainable Development Goals through switching cost. *J. Clean. Prod.*, 232, 1430–1441. https://doi.org/https://doi.org/10.1016/j.jclepro.2019.06.033
5 Onwukamike, K.N., Grelier, S., Grau, E. et al. (2018). Critical review on sustainable homogeneous cellulose modification: why renewability is not enough. *ACS Sustainable Chem. Eng.* 7 (2): 1826–1840.
6 Sheldon, R.A. (2007). The E factor: fifteen years on. *Green Chem.* 9 (12): 1273–1283.
7 Koller, G., Fischer, U., and Hungerbühler, K. (2000). Assessing safety, health, and environmental impact early during process development. *Ind. Eng. Chem. Res.* 39 (4): 960–972.
8 Alnur, A., Sakinah, M., Mustafa, A.B. et al. (2013). Simulation of xylitol production: a review. *Aust. J. Basic Appl. Sci.* 7 (5): 366–372.
9 Ardiansyah, F., Gunningham, N., and Drahos, P. (2012). An environmental perspective on energy development in Indonesia. In: *Energy and Non-Traditional Security (NTS) in Asia* (ed. M. Caballero-Anthony, Y. Chang, and N.A. Putra), 89–117. Springer.

10 Yana, S., Nizar, M., and Mulyati, D. (2022). Biomass waste as a renewable energy in developing bio-based economies in Indonesia: a review. *Renewable Sustainable Energy Rev.* 160: 112268.

11 Buana EC (2022). Indonesia's medium-term development opportunities and challenges. National workshop "building forward better: securing inclusive, resilient, and green development in Indonesia," Jakarta (7 March 2022) https://www.unescap.org/sites/default/d8files/event-documents/Green%20Economy%20-%20UNESCAP%20Dit.%20PMAS.pdf (accessed on 23 October 2022).

12 Ronaldo, R. and Suryanto, T. (2022). Green finance and sustainability development goals in Indonesian Fund Village. *Res. Policy* 78: 102839.

13 Sadhukhan, J., Martinez-Hernandez, E., Murphy, R.J. et al. (2018). Role of bioenergy, biorefinery and bioeconomy in sustainable development: strategic pathways for Malaysia. *Renewable Sustainable Energy Rev.* 81: 1966–1987.

14 Ozturk, M., Saba, N., Altay, V. et al. (2017). Biomass and bioenergy: an overview of the development potential in Turkey and Malaysia. *Renewable Sustainable Energy Rev.* 79: 1285–1302.

15 Ministry of Finance Malaysia (2022). MySDG foundation a platform for better coordination of grants. Press release (27 January).

16 Yang, W., Meng, F., Peng, J. et al. (2014). Isolation and identification of a cellulolytic bacterium from the Tibetan pig's intestine and investigation of its cellulase production. *Electron. J. Biotechnol.* 17 (6): 262–267.

17 Piarpuzán, D., Quintero, J.A., and Cardona, C.A. (2011). Empty fruit bunches from oil palm as a potential raw material for fuel ethanol production. *Biomass Bioenergy* 35 (3): 1130–1137.

18 Chiew, Y.L. and Shimada, S. (2013). Current state and environmental impact assessment for utilizing oil palm empty fruit bunches for fuel, fiber and fertilizer – a case study of Malaysia. *Biomass Bioenergy* 51: 109–124.

19 Juturu, V. and Wu, J.C. (2014). Microbial cellulases: engineering, production and applications. *Renewable Sustainable Energy Rev.* 33: 188–203.

20 Hasunuma, T., Okazaki, F., Okai, N. et al. (2013). A review of enzymes and microbes for lignocellulosic biorefinery and the possibility of their application to consolidated bioprocessing technology. *Bioresour. Technol.* 135: 513–522.

21 Gozan, M., Harahap, A.F., Bakti, C.P., and Setyahadi, S. (2018). Optimization of cellulase production by *Bacillus* sp. BPPT CC RK2 with pH and temperature variation using response surface methodology. *E3S Web Conf.* 67: 02051.

22 Surya, E.A., Rahman, S.F., Zulamraini, S., and Gozan, M. (2018). Preliminary plant design of *Escherichia coli* BPPTCC-EgRK2 cell culture for recombinant cellulase production using oil palm empty fruit bunch (OPEFB) as substrate. *IOP Conf. Ser.: Earth Environ. Sci.* 141 (1): 012030. IOP Publishing.

23 Kim, S., Park, J.M., Seo, J.W., and Kim, C.H. (2012). Sequential acid-/alkali-pretreatment of empty palm fruit bunch fiber. *Bioresour. Technol.* 109: 229–233.

24 Medina, J.D., Woiciechowski, A., Zandona Filho, A. et al. (2016). Steam explosion pretreatment of oil palm empty fruit bunches (EFB) using autocatalytic hydrolysis: a biorefinery approach. *Bioresour. Technol.* 199: 173–180.

25 Duangwang, S., Ruengpeerakul, T., Cheirsilp, B. et al. (2016). Pilot-scale steam explosion for xylose production from oil palm empty fruit bunches and the use of xylose for ethanol production. *Bioresour. Technol.* 203: 252–258.

26 Halliwell, D.J., Barlow, K.M., and Nash, D.M. (2001). A review of the effects of wastewater sodium on soil physical properties and their implications for irrigation systems. *Soil Res.* 39 (6): 1259–1267.

27 Laurenson, S., Bolan, N.S., Smith, E., and McCarthy, M. (2012). Use of recycled wastewater for irrigating grapevines. *Aust. J. Grape Wine Res.* 18 (1): 1–10.

28 Zhu, J.K. (2003). Regulation of ion homeostasis under salt stress. *Curr. Opin. Plant Biol.* 6 (5): 441–445.

29 Akhdiya, A. (2003). Isolasi bakteri penghasil enzim protease alkalin termostabil. Indonesian Ministry of Agriculture. *Buletin Plasma Nutfah* 9 (2): 38–44. [in *Bahasa*]. https://media.neliti.com/media/publications/55494-ID-isolasi-bakteri-penghasil-enzim-protease.pdf.

30 Abd-Aziz, S., Gozan, M., Ibrahim, M.F., and Phang, L.Y. (2022). Demand and sustainability of palm oil plantation. In: *Biorefinery of Oil Producing Plants for Value-Added Products*, vol. 1 (ed. S. Abd-Aziz, M. Gozan, M.F. Ibrahim, and L.Y. Phang), 11–28. Wiley-VCH.

31 Erman, E. (2018). Dibalik Keberlanjutan Sawit: Aktor, Aliansi Dalam Ekonomi Politik Sertifikasi Uni Eropa. *Masyarakat Indonesia* 43 (1): 1–13.

32 Varkkey, H. (2015). *The Haze Problem in Southeast Asia: Palm Oil and Patronage*, 1e. London: Routledge https://doi.org/10.4324/9781315717814.

33 Banten Province (2022). One-stop integrated service investment service (DPMPTSP), Banten Province. https://dpmptsp.bantenprov.go.id/home (accessed 13 December 2022).

34 Prabavathy VR, Rengalakshmi R, Nair S (2007). Decentralised production of biofertilisers–*Azospirillum* and Phosphobacteria. JRD Tata Ecotechnology Centre, Chennai, India, p. 36. http://59.160.153.188/library/sites/default/files/Ecoenterprises%20for%20Sustainable%20Livelihood.pdf (accessed on 12 August 2022).

35 Okon, Y. (1985). *Azospirillum* as a potential inoculant for agriculture. *Trends Biotechnol.* 3 (9): 223–228.

36 Spaepen, S., Vanderleyden, J., and Okon, Y. (2009). Plant growth-promoting actions of rhizobacteria. *Adv. Bot. Res.* 51: 283–320.

37 Li Yangrui (2008). Method for using molasses alcohol fermentation liquid as sugarcane liquid fertilizer. China Patent CN101439994B, Worldwide applications. https://patents.google.com/patent/CN101439994B/en.

38 Biwer, A. and Heinzle, E. (2004). Process modeling and simulation can guide process development: case study α-cyclodextrin. *Enzym. Microb. Technol.* 34 (7): 642–650.

39 Farid, S.S. (2007). Process economics of industrial monoclonal antibody manufacture. *J. Chromatogr. B* 848 (1): 8–18.

40 Krajnc, D., Mele, M., and Glavič, P. (2007). Improving the economic and environmental performances of the beet sugar industry in Slovenia: increasing

fuel efficiency and using by-products for ethanol. *J. Clean. Prod.* 15 (13–14): 1240–1252.

41 Kwiatkowski, J.R., McAloon, A.J., Taylor, F., and Johnston, D.B. (2006). Modeling the process and costs of fuel ethanol production by the corn dry-grind process. *Ind. Crop. Prod.* 23 (3): 288–296.

42 Pérez Sánchez, A., Singh, S., Pérez Sánchez, E.J., and Segura Silva, R.M. (2018). Techno-economic evaluation and conceptual design of a liquid biofertilizer plant. *Rev. Colomb. Biotecnol.* 20 (2): 6–18.

43 Kour, D., Rana, K.L., Yadav, A.N. et al. (2020). Microbial biofertilizers: bioresources and eco-friendly technologies for agricultural and environmental sustainability. *Biocatal. Agric. Biotechnol.* 23: 101487.

44 Šarić, L.Ć., Filipčev, B.V., Šimurina, O.D. et al. (2016). Sugar beet molasses: properties and applications in osmotic dehydration of fruits and vegetables. *Food Feed Res.* 43 (2): 135–144.

45 Sebayang, F. (2006). Pembuatan etanol dari molase secara fermentasi menggunakan sel Saccharomyces cerevisiae yang terimobilisasi pada kalsium alginat. *Jurnal Teknologi Proses.* 5 (2): 75–80.

46 Sinsuw, A.A., Wuisang, C.E., and Chu, C.Y. (2021). Assessment of environmental and social impacts on rural community by two-stage biogas production pilot plant from slaughterhouse wastewater. *J. Water Process. Eng.* 40: 101796.

47 Rosyidi, S.A., Bole-Rentel, T., Lesmana, S.B., and Ikhsan, J. (2014). Lessons learnt from the energy needs assessment carried out for the biogas program for rural development in Yogyakarta, Indonesia. *Procedia Environ. Sci.* 20: 20–29.

48 Singh, N., Singhania, R.R., Nigam, P.S. et al. (2022). Global status of lignocellulosic biorefinery: challenges and perspectives. *Bioresour. Technol.* 344: 126415.

49 Awasthi, M.K., Sindhu, R., Sirohi, R. et al. (2022). Agricultural waste biorefinery development towards circular bioeconomy. *Renewable Sustainable Energy Rev.* 158: 112122.

50 Kratky, L. (2022). Lignocellulosic waste treatment in biorefinery concept: challenges and opportunities. In: *Zero Waste Biorefinery* (ed. Y.K. Nandabalan, V.K. Garg, N.K. Labhsetwar, and A. Singh), 59–94. Springer Singapore.

51 Ioannidou, S.M., Filippi, K., Kookos, I.K. et al. (2022). Techno-economic evaluation and life cycle assessment of a biorefinery using winery waste streams for the production of succinic acid and value-added co-products. *Bioresour. Technol.* 348: 126295.

52 Solarte-Toro, J.C., Laghezza, M., Fiore, S. et al. (2022). Review of the impact of socio-economic conditions on the development and implementation of biorefineries. *Fuel* 328: 125169.

Index

a

accelerated solvent extraction   26
acetone–butanol–ethanol pathway   342
acid–base titration procedure   239
acid-catalyzed hydrolysis   225
acid hydrolysis   295
   advantages of   290
   disadvantages of   290
acoustic cavitation   28
activated carbon, from seaweed   64–66
aerobic microorganisms   323
agricultural biomass   4, 6, 146, 269, 271
agricultural by-products
   availability and reliability of   368
   market opportunity   369
   production cost   368
   technical challenges   368–369
agricultural sector   1, 4, 11, 34, 87, 301
agricultural wastes
   antioxidant sources from   22
   categories of   3
   chemical substitutes   3
   defined   1
   industrial by-products
      agriculture, horticulture, and landscaping   8
      raw material or additive   8–9
   sources of   2
   types of   4–5
   waste utilization routes

fertilizer application   5
industrial enzymes   7–8
mushroom cultivation   6–7
organic acids   7
textile industry fibers   5–6
agro-industry wastes   145, 216
algae   59, 63, 77
alginate-based nano-Si electrodes   81
alginic acid   63, 76, 81, 83
alkaline delignification   309
amino acid composition, of chicken feathers   126
anaerobic microorganisms   323–325
antioxidant enzymes   20, 33
antioxidant extracts   25, 33
antioxidants   19
   from agricultural wastes   22
   future directions of   34–35
   extraction from agricultural wastes   22–30
   cosmetic industry   32–33
   food industry   30–32
   green extraction   23
   maceration   22–25
   microwave-assisted extraction   27–28
   pressurized liquid extraction   26–27
   supercritical fluid extraction   29–30
   therapeutic industries   33–34

Chemical Substitutes from Agricultural and Industrial By-Products: Bioconversion, Bioprocessing, and Biorefining,
First Edition. Edited by Suraini Abd-Aziz, Misri Gozan, Mohamad Faizal Ibrahim, and Lai-Yee Phang.
© 2024 WILEY-VCH GmbH. Published 2024 by WILEY-VCH GmbH.

antioxidants (*contd.*)
  ultrasound-assisted extraction 28–29
  mechanisms of action 20
  oxidative homeostasis 20
  physiological effects of 21
  sources of 21–22
aqueous two-phase extraction 182
aromatherapy 43, 45, 46, 48, 103, 105, 114, 115, 117, 118
Aspen Plus 13, 358
Aspergillus niger 7, 217
Auricularia auricula. *see* wood ear mushroom
Azospirillum brasilense 365, 367

b

Bacillus cereus-immobilized sugarcane bagasse 220
Bacillus sp. 135, 136, 360
battery component, seaweed derived-chemical materials for 79–82
bioactive compounds 22, 23, 26, 29, 34, 35, 46, 49–51, 53, 63
  with antioxidant-related effects 23
bioactive molecules 22
bio-bleaching 146
bio-bleaching enzymes 10–11
  agricultural waste bioprocessing into
    block flow diagram 154–157
    downstream processing 158–159
    upstream processing 157–158
  biomass substrate characteristics 146–148
  challenges and future research 165–167
  economic analysis 162–165
  microbial sources of 148–150
  technical analysis 159–162
  usage in pulp and paper industry
    cellulases 151–152
    laccases 152–154

xylanases 150–151
bioconversion of sago wastes 194
  into biosugars 200–202
biodiesel 13, 337–350
Bioeconomy Transformation Program 359
biofertilizers 10, 11, 13, 59, 63, 95, 124, 129–131, 213–230, 364–367
bio-industry, sustainable society for 357
biological delignification 309
biomass 76, 147
  compositional attributes of 258
  hydrolysis 259–260
  pretreatment 259
  removal 343
  sago wastes 196–200
  size reduction 258–259
bioplastics 10, 124, 130
bioprocessing of chicken feathers 124, 128–132
biorefinery 1–13, 209, 228, 229, 257, 258, 295, 313, 330, 332
bioremediation process, SMS-derived enzymes for 91–92
bio-succinic acid
  additives 264
  application of 272–273
  bacteria 260–261
  carbon dioxide 264
  carbon sources 263
  cell immobilization 266–267
  consolidated bioprocessing 265–266
  fermentation conditions 261–263
  fungi and yeasts 261
  global market and demands for 256
  microbial 260
  nitrogen sources 264
  operational modes 266
  separate hydrolysis and fermentation 265

simultaneous saccharification and fermentation/co-fermentation 265
  valorization of biomass to 257–260
biosugars 194
  sago wastes bioconversion into 200–202
black liquor 223
bleached red seaweed, as supercapacitor electrode 77
Borneo tallow 235, 236
brown seaweed 22, 61–64
Brunair–Emmet–Teller equation 220
B-serum 186, 187
butanol
  applications of 346–348
  in situ gas stripping 345
  in situ pervaporation 346
  and 1,3-propanediol 340–342
butanol dehydrogenase 342
butyryl-CoA dehydrogenase 341

c

Calocybe indica 6
Candida sp. 149
Carbohydrate-Active Enzyme database 321
carbohydrate-binding modules 321, 323, 324
carbonization process 65, 66
Caulerpa lentillifera 63
cavitation phenomenon 29
cellobiohydrolases (CBHs) 151, 152, 320
cellobiose 201–203
cellulases 12, 151–152
  aerobic microorganisms 323
  agricultural waste, utilization of 326–328
  anaerobic microorganisms 323–325
  catalytic mechanisms of 322
  consolidated bioprocessing 330–331
  diversity 320–322
  production 327–330, 360
  properties 325–326
  structures of 321–322
  techno-economic analysis 331–332
cellulose 88
  chemical structure of 198
  hydrolysis reaction 289
  in sago bark 197–199
cellulosomes 323–325
chemical delignification 309
chemical-derived seaweed, for battery component 79, 82
chemical extraction 46, 48–49
chemical process industry 365
chicken feathers
  biological treatment 127
  into chemical substitutes, bioprocessing of 128–132
  chemical treatment 127
  composition of 125
  keratin in 130
  physical treatment methods 126
  single-stage combined treatment 127
  two-stage combined treatment 127–128
  valorization of 124–128, 140
  waste 123
chitinases 178
  lysozymes 180–181
  plant, microbe and animal 179
  purification methods for 184–185
  structure 178
chitooligosaccharides 178, 182
5-chloromethylfurfural 280–281
circular bioeconomy 125, 175, 357
circular economy 3, 23, 34, 90, 95, 118, 125, 140, 216, 229, 358, 359
closed-chain dehydration 291
cocoa butter equivalent 11, 237, 238, 240–242, 249, 250

cocoa butter substitute 249
co-culture system 128, 129
cohesin modules 323
commercial mushrooms 10, 87, 88, 94
compound annual growth rate 116, 320
consolidated bioprocessing 265–266, 330–331
corn cob 147, 286
    lignocellulosic content in 148
corn stover 286
COVID-19 pandemic
    lactic acid demand during 202
    mushroom supply during 87
crude glycerol 337
    into butanol 340–342
    characteristics 338–340
    impurities 338–340
    into 1,3-propanediol 340–342
crystallization process 271–272
C-serum 175, 186, 187
Cymbopogon sp.
    C. citratus 9, 39–41, 51
    C. exuosus 40
    C. flexuosus 46

d

dark chocolate, health benefits 22
deashing process 65
deep eutectic solvent 70, 71, 294
dehydration process 12, 279, 289, 291, 292
deionized water 343
dextrose 40, 201
Dipterocarpaceae family 235
directed evolution, of recombinant keratinase 137
distillation method 106
    hydro-distillation 107
    Soxhlet extraction 107–108
dividing wall column (DWC) 309, 312
dockerin (Doc) modules 323

e

effluent 151, 193, 196
E-isomer 42
electrophilic substitution reaction 284
empty fruit bunches 88, 286, 313, 360
endoglucanases 93, 151, 152, 260, 320
end-product inhibition 342
Enteromorpha prolifera macroalgal pollutants 77
enzymatic antioxidant 19
enzymatic hydrolysis 92, 129, 194, 197, 201, 202, 259–260, 265, 290, 319
enzyme-assisted extraction 104, 106, 109, 110, 118
Escherichia coli 7, 112, 135, 136, 325, 360
essential oils 39, 42, 45, 103
    distillation 106–107
    extracted essential oil compounds 112–113
    extraction of 106
        enzyme-assisted extraction 109
        hydro-distillation 107
        supercritical fluid extraction 109–112
    and hydrosols 114
        applications of 114–116
        market analysis 116–117
ester and HMFCA 281, 282
ethylene glycol 226
ethyl lactate 113
ethyl propionate 112
Euchema cottoni 63
Eucheuma denticulatum 62
Eucheuma Spinosum 63
exoglucanases 151, 260, 320–322
extractable proteins 173

f

feather meal 10, 130, 132, 140
    as animal feed 129

feather waste disposal methods   124
feedback inhibition   320
Fe$_3$O$_4$/Fe-doped graphene nanosheets   70, 71
Fe$_2$O$_3$ hollow nanoparticles/graphene aerogel (Fe$_2$O$_3$-HNPs/N-GAs)   71, 73
food flavoring   39–54, 116
forest/wood biomass   287, 288
formic acid   12
  biological pretreatment   307–308
  from biomass   302
  case studies   313–314
  delignification   308–309
  economic feasibilities   312–313
  formation   306
  pretreatment and production   306–309
  purification technologies   309, 312
  rice straw   304
  size reduction   306–307
  sugarcane bagasse   304–306
  supplying countries and companies   303
  utilization   302
forward osmosis   271
free radicals   19
  examples of   20
  types of   20
Frey–Wyssling complexes   187
fungal laccases   153
furaldehyde   279
furanaldehyde   279
2-furancarboxaldehyde   279
furan molecule   283, 284
furfural
  as building block material   280
  5-chloromethylfurfural (CMF)   12, 280, 281
  dehydration process   291, 292
  extraction process   293
  heterogeneous catalysts   293, 294
  homogeneous catalysts   293, 294
  hydrogenation   283
  hydrolysis process   289, 290
  lignocellulosic biomass   285, 286
  pretreatment process   288, 289
  purification process   292, 293
  solvent   294
  techno-economic analysis   295, 296
2-furfuraldehyde   279
furfuryl alcohol   12, 230, 279, 281–285, 296
furoic acid   279, 284, 285

g

gas stripping–pervaporation   349
generally recognized as safe   112
genetic modification   149, 167, 208, 331
genome editing   134, 135
Glacilaria sp.   63
glucose   90, 151, 152, 195, 200–203, 225, 226, 258, 260, 263, 289, 291, 320, 324, 328, 329
  degradation   290
β-glucosidases   152
glycerine   337
glycosyl hydrolase enzyme   319
graphene, from seaweed   66–75, 82
green butter   236
green fuel feedstock production, SMS-derived enzymes for   92–94
green seaweed   63
Green Technology Policy   359

h

hemicellulose   4, 5, 10, 11, 88–90, 197, 198, 200, 201, 204, 215, 224, 225, 258, 259, 263, 279, 285, 287–290, 304, 306–309
  hydrolysis   290
heterogeneous catalysts   293, 306
high ammonia natural rubber latex   175
highly-ammoniated latex (HA latex)   176
high-redox potential laccases (HRPLs)   153
homogeneous catalysts   293, 306
Hummers and Offeman's method   69
hydro-distillation   104, 107, 108
hydrodistillation-based separation   343–345
hydrogen bond acceptor (HBA)   294
hydrogen bond donor (HBD)   294
hydrolysis process   239, 289–290, 306–308
hydrolytic enzymes   10, 97, 148, 259, 265
hydrosols   114–117
5-hydroxymethyl-2-furan carboxylic acid (HMFCA)   281
5-hydroxymethylfurfural (HMF)   12, 263, 280

i

illipe butter   236
illipe tallow   236
Indonesian Government's National Economic Recovery   358
industrial by-products   1–13, 355–370
industrial enzymes   7–8, 145–146, 148, 158, 162, 164, 165, 319, 320, 332
in situ gas stripping   345
in situ pervaporation   346
integrated nutrient management (INM)   5

ion exchange   271
  resin-based separation   343, 344

k

Kappaphycus sp.   63
  K. alvarezii   62
  K. striatus   62
keratinase   10, 124, 128–129, 131–133, 135–140
keratinase production, molecular approach to   132, 133
  enzymatic consortium   134–135
  keratinase overexpression   132, 133
  strain engineering   134
keratin-based plastics   130
keratin degradation performance   139
keratinolytic microbes   127, 128
KOH-activation method   82
kojic acid   206–207
Krebs cycle   260

l

laccase-mediator system (LMS)   153
laccases   10, 89, 91, 92, 94, 149, 152–154, 205
laccase substrates   153
lactic acid   113, 186, 202–204, 226–227
lactic acid ethyl ester   113
Lactococcus lactis IO-1   202, 204
large rubber particles (LRP)   176
latex fruit syndrome   174
lemongrass   39
  chemical substitutes
    essential oils   42
    oleoresins   43
    phytoconstituents   43
  types of   40
lemongrass oleoresins   40
  applications of   52
  characteristics and properties   44
  composition and function   44
  extraction technique of   46

chemical extraction method   46
food flavoring   51
pressurized liquid extraction   50
steam distillation   49
Lentinula edodes   6, 87
Lessonia nigrescens based carbon materials for supercapacitor electrodes   76–79
levulinic acid
  biological pretreatment   307–308
  from biomass   302
  case studies   313–314
  delignification   308–309
  economic feasibilities   312–313
  formation   306
  pretreatment and production   306–309
  purification technologies   309, 312
  rice straw   304
  size reduction   306–307
  sugarcane bagasse   304–306
  supplying countries and companies   303
  utilization   302
life cycle analysis (LCA)   357
lignin   88, 285
  deconstruction   148
  in sago bark   198
lignin-degrading enzymes   90, 91, 146
lignin-modifying enzymes   89, 90
lignocellulose
  biomass   12, 286
  saccharification of   290
lignocellulosic biomass   197, 285, 286, 293, 357
  composition of   288
  corn cob and stover   286
  forest/wood biomass   287
  oil palm biomass   287
  rice straw   287
  sugarcane bagasse   286
  waste   145

lignocellulosic composition, of sago fronds   200
lignocellulosic conversion   223
L-lactic acid   202–204
London dispersion forces   27
lutoid layer   187
lysozymes   175
  activity assays for   178, 180–182
  assays for   178, 182
  plant-derived   177–178, 182, 183
  recovery and purification methods   182, 184–187
  structure   177

m

maceration   22–23, 48
maceration extraction   23, 25, 49
macroalgae   59, 62, 70
  as biofuel feedstock   64
Malaysia Sustainable Development Goal Foundation   359
Malaysia Sustainable Development Goals Trust Fund   359
medium-ammoniated latex   176
methyl isobutyl ketone   293, 295
Metroxylon sago   193
microbial sources, of bio-bleaching enzymes   148
  bacteria   149, 150
  fungi   148, 149
  yeast   149
Micrococcus lysodeikticus   178, 182
microwave-assisted extraction   27–28, 104, 107
modifiers   30
Monostroma nitidum   63
mushroom cultivation   6–7, 10, 87, 88, 90, 95
mushroom-derived enzymes   90

n

nanocarbon materials 9, 59–83
nanotechnology 208
National Biotechnology Policy 2.0 359
natural rubber latex 11, 173
  deproteinization of 176
  lysozymes and chitinases 178
  overview 175
  plant-derived lysozymes and chitinases 177, 182, 183
  potential strategy for lysozymes and chitinases 182
  preservation 176
  structure 176
net present value (NPV) 162, 164, 247–249, 296, 329, 362, 366
nitrogen stripping 292
nonenzymatic proteins 19
nonsolvent-induced phase separation procedure 82

o

oil palm biomass 6, 33, 280, 287
oil palm empty fruit bunches 7, 287, 294, 296, 309, 358, 360, 363
oil palm trunk 287, 328
oleoresins 39, 40, 43
  characteristics and properties 44
  prospect 53
  solvent extraction for 48
one factor at a time 46, 139
OPEFB. *see* oil palm empty fruit bunches (OPEFB)
open-chain dehydration 291
open reading frame 2 (Orf2p) 323
orange oil 117
organic acids 7, 90, 103, 104, 187, 202, 206, 216, 217, 269, 293, 342–344
organic solvents 23, 293, 294
outer layer protein B (OlpB) 323, 324
oxidation rates, of laccases 153
oxidative enzymes 148
oxidative homeostasis and antioxidants 20, 21
oxidative stress 20, 21, 32, 33, 198
oxolane 284
oyster mushroom *(Pleurotus ostreatus)* 6, 88, 91, 92, 94–96, 307

p

palm mid-fraction 249–250
palm oil mill effluent (POME) hydrolysis 92, 93
PANI-coated seaweed graphene's manufacturing process 74
phenol degradation 218–222
phenolic acid 21, 43
phytoconstituents 39, 43
Pichia pastoris 132, 135, 136, 149, 205
pineapple essential oils 10, 104–106, 112, 114, 118
pineapple leaves fiber (PALF) 6
pineapple wastes 104
  chemical composition of 105
  recycling of 103
plant growth-promoting bacteria (PGPB) 364–365
plant growth-promoting rhizobacteria (PGPR) 367
plasmid overexpression, in native keratinolytic host 132, 133
plastics 90, 130, 226, 272, 297, 346, 347, 369
plate and frame microfiltration 158
polybutylene succinate 272, 273
polyphenols 22, 23, 31, 43
POME hydrolysis 92, 93, 97
porous carbon 65, 82, 83
potassium hydroxide (KOH) 127, 309, 338
poultry meat consumption 123, 124
poultry waste (feathers), recycling of 125
press mud cake 213
pressurized fluid extraction 26

pressurized hot-solvent extraction 26
pressurized hot-water extraction 26
pressurized liquid extraction 26–27, 50–51
Prohevein 174
promoter engineering 136–137
1,3-propanediol
  applications of 346–348
  butanol and 340–342
  hydrodistillation-based separation 343–345
  ion-exchange resin-based separation 343, 344
propanoic acid ethyl ester 112, 113
propeptide engineering 136
pulp and paper industry 10, 146, 147, 150–154, 165, 369
pulp bleaching 146, 151

r

radiometric assays 178
random mutagenesis 134
raw sago hampas, chemical composition of 199
reactive extraction 270, 304
reactive oxygen species (ROS) 19–21
recombinant keratinase production, molecular approach 135
  alteration of protein domains 138
  directed evolution 137
  promoter engineering 136, 137
  propeptide engineering 136
  signal peptide engineering 137
red seaweed 22, 61–63, 69, 77
renewable energy (RE) 358
response surface methodology (RSM) 46
rice straw 6, 7, 12, 287, 301–314, 328
rubber elongation factor 174
rubber tree *(Hevea brasiliensis)* 11, 173–175, 186

s

sago bark 193, 194, 196–199
sago frond 193, 200
  antimicrobial and prebiotic sugar (cellobiose) from 203
  lignocellulosic composition of 200
  pre-sterilized sap 203
  sap 200
  silage 204
sago hampas 193, 199
  advantages 199
  raw, chemical composition of 199
sago industrial wastes 193
sago logs, debarking of 196
sago palm
  cultivation 194
  plantation 193
  resilience against wildfires 194
sago pith residue 199
sago starch 201
  enzymatic hydrolysis of 202
  industry
    limitation 207
    in Malaysia 195
    processing in Sarawak 195–196
sago starch industry, in Malaysia 195
sago wastes
  bioconversion into biosugars 200, 202
  bioprocessing fermentable sugar for chemicals substitute 202, 207
sago wastes biorefinery
  challenges 207
  future direction of 207, 208
sago wastewater 196, 197
Salmonella enterica 112
Sargassum sp. 62, 63
Sargassum tenerrimum functionalized graphene sheets 71

scaffold in dockerin-binding protein A   323
seaweeds
  activated carbon from   64, 66
  aquaculture production per country   60
  benefits of   59
  biomass
    future research and challenges   82, 83
    proximate analysis result of species   65
  brown   62, 63
  categories of   62
  graphene from   66, 75
  green   63
  hydrothermal carbonization stage   66
  red   62
  role in global aquaculture   59
  taxonomic groups   61
  utilization for energy storage component   76, 82
  world production   60, 61
separate hydrolysis and fermentation (SHF)   265, 330, 331
Shiitake *(Lentinula edodes)*   87
Shorea stenoptera   11, 235, 236
signal peptide engineering   137
simultaneous saccharification and fermentation (SSF)   330
simultaneous saccharification and fermentation/co-fermentation   264, 265
site-specific mutagenesis   134
size reduction, levulinic and formic acid   306–307
small rubber particles   176
sodium dodecyl sulphate   175
solid agricultural waste   146
solid fat content   250
solid fat content, of Tengkawang butter   239–240

solid-state fermentation   7, 96, 97, 155, 216, 328–330
solvent production process   342
Soxhlet extraction   49, 107–108
spent mushroom compost   89, 97
spent mushroom substrate   10, 87, 89, 90
  beneficial activities   89
  challenges using   94, 95
  enzymes extracted from
    for bioremediation   91, 92
    for green fuel feedstock production   92, 94
  formation process   89
  future prospects using   95–-97
spent wash   213
Sphaerotheca humuli Burrill   182
spray drying   158–159
Staphylococcus aureus   34, 112, 203, 206
steam distillation   10, 44, 48–50, 107, 244
steam stripping process   292
subcritical water extraction   26
submerged fermentation   7, 155, 157, 163, 167, 197, 206, 216, 328, 329
succinic acid   7
  crystallization process   271, 272
  downstream processing method   269, 270
  fermentation broth constituents   267, 269
  forward osmosis   271
  ion exchange   271
  production   255, 256
  reactive extraction   270
sugarcane bagasse   7, 11, 145, 147, 155, 213, 214, 286, 295, 301–314, 328, 332
  levulinic and formic acid   304–306
  lignocellulosic content in   148
  production   215

as raw material for soil improver 218–222
sugarcane by-products
  chemical composition of 214
  conversion into biofertilizer 216, 218
  conversion into chemical 223, 227
  industrial cycle 228, 229
  as material for biocomposites 227, 228
  as raw materials 218, 222
  usage and applications of 229, 230
sugarcane industrial process 214
sugarcane molasses 216
sugarcane refining process 215
sugarcane vinasse 217, 218
supercapacitor, seaweed derived-carbon material for 76–79
supercritical fluid extraction 29–30, 104, 109–112
superheated water extraction 26
superoxide dismutase 19, 20
SuperPro Designer 146, 159, 162, 358, 360–367
sustainable development goals 230, 357
sustainable development strategy
  Indonesia's nation 358–359
  Malaysia's policies and regulations 359
  market opportunities 368–369

t

technically specified rubber (TSR) 187
techno-economic analysis (TEA) 331–332, 360–367
techno-economic biofertilizer
  environmental impact analysis 367
  social aspect 367
  SuperPro Designer 364–367
techno-economic cellulase
  environmental impact analysis 363–364
  social aspect 364
  SuperPro Designer 360–363
Tengkawang *(Shorea stenoptera)* 11, 235–250
Tengkawang butter 236
  benefits of 249, 250
  chemical purification of 243
  economic feasibility 244, 249
  fatty acids profile of 237, 238
  physical purification of 244
  physicochemical properties of 238
  quality parameters of 238, 239
  solid fat content of 239, 254
  thermal properties of 240, 254
  treatment process 241, 242
Tengkawang fat 236
tetrahydrofuran 284
tetrahydrofurfuryl alcohol 279, 281
tetraphenol 283
thermostable Xyl-11 150
3D stratified macro-meso-microporous sulfur-doped carbon aerogel 77
total capital investment 331, 362
Total Plant Direct Cost 163, 365
Total Plant Indirect Cost 163, 365
tricarboxylic acid cycle 256, 260, 264

u

ultrasound-assisted extraction 28–29, 107
Undaria pinnatifida 61, 63

v

vinasse 213, 215–218
Volvallella volvacea 6

w

waste utilization routes
  fertilizer application 5
  industrial enzymes 7–8
  mushroom cultivation 6–7
  organic acids 7

waste utilization routes (*contd.*)
  textile industry fibers 5–6
waste valorization 11, 145
wheat straw 145–167
white rot basidiomycetes 153
wood ear mushroom 88

x

xylanase 7, 10, 149–152, 157, 167, 324, 327, 328, 332
  producers for pulp bleaching 151
xylose 12, 90, 198, 201, 204, 263, 270, 279, 285, 287, 289–291, 293, 294
  degradation 290
  dehydration 291

y

Yarrowia lopolytica 149